AF356148

Landscape Governance in the Age of Social Media

Landscape Governance in the Age of Social Media

Editors

Cecilia Arnaiz Schmitz
Nicolas Marine
María F. Schmitz

Basel • Beijing • Wuhan • Barcelona • Belgrade • Novi Sad • Cluj • Manchester

Editors

Cecilia Arnaiz Schmitz
Universidad Complutense
de Madrid
Madrid
Spain

Nicolas Marine
Universidad Politécnica
de Madrid
Madrid
Spain

María F. Schmitz
Universidad Complutense
de Madrid
Madrid
Spain

Editorial Office
MDPI
St. Alban-Anlage 66
4052 Basel, Switzerland

This is a reprint of articles from the Special Issue published online in the open access journal *Land* (ISSN 2073-445X) (available at: https://www.mdpi.com/journal/land/special_issues/landscape_social_media).

For citation purposes, cite each article independently as indicated on the article page online and as indicated below:

Lastname, A.A.; Lastname, B.B. Article Title. *Journal Name* **Year**, *Volume Number*, Page Range.

ISBN 978-3-7258-0141-1 (Hbk)
ISBN 978-3-7258-0142-8 (PDF)
doi.org/10.3390/books978-3-7258-0142-8

Cover image courtesy of Amaury Laporte

Contents

Preface . . . . **vii**

Yiwei Huang, Zhixin Li and Yuhan Huang
User Perception of Public Parks: A Pilot Study Integrating Spatial Social Media Data with Park
Management in the City of Chicago
Reprinted from: *Land* **2022**, *11*, 211, doi:10.3390/land11020211 . . . . **1**

Aireona B. Raschke, Jeny Davis and Annia Quiroz
The Central Arizona Conservation Alliance Programs: Use of Social Media and
App-Supported Community Science for Landscape-Scale Habitat Restoration, Governance
Support, and Community Resilience-Building
Reprinted from: *Land* **2022**, *11*, 137, doi:10.3390/land11010137 . . . . **22**

Fan Tang, Jianping Yang, Yanxia Wang and Qiuling Ge
Analysis of the Image of Global Glacier Tourism Destinations from the Perspective of Tourists
Reprinted from: *Land* **2022**, *11*, 1853, doi:10.3390/land11101853 . . . . **39**

Jiani Zhang, Xun Zhu and Ming Gao
The Relationship between Habitat Diversity and Tourists' Visual Preference in Urban Wetland
Park
Reprinted from: *Land* **2022**, *11*, 2284, doi:10.3390/land11122284 . . . . **60**

Keying Ding, Mian Yang and Shixian Luo
Mountain Landscape Preferences of Millennials Based on Social Media Data: A Case Study on
Western Sichuan
Reprinted from: *Land* **2021**, *10*, 1246, doi:10.3390/land10111246 . . . . **79**

Tingting Ding, Wenzhuo Sun, Yuan Wang, Rui Yu and Xiaoyu Ge
Comparative Evaluation of Mountain Landscapes in Beijing Based on Social Media Data
Reprinted from: *Land* **2022**, *11*, 1841, doi:10.3390/land11101841 . . . . **96**

Siya Cheng, Zheran Zhai, Wenzhuo Sun, Yuan Wang, Rui Yu and Xiaoyu Ge
Research on the Satisfaction of Beijing Waterfront Green Space Landscape Based on Social
Media Data
Reprinted from: *Land* **2022**, *11*, 1849, doi:10.3390/land11101849 . . . . **126**

Diego Martín Sánchez and Noemí Gómez Lobo
Urban Forest Tweeting: Social Media as More-Than-Human Communication in Tokyo's
Rinshinomori Park
Reprinted from: *Land* **2023**, *12*, 727, doi:10.3390/land12040727 . . . . **151**

Ángeles Layuno Rosas and Jorge Magaz-Molina
Contributions of Social Media to the Recognition, Assessment, Conservation, and
Communication of Spanish Post-Industrial Landscapes
Reprinted from: *Land* **2023**, *12*, 374, doi:10.3390/land12020374 . . . . **166**

Graziella Trovato
Postproduction in the Research on the Urban Cultural Landscape: From the Transfer of Results
to the Exchange of Knowledge on Digital Platforms and Social Networks—The TRAHERE
Project in Madrid
Reprinted from: *Land* **2023**, *12*, 31, doi:10.3390/land12010031 . . . . **185**

Nicolas Marine, Cecilia Arnaiz-Schmitz, Luis Santos-Cid and María F. Schmitz
Can We Foresee Landscape Interest? Maximum Entropy Applied to Social Media Photographs:
A Case Study in Madrid
Reprinted from: *Land* **2023**, *11*, 715, doi:10.3390/land11050715 . . . . **207**

Preface

At the end of the 20th century, documents such as the World Heritage Guidelines or the European Landscape Convention proposed new and challenging ways of conceptualizing landscape assessment and governance. Consequently, in the recent two decades, numerous countries have re-evaluated their national planning systems and landscape conservation policies.

Simultaneously, social media has grown into an extensive source of data with a certain influence on how we regard spaces. Currently, numerous researchers are advocating the value of social media data to better comprehend ecosystem services provision, use, and intensity. Through this, we may also be in a more optimal position to understand how to recognize people's patterns of behavior or how they perceive the landscape.

The relation between social media and the latest conceptualizations of landscape allows us to ask pertinent questions that are the base of this collection of chapters: Is social media useful for administrations to recognize and adapt to changes in land use, patterns of mobility, or landscape meaning? Is it bringing a more democratic understanding of the landscape and its conservation? Does it serve local communities to express their feelings towards governance policies? Do any of these factors align with the concepts laid out by international organizations, such as the IUCN, UNESCO, or the European Council?

The aim of this Special Issue is to delve into the relationship between the contemporary forms of landscape valuation and governance and present-day social media. The lines of research outlined by the included papers focus on the current connections of social media with the following:

Traditional ecological knowledge (TEK) and governance decentralization;
Participatory scenarios and land planning based on online technologies;
Multi-scale processes and social–ecological resilience;
Spatio-temporal patterns for the maintenance of the living landscape;
Sustainable governance and rural landscape stewardship;
User-generated content (UGC) as a tool for landscape studies;
Inclusive and participatory land governance—a cross-country comparison.

This Special Issue comprises articles written by authors from a diverse geographic distribution and different academic backgrounds. In total, the 11 papers collected here provide a multifaceted collection of approaches to the afore-mentioned problem. First, we find research studies directly focused on management—such as that of Yiwei Huang, Zhixin Li, and Yuhan Huang—which are centered around Chicago's public parks. Similarly, Aireona B. Raschke, Jeny Davis, and Annia Quiroz explain the interesting and innovative proposals of the Central Arizona Conservation Alliance.

On the other hand, several papers focus on the management and appreciation of the tourist population, for example, in relation to glacier destinations—authored by Fan Tang, Jianping Yang, Yanxia Wang, and Qiuling Ge; urban wetland parks—by Jiani Zhang, Xun Zhu, and Ming Gao; mountain landscapes in Beijing—by Tingting Ding, Wenzhuo Sun, Yuan Wang, Rui Yu, and Xiaoyu Ge; and mountain landscapes in Western Sichuan—by Keying Ding, Mian Yang, and Shixian Luo. In addition, some papers address the issues of user preference and the relationship between public spaces and social networks more generally. This is the case of a paper on citizen satisfaction on the Beijing waterfront—by Siya Cheng, Zheran Zhai, Wenzhuo Sun, Yuan Wang, Rui Yu, and Xiaoyu Ge—and an interesting approach to social media content in an article by Diego Martín Sánchez and Noemí Gómez Lobo, who work on a case study of Tokyo's Rinshinomori Park.

In addition, this Special Issue is completed with two papers related to research projects that employ social networks to the recognition, assessment, conservation, and communication of Spanish post-industrial landscapes. One is authored by Ángeles Layuno Rosas and Jorge Magaz-Molina, and the other by Graziella Trovato.

This Special Issue closes with a paper that critically evaluates the use of predictive tools to determine the future interest in the landscapes around us, authorized by Nicolas Marine, Cecilia Arnaiz-Schmitz, Luis Santos-Cid, and María F. Schmitz.

Cecilia Arnaiz Schmitz, Nicolas Marine, and María F. Schmitz
Editors

 *land*

Article

User Perception of Public Parks: A Pilot Study Integrating Spatial Social Media Data with Park Management in the City of Chicago

Yiwei Huang [1,*], Zhixin Li [2] and Yuhan Huang [3]

1   Department of Horticulture and Landscape Architecture, Purdue University, West Lafayette, IN 47907, USA
2   Geospatial Sensing and Modeling Lab, Lyles School of Civil Engineering, Purdue University, West Lafayette, IN 47907, USA; li2887@purdue.edu
3   Department of Land, Air, and Water Resources, University of California Davis, Davis, CA 95616, USA; yuhhuang@ucdavis.edu
*   Correspondence: huan1655@purdue.edu

Abstract: User-generated content (UGC) is a relatively young field of research; however, it has been proven useful in disciplines such as hospitality and tourism, to elicit public opinions of place usage. In landscape architecture and urban planning, UGC has been used to understand people's emotions and movement in a space, while other areas and additional functions are yet to be discovered. This paper explores the capability of UGC in revealing city-scale park management problems and the applicability of social media as a future tool in bridging visitor feedback to city parks and recreation department staff. This research analyzed the spatial characteristics and patterns of Google Maps review quantity, rating score, and review comments. The results of this pilot study indicate the spatial and structural features of the Chicago parks and demonstrate distribution problems, financial investment priority concerns, park usage characteristics, and user preferences of the park attributes. Findings affirm that user-generated online reviews can be used as an alternative and self-reporting data source to effectively assess the natural performance and users' experience of city parks and can potentially serve as an evaluative tool for public park management.

Keywords: user-generated content (UGC); park and recreation; Google Maps; online views; park experience

Citation: Huang, Y.; Li, Z.; Huang, Y. User Perception of Public Parks: A Pilot Study Integrating Spatial Social Media Data with Park Management in the City of Chicago. *Land* **2022**, *11*, 211. https://doi.org/10.3390/land11020211

Academic Editors: María Fe Schmitz, Nicolas Marine and Cecilia Arnaiz Schmitz

Received: 24 December 2021
Accepted: 27 January 2022
Published: 29 January 2022

Publisher's Note: MDPI stays neutral with regard to jurisdictional claims in published maps and institutional affiliations.

1. Introduction

1.1. Urban Parks and Parks-and-Recreation in Cities

Urban parks are defined as delineated open space areas, which are mostly dominated by vegetation and water, and generally reserved for public use [1]. Parks vary in size; while most urban parks are large, some can be small and are called "pocket parks". Parks are usually defined by authorities, are typically owned and managed by their local municipality and/or government agencies, and aim to provide sports, physical activities, cultural and environmental programs to local residents and visitors [2]. Urban parks and public open spaces are crucial to livable and sustainable cities and towns. The experience of nature in urban environments can elicit positive feelings and beneficial services that satisfy the social functions and psychological needs of its users [3]. They are important assets to cities and have been shown to provide tremendous benefits to urban dwellers' wellbeing. For example, the presence of natural assets and components in an urban context can reduce stress [4] and provide a sense of peacefulness [5]. Live plantings, such as trees and grass in outdoor spaces, may promote social connectedness [6]. Trees, water, and open spaces, especially attractive ones, are also associated with higher house prices and have the potential to bring economic benefits to the surrounding neighborhood [7].

Urban parks should be inclusive for urban dwellers and visitors as the accessibility, quality, and availability of urban parks impact life in cities [8]. In the United States, the municipality's parks' management agency, typically its parks and recreation department, plays an important role in monitoring an accessible and equitable distribution of urban parks to visitors. According to the Park and Recreation Professional's Handbook, with the current challenges the world faces, including inequity, obesity, polities, and technology development among others, park and recreation professionals have many opportunities to engage with people's leisure time, understand people's needs, and continue to play a stronger role in improving lives [2]. As public amenities and a form of public investment, urban parks should serve communities fairly, especially for those with inadequate access to private recreational activities, such as low-income populations, older adults, youth, and ethnic minorities [9].

1.2. Current Park Management Strategy Challenges

A city's parks and recreation department is responsible to provide places and programming that help residents and visitors of all ages, backgrounds, and economic and social status stay healthy and learn new skills. With large quantities of facilities and huge areas to manage, doing this job likely faces many challenges. Some frequently mentioned challenges include:

(a) New methods are required to keep information up to date.

Traditionally, and even today, the standard practices to monitor park usage and performance are surveys, questionnaires, and observations. For example, Cohen et al.'s two-year study looked at the relationship between park usage, park characteristics and demographic factors. They surveyed 51 park directors and more than 4000 park users and residents, and conducted observations on 30 parks in a Southern California metropolitan area [10]. Chiesura distributed 750 questionnaires to understand the significance of nature in citizens' well-being and the contribution to the sustainability of the city [1]. In a case of neighborhood parks performance assessment in New Orleans, the researchers conducted observations on a total of 39 neighborhood parks with more than 170 activity areas. To maintain research rigidity, the observations were conducted six times per day, with a half-hour interval over a three-hour period [11]. Those methods provide a sufficient dataset if conducted right. However, they are typically limited by staff capacity in terms of time and number of employees, can be potentially costly to conduct, especially regularly, and are not always spatially explicit [12].

Many internal and external factors and changes may impact park visitation and usage, and it may be difficult to maintain up-to-date information under current monitoring methods. For example, Zhang and Zhou found that transportation accessibility is a significant factor in park usage [13]; however, city administrators normally do not conduct a survey of park usage before and after the construction of every new bus stop. The same story applies to the COVID-19 pandemic, that typical information generating methods, such as questionnaires and surveys, are not sufficient to draw any meaningful conclusions on how the usage pattern or visiting groups change over time, or under particular circumstances. Adopting and implementing new methods to generate data about gathering patterns, popular programs, and immediate concerns is an urgent task for parks and recreation departments to undertake.

(b) Data collection and park performance measurements require improvement.

As the recreational division owns or maintains a large and complicated array of programs located in different places that are aligned with different operating models, and target various customer groups, maintaining a simple and consistent way of data collection and performance measurement is challenging. Previously and currently, as aforementioned, data is often collected via surveys and questionnaires to solicit community feedback. However, those methods are often lacking in fidelity due to the lack of participation of certain populations, especially those who are marginalized. For instance, Scott and Munson's study revealed that low-income family members' park usage were limited by many reasons, including fear of crime, health conditions, transportation, and costs. Moreover, members of low-income groups have always been under-sampled, due to reasons such as busy work schedule or family duties [14]. The city of Seattle's Recreational Evaluation Plan pointed out that to prioritize recreational services for underserved communities, additional data collection and reporting is needed [15].

All data collection, analysis, and performance measurement require a certain level of educational background, professional accountability, continued training and professional development [16]. Who has the skillsets to do these analyses? How many times and how often do park and recreational staff members normally conduct a survey? How are data interpreted and how are those interpretations used for future park development and planning? How can community members be involved in data contribution and monitoring performance evaluation processes? All these questions are important but remain unanswered and ripe for novel solutions.

(c)   Environmental justice problems and efforts from planning.

Research has found that green spaces are inequitably distributed within cities. Cities in UK, Australia, Turkey, and the US, have reported that the so-called minority groups are often disproportionally displaced to areas with less access to urban open spaces, and may consequentially be exposed to greater health-related issues [17]. Byrne, Wolch, and Zhang argue in their systematic review that although many recent park usage articles attempt to explain differences in park visits based on factors such as race, gender, and age, they ignore important social-spatial factors that may support park use. Geographical variables such as residential location, park distribution, and facility supply must also be considered as potentially relevant factors for park use and as such, require more in-depth investigation [17].

To battle with existing environmental justice problems, an equity-oriented approach to landscape planning that better articulates park needs, recreational and health disparities, and park resources distribution is required [18]. Some previous research, most of which utilized ArcGIS and open city data, has begun to shed light on future planning efforts. Previous discussed topics mostly include park proximity, acreage, and park qualities [19]. These quantitative analyses show multifaceted patterns of environmental injustice. However, to retrieve feedback and perceptions from residents and affected groups, additional qualitative data needs to be acquired for further analysis.

(d)   Insufficient budget and financial investment.

Another factor is sometimes drawn from insufficient funding from public entities. Takyi and Seidel showcased a case study of parks in the city of Vancouver to illustrate the fact that the indirect economic values of urban parks make it difficult to represent their financial benefits. This affects the ability to assess the true costs and benefits for decision makers. This adversely impacts the level of investment in the ongoing development of the park, thereby limiting sustainable management of the entire park system [20]. To become more effective in dealing with rising costs associated with providing basic services, park and recreation agencies have had to become more business-like [21], which may lead to uneven attention to all the parks in city.

1.3. User-Generated Content and Its Potential to Contribute to Landscape Governance

With the development of science and technology, as well as the proliferation of electronic device usages in daily life, the forms of information we can gather have also changed. There has been a shift in mindset about how to collect and analyze data in public re-digital formats to obtain better and more innovative results. User-generated content, or UGC, is one form of data that has effects on society, economy, and individuals [22]. According to Wyrwoll, UGC is content that is published on online platforms by users, through a process that does not require users to be equipped with programming skills. Social media then comprises platforms that contain user-generated content [22].

The innovation of UGC is that it consists of different forms of data, which enlarges the scope and aspects of the data characteristics. It may help researchers further examine the correlations of data content with other information, for example, demographic records. A traditional UGC unit consists of core data, or the content, and metadata, or the information about the given piece of information, such as the date and time of publication, the associated author information, and the number of views [22]. Moreover, one of the benefits of UGC is that almost all the content is voluntarily uploaded by users, so the content itself is unobtrusive, and reduces the researchers' need to be directly involved in data collection.

In terms of disciplines, journalism, computer science, media and culture, marketing, hospitality, and tourism are employing UGC research. The most researched social media platforms are Twitter, Flickr, YouTube, Facebook, and Instagram, to name a few. User-generated content is being used to understand customer needs [23], such as how it may change users' behavior and travel habits [24], and how it provides first-time users the opportunity to understand a place by exploring the descriptions and opinions from others [25].

The authors believe UGC has potential to contribute to a better understanding of environmental experiences and landscape governance, and in turn with the generated information, to help parks and recreation staff more efficiently manage entities within their city scale. In the field of landscape architecture, several social media platforms have been studied and have contributed to the understanding of user movements, perceptions, and feature popularities within or outside parks. Examples include Flickr and Twitter data which can showcase human visitation dynamics and indicate the equitability of park access [26]. Flickr images can be analyzed to explore people's perception and attitudes towards a phenomenon in city parks [27]. Instagram posts can be collected and coded to understand users' emotions and activities associated with specific park features [28].

There is currently limited research utilizing Google Maps user reviews to understand park management deficiencies or visitor feedback on park conditions. However, Google reviews have been utilized in other fields, to examine airport service quality [29], restaurant service and customer's eating experience [30], students' educational experience and their attitudes towards quality of teaching, course design, learning environment and support received [31], and so on. Google Maps user reviews have also been used for branding tourist destinations and to predict public perceptions of visiting places [32]. However, as reported, few of these studies have focused on using the core data, the content metadata, spatially explicit information, or the other values associated with the core data.

This project is a pilot study intending to make breakthroughs in this area, through further use of metadata, especially the spatial attributes of core data, and to analyze the relationship between core data, metadata, and other data that exists within the city boundary on websites. This analysis may help to support the use of UCG as a relevant assessment tool for future park management of any city. The objective of this paper is to use Google Map reviews of Chicago public parks as an example of UCG to determine a relationship between popularities of reviews together with their spatial pattern, most-discussed topics, and the corresponding relationship with household incomes, population distribution, and the equality of regional development.

2. Data and Methods

2.1. Study Area

The authors chose the city of Chicago to conduct a pilot study for several reasons. Chicago has long been an experimental mecca for urban design, planning, and landscape architecture. In Dreaming the Metropolis [33], Cronon described the importance of land geography in Chicago, and how the location of resources, transportation routes, and culture shaped the city to what we know today. Chicago also has geographically related inequities including health disparities [34], healthy food access [35], and transportation and mobility issues [36]. Chicago also has one of the largest urban public park districts in the world staffed by 3000 full-time and 3000 seasonal employees in the 2000s [16]. The Chicago Park District now has the stewardship of more than 8000 acres of open spaces, more than 570 parks, 30 plus beaches, and 50 nature areas [37]. Moreover, the city of Chicago has public demographic data that are accessible to researchers, and Chicago's parks also have received large quantities of reviews on Google Maps which made the quantitative analysis abundant in samples.

As aforementioned, the social media platform for this research is Google Maps. All the parks analyzed in this research were registered as parks under the official Chicago Park District website [38]. The core data reviewed in this paper are the review content, including the comments and review scores, which range from 1 to 5, with 1 being the lowest and 5 being the highest for their satisfaction with the Chicago urban parks. The boundary of this project follows the city boundary set by the planning and development department, as shown by Figure 1. The reviews of the parks that were studied are drawn from the amenity lists from the Chicago Park District. Pertinent to this paper, the Google review average score, numbers of ratings, as well as the first five reviews are public data that can be collected by anyone (see Figure 2).

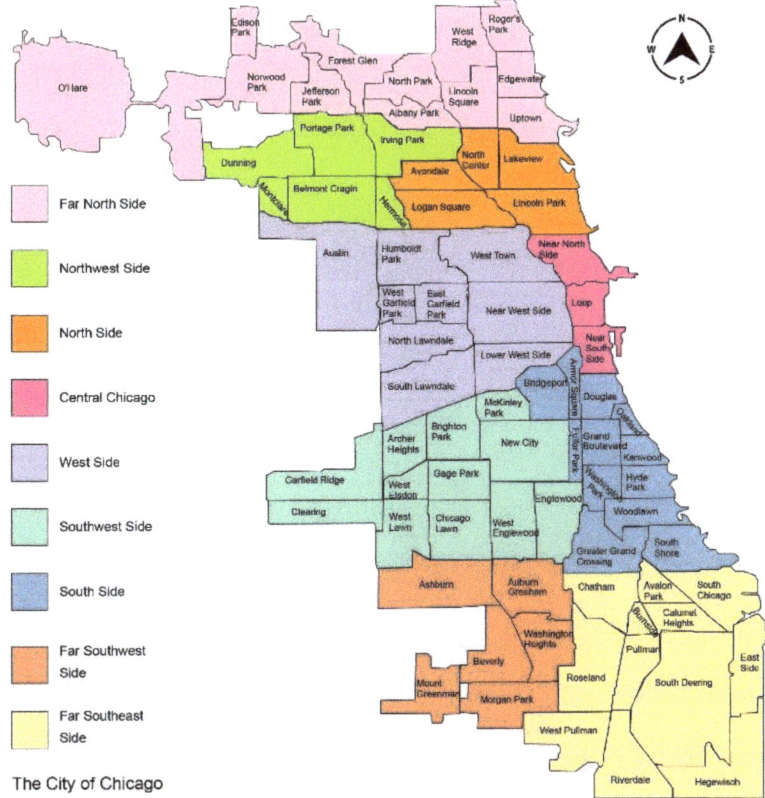

Figure 1. Chicago neighborhood maps. Adapted from Wiki commons.

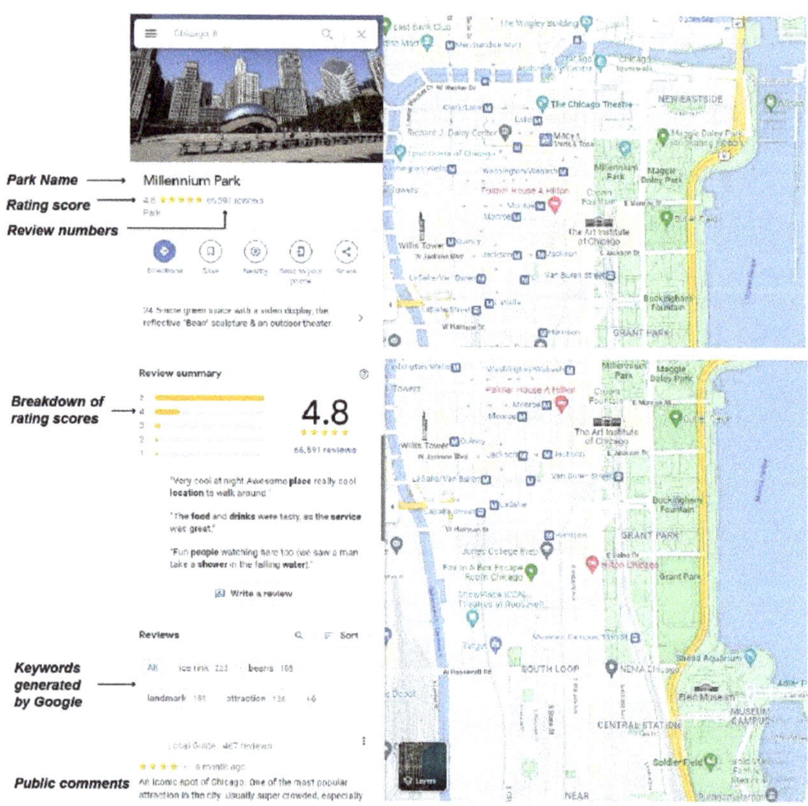

Figure 2. Google Map Reviews interface example.

2.2. Data Collection

As a web-mapping service, user-generated reviews from Google Maps have previously been used to analyze customer perceptions of theme parks, restaurants, and libraries [39,40]. We suggest that these reviews may also help understand the spatial patterns and the user experiences of the public parks under study.

There has been a dramatic increase in the number of Google Map reviews received since 2015 as compared with other review platforms [41]. Pertinent to this research, when compared with other platforms such as Yelp or TripAdvisor, Google Maps also received more reviews for public parks, especially for community parks that are relatively small, providing inclusive information and samples for further analysis.

We retrieved records of 605 public parks in Chicago through Google Maps Application Programming Interface (API). Based on the names and addresses of the public parks from the Chicago Park District website (https://www.chicagoparkdistrict.com/ (accessed on 6 July 2021)), as illustrated in Table 1, we collected the park attributes, geolocations, park ratings, as well as the frequency and the content of reviews for each park up from the beginning time that the review is available to August 2021. The collected attributes of the public parks were further converted into spatial points with their corresponding structural and non-spatial information in the Environmental Systems Research Institute (ESRI) Shapefile format. We further extracted keywords of the collected reviews for each individual park along with the total number of reviewers who mentioned the keywords in their reviews (Table 2).

Table 1. Park information collected through the Google Maps API.

Attributes	Description
name	Park name.
formatted_address	A string containing the human-readable address of this park.
place_id	A unique identifier of this park, which can be used with other Google APIs.
rating	Park rating, from 1.0 to 5.0, based on aggregated user reviews.
user_ratings_total	The total number of reviews of this park, with or without text.
lat	Latitude of this park in decimal degrees.
lng	Longitude of this park in decimal degrees.
url	The URL (Uniform Resource Locator) of the official Google page for this park. This is the Google-owned page that contains the best available information about the place.

Table 2. Key words from the Google Maps Reviews.

Attributes	Description
place_id	A unique identifier of this park, which can be used with other Google APIs.
key_words	Key words mentioned by multiple reviews that can label the features of a park.
Kw_mentioned	Total number of each keyword for all reviews.

To explore the relationship between the pattern of public park distribution and the socioeconomic conditions of the surrounding communities, we also collected the 2019 household income information of Chicago residents at the census tract level (https:// datausa.io/profile/geo/chicago-il (accessed on 6 July 2021)), 100 m gridded population structure data in 2020 from WorldPop (https://www.worldpop.org/ (accessed on 6 July 2021)), 38 m human settlement history layer showing the presence of built-up in different epochs [42], and the human modification layer in 2016 which shows the percentage of human activities, such as urban infrastructure, agriculture, mining, or transportation, in each 1 km pixel [43]. We further summarized the map of the percentage of children under the age of 15 from the WorldPop population structure data.

2.3. Methods

2.3.1. Web Crawler and Web Content Parsing

The web crawler, also known as a spider [44] or an automatic indexer [45], is a powerful technology that collects data from web sources by iteratively extracting web contents from a list of URLs, which are also called seeds. In our data collection process, the URLs for all parks shown in Table 1 are considered as the seeds of the web crawler and corresponding web pages are stored by accessing these seeds. Since web pages are built using text-based mark-up languages (HTML and XHTML), with data distributed in the contents, same class information is typically encoded into similar pages by a common script or pattern. After crawling the web pages, we identified and scraped web elements with the targeting information using the Selenium package in Python. All data stored in web elements found by Selenium were saved and cleaned using regular expression [46] to remove the redundant and noisy records. Figure 3 shows a subset of the data collected and cleaned:

place_id	key_words
ChIJvX8aF-ElDogRws8sWkRmL60	All kids:18 house:9 play:8 walk:8 pool:5 basketball:5 peaceful:4 community:3 baseball:3 restaurants:3
ChIJqXs7PlckDogRS2CeQDq4BAY	All kids:18 play:6 basketball:6 football:4 swimming:4 house:3
ChIJb2Z0dB_TD4gRIOp_StKNk80	All sandbox:15 splash pad:11 toddler:8 swings:6 clean:6 water feature:5 structures:3 parents:3 slides:3 benches:3
ChIJOeKUdVMrDogRmMZVEmCCSdE	All beach:10 walk:6 benches:5 lake:4 trees:4 bike path:3 water:3

Figure 3. A subset of the collected and cleaned data. The left column represents the unique park identifier and the right column contains the key word information.

2.3.2. Kernel Density Estimation

As an important nonparametric technique in statistical analysis, kernel density estimation (KDE) is used to estimate the probability density function of a random variable [47]. Kernel density estimation has been widely used for multiple purposes such as spatial data smoothing, hot spot detection, and risk prediction [48–50]. When dealing with geospatial information, KDE generates a density surface where each cell is rendered based on the kernel density at the pixel center. For each observed geographic point, KDE fit a kernel function, assuming that each observation is continuously spread within its kernel window. Given by n observed points p_i, the predicted density $\rho(x)$ at a new location x is determined by the following formula:

$$\rho(x) = \frac{1}{r^2} \sum_{i=1}^{n} \frac{3}{\pi} pop(p_i) \left(1 - \left(\frac{dist(p_i, x)}{r} \right)^2 \right)^2, \, for \, dist(p_i, x) < r \tag{1}$$

where r represents the search radius, function $pop(p_i)$ represents the population field, which serves as the weight on each observation, and $dist(p_i, x)$ computes the distance from each location x to the observation p_i.

Following Equation (1) above, we further generated the equally weighted kernel density map in ArcGIS Pro to show the spatial pattern of the public park distribution in Chicago, with each pixel on the resulting image indicating the number of public parks per square meter.

2.3.3. Global Moran's I and Getis–Ord Gi* Statistics

To understand the spatial patterns and to validate the significance of the park ratings and of the number of ratings of the different parks reviewed, we conducted spatial autocorrelation analysis and calculated both the global Moran's I and Getis–Ord Gi*. In general, Moran's I compares the similarity of the value at the current location with its adjacent locations [51]. Getis–Ord G* identifies spatial clusters where high or low values are observed [52].

We used global Moran's I to measure the spatial autocorrelation among ratings and the number of ratings received for each park based on its location and the value simultaneously. This spatial autocorrelation analysis measures the overall pattern of the Chicago public park distribution, which ranges from clustered, random, to dispersed [53]. The Moran's I statistic for spatial autocorrelation is calculated as:

$$I = \frac{n}{S_0} \frac{\sum_{i=1}^{n} \sum_{j=1}^{n} w_{i,j}(x_i - \mu)(x_j - \mu)}{\sum_{i=1}^{n} (x_i - \mu)^2} \tag{2}$$

where x_i is the value of the observation i, n is the number of the observations, μ is the mean of the observation, w_{ij} represents the spatial weight between i and j, and S_0 is the aggregate of all weights.

A large and positive Moran's I indicates a high similarity between the parks and their adjacent parks in terms of the rating or number of ratings received, and a negative value represents the dissimilarity when compared with adjacent parks.

The Getis–Ord Gi* Statistic evaluates the significance of local pattern and clusters of public parks [54,55]. By testing each park within the context of neighboring features, Getis–Ord Gi* identifies the clusters. An observation with a high value and surrounded by high-value points can be called a statistically significant hot spot and can be identified by the Getis–Ord Gi* statistic. The Getis–Ord Gi* can be calculated as following:

$$G_i^* = \frac{\sum_{j=1}^{n} w_{i,j}x_j - \sum_{j=1}^{n} w_{i,j}x_i}{S\sqrt{\frac{n\sum_{j=1}^{n} w_{i,j}^2 - \left(\sum_{j=1}^{n} w_{i,j}\right)^2}{n-1}}} \tag{3}$$

where x_i is the value of the observation i, n is the number of the observations, w_{ij} represents the spatial weight between i and j, and S is calculated as

$$S = \sqrt{\frac{\sum_{j=1}^{n} x_j^2}{n} - \left(\frac{\sum_{j=1}^{n} x_j}{n}\right)^2} \tag{4}$$

In our analysis, a high value of Getis–Ord Gi* indicates that the total park ratings or the total number of people evaluating parks in the neighborhood is high relative to the average of all public parks in Chicago. Likewise, a negative value indicates a low value cluster and a value approaching to 0 means the intermediate condition.

2.3.4. Review Keywords Analysis

To reduce the dimensionality of the scraped park reviews and explore the features of the public parks in Chicago in terms of management improvement, keywords of park reviews and their corresponding frequency of being mentioned were exploited to explore what features are of most concern to visitors. First, the keyword frequencies for all parks were aggregated together and the top 10 mentioned keywords were visualized to give a big picture of the park features of most concern. Second, we stratified park reviews according to their ratings to illustrate the potential differences in park attributes, conditions, and environments that lead to differences in park ratings. The pie charts of keywords are then generated for parks with ratings ranging from 1 to 2 (1 park), 3 to 4 (22 parks), and 4 to 5 (329 parks) stars (scores). When we scraped the review data, there were no parks in Chicago rated between 2 to 3 stars on Google Maps and some parks do not have reviews. Finally, 6 parks in different geographical locations are manually selected as examples to illustrate differences of features that visitors mentioned.

3. Results and Conclusions

3.1. Spatial Patterns of Public Parks in Chicago

Overall, both park location and the most frequently rated parks are significantly clustered by Lake Michigan, with a densely populated zone of public parks extending from North Side to Far Southeast Side of the city. The spatial patterns seem be controlled by the distance to the urban infrastructure, local socioeconomic conditions, and park users' behaviors. Figure 4 shows the kernel density map of the public parks in Chicago. Central Chicago and west of the West Side have the most observed dense park distribution in the city.

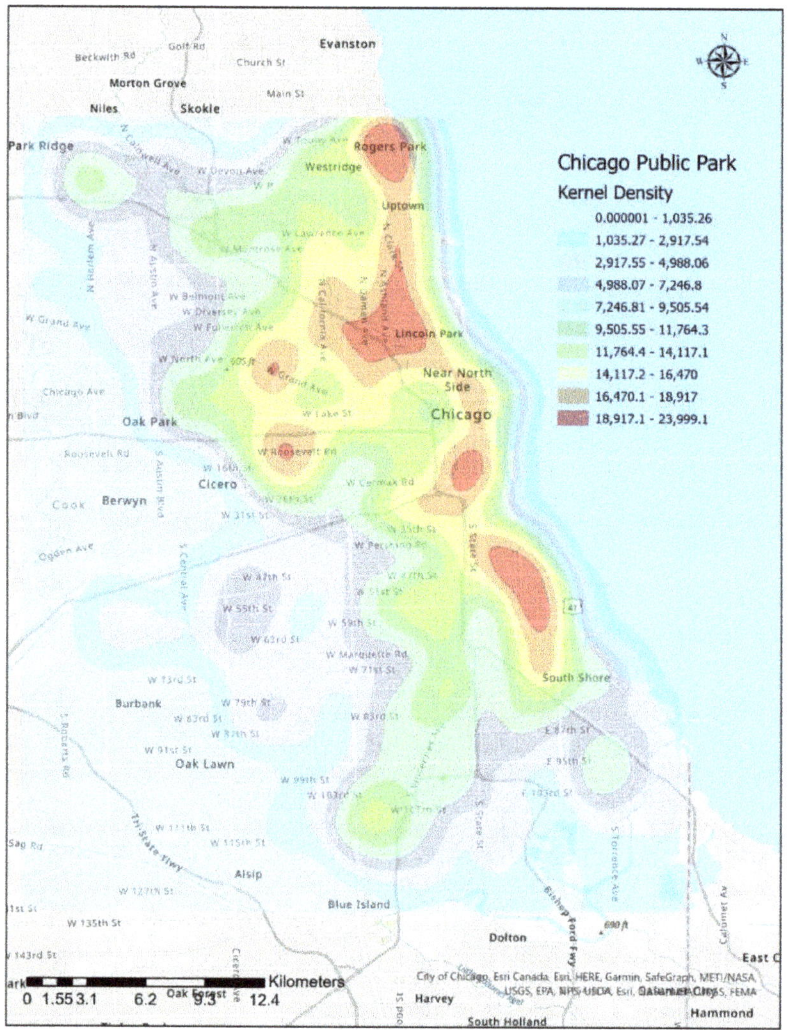

Figure 4. Kernel density map of public parks in Chicago.

The results of the spatial autocorrelation analysis indicate the patterns of park ratings and rating numbers at both the global and local scale. In terms of the global spatial autocorrelation, as summarized in Table 3, there are clustered distributed patterns for park ratings and rating numbers for all public parks in Chicago. This finding implies that high-rating parks are located close to each other in space and the most-visited parks are also spatially clustered together. For the patterns of park rating, an extremely high critical score (z-score) of 4.4 and a very small p-value of 0.000011 were received from the statistical test, representing that it is statistically significant and allowing us to reject the null hypothesis that the pattern is randomly distributed. Parks with similar ratings are thus highly clustered across the space and there is less than 1% likelihood that this clustered pattern is a result of random chance. As for the number of reviews made for each park, a z-score of 2.13 and p-value of 0.033 indicates a less than 5% likelihood that this clustered pattern occurred by chance.

Table 3. Global spatial autocorrelation results.

Attributes	Moran's I	z-Score	*p*-Value
Park rating	0.089	4.40	0.000011
Park rating numbers	0.035	2.13	0.033

In terms of the local distribution, the hot and cold spots with respect to the park ratings and rating numbers were detected through the Getis–Ord Gi* statistic. A hot spot on the map represents a cluster of parks with high ratings and a cold spot refers to low-rated parks. We further overlaid the quantified hotspots with the 2019 household income information of Chicago at the census tract level shown as the grayscale color scheme base map in Figure 5, to illustrate the relationship between socioeconomic conditions in the neighborhood and the park rating ranking distribution. The detected hot and cold spots with respect to the park ratings and rating numbers are displayed as red and blue colors, respectively. The identified hot and cold spots measure the relative degree of parks being high or low in park ratings. Across all the public parks in Chicago, there is an average score of 4.36 on Google Maps Reviews, indicating there are a higher proportion of parks falling into a high-score range in visitor perceptions (see Figure 5).

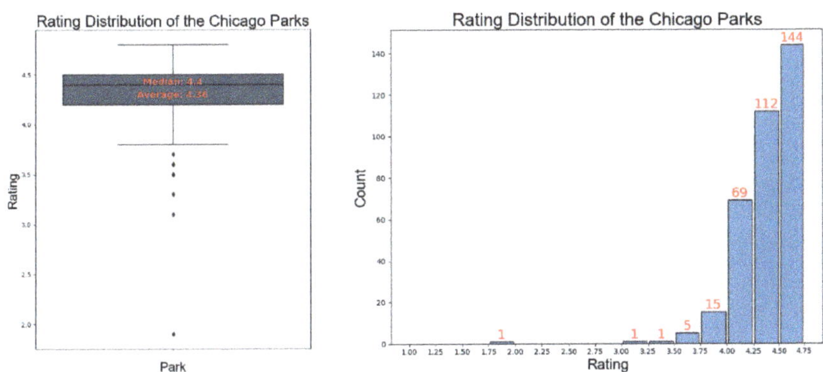

Figure 5. Boxplot and histogram of Chicago parks' ratings.

According to the U.S. census tract household income information, the level of the household income is directly identified in the base map of Figure 6a,b, showing that people with a high level of income (USD 100 k~150 k annually) are more likely to live in several neighborhoods, including Central Chicago, the North Side, the West End of the Far North Side, Hyde Park in the South Side, and several neighborhoods in the Far Southwest Side. As seen in Figure 6a for park rating clusters, which indicate the spatial autocorrelations of park ratings, there are two significant hot clusters of high-rated parks by the Hyde Park area and the North Side, and three significant cold clusters in the inner city, which are in the West Side, and the Southwest Side, and some in the Far Southeast Side. While it is intuitive to assume people live in a neighborhood that has both positively and negatively rated parks, the distribution that we found implies that people living in hot clusters have a much higher possibility of visiting parks that are all highly rated (with parks rated on an average of 4.51 out of 5). On the other hand, people living in the cold clusters are less likely to have opportunities and access to good-quality parks that are found in hot clusters if relying on walking or biking distances, and by virtue of location, are more frequently proximal to parks with low scores (with an average rating of 3.06 and 2.45 respectively out of a 5-score system). A comparison of the average scores among these cluster indicates that, although most parks in Chicago received a relatively high score in this Google rating system, those cold clusters identified from our analysis were still, under common sense, poorly rated online and had a relatively large difference from parks in the cluster of highly rated parks.

The clusters of high-rating score parks on the map also correlate with income levels in obvious way, which reaffirms the environmental justice issues discussed in the Introduction section of this paper. Hence, some of the local socio-economic neighborhoods with lower household incomes have lower quality parks and thus have a reduced chance to reap the benefits of nature, which reaffirm an urban environmental and social injustice issue.

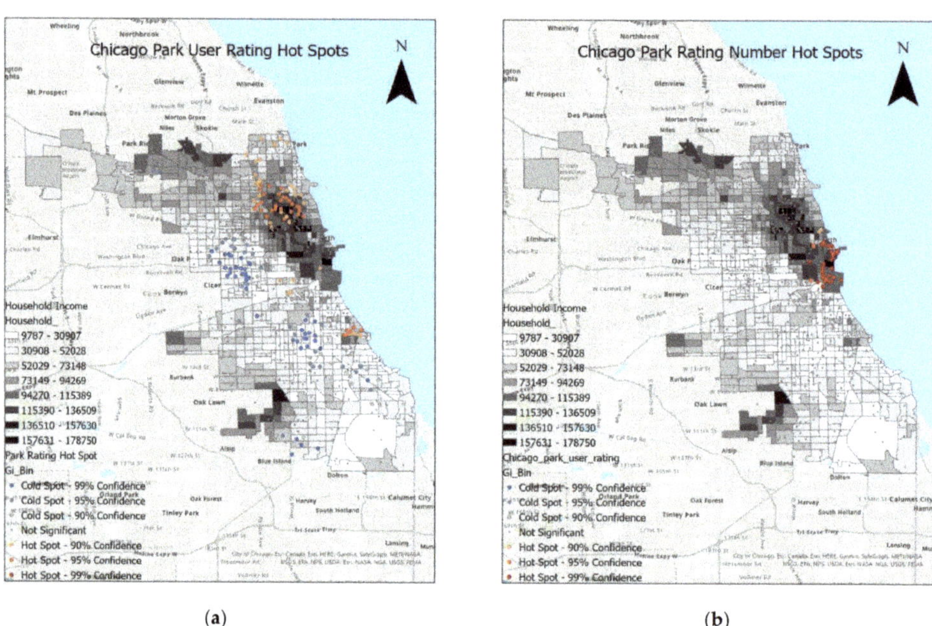

(**a**)   (**b**)

Figure 6. The hot and cold spots detected through the Getis–Ord Gi* statistic with respect to the park rating scores (**a**) and rating numbers (**b**). The household income of census tract in 2019 is colored by grayscale.

In terms of the review and rating quantities, as shown Figure 6b, only one hot spot for rating numbers is identified, which is in the downtown area, in the Central Chicago neighborhood close by Lake Michigan. This finding indicates that public parks located in downtown Chicago have been visited and reviewed the most. This makes sense, as they are likely more accessible to greater numbers of people by virtue of their central location, including tourists, who tend to cluster in downtown Chicago to visit its myriad attractions. In addition, the spatial patterns of park rating and rating numbers also indicate that areas with the most visited parks are not necessarily the places with more high-rating parks. The downtown area in Central Chicago seen Figure 6b has a cluster of hot spots of reviews with high statistical confidence but as mentioned above, has one hot spot of high-rating parks on Figure 6a, which means the downtown area has parks with varying levels of ratings. Conversely, neighborhoods with clusters of high- or low-ratings on Figure 6a, for example the neighborhoods in the North Side and the West Side, had no significant hot spots based on the number of reviews on Figure 6b.

We further examine the spatial relationship between the hot spots and cold spots of park ratings with the population, the percentage of children, the percentage of human modification, and the age of the urban built-up, as shown in Figure 7. Overall, clusters of high-rating parks are within the zones of higher populated areas when compared with those clusters of low-rating parks (Figure 7a). Clusters of high-rating parks are also more likely to be in the regions where urban built-ups were constructed before 1975 while most low-rating parks are in urban regions built in between 1975–1990 (Figure 7c). Although there is little difference between the hot and cold spots in terms of their spatial distribution across the maps of child percentage and the degree of human modifications (how much human activity has changed the wilderness), all of these park clusters are within regions where there are higher percentages of children and higher human modification than all other parks that are sparsely distributed (Figure 7b,d).

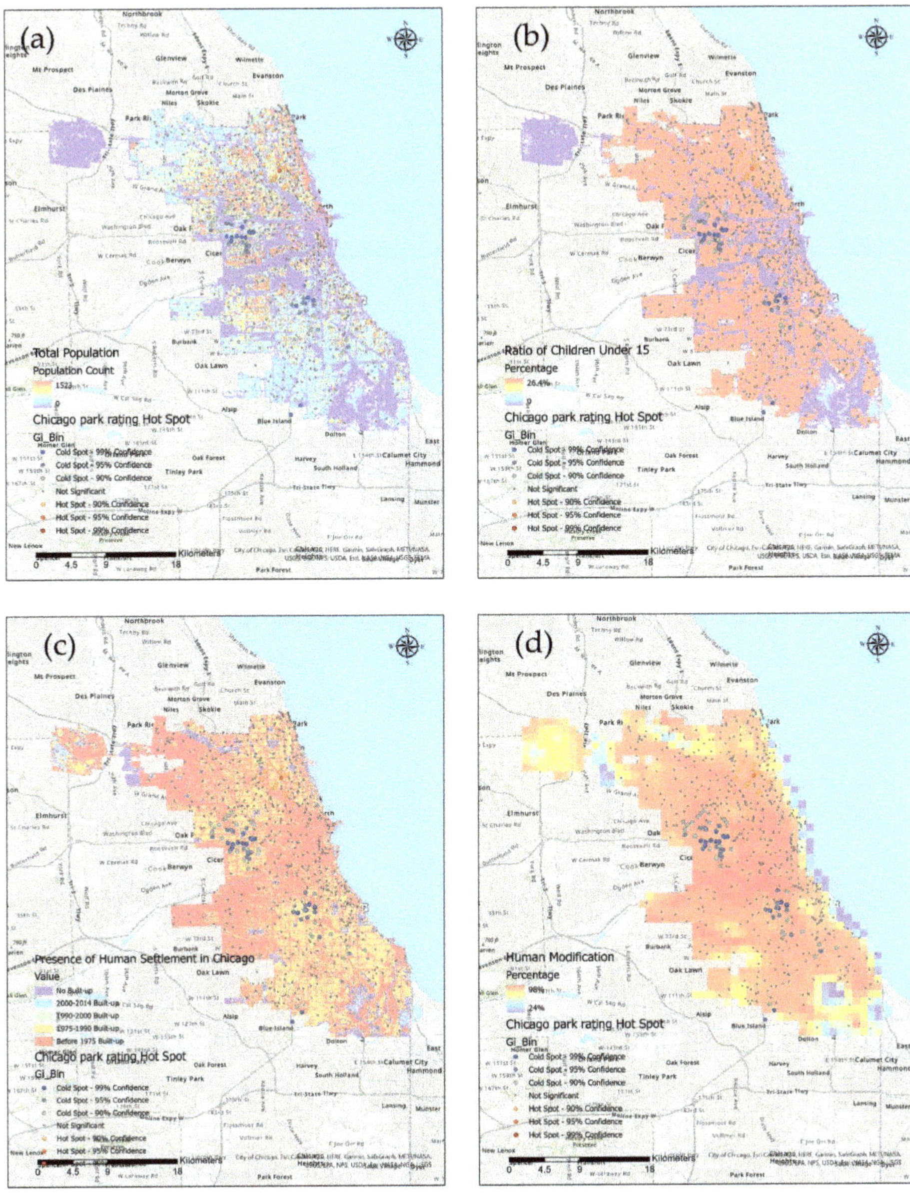

Figure 7. Maps of the relationship between park ratings of Chicago and the spatial patterns of (**a**) the total population of Chicago, (**b**) the percentage of children in the population who are under 15 years old, (**c**) the epoch (before 1975, from 1975–1990, from 1990–2000, and from 2000–2014) of presence of the human built-up, and (**d**) the percentage of human modification (how much human activity has changed the wilderness).

3.2. Keywords of Park Reviews by Different Ratings

As shown in previous figures, the review content can be categorized by its rating scores, and further analysis can be done to understand the reasons behind higher or lower reviews. To develop the overall picture of the park characteristics, the top 10 most-mentioned keywords with respect to frequency are visualized in Figure 8. Referencing Figure 9, the pie charts of the keywords were generated for parks with the star ratings ranging from 1–2 (low-rating), 3–4 (medium-rating), and 4–5 (high-rating). Although there are more parks falling into the star rating range of 4–5 and fewer in the range of 1–2, we still use these numeric splits instead of the relative values of even splits based on the differences of park ratings, to partition the keywords of the parks, in order to better simulate the actual opinion of park users under the general intuition that a high rating represents a good quality while a low rating implies dissatisfaction from previous visitors. The larger slice of the pie chart represents a higher frequency of the word mentioned by park reviewers.

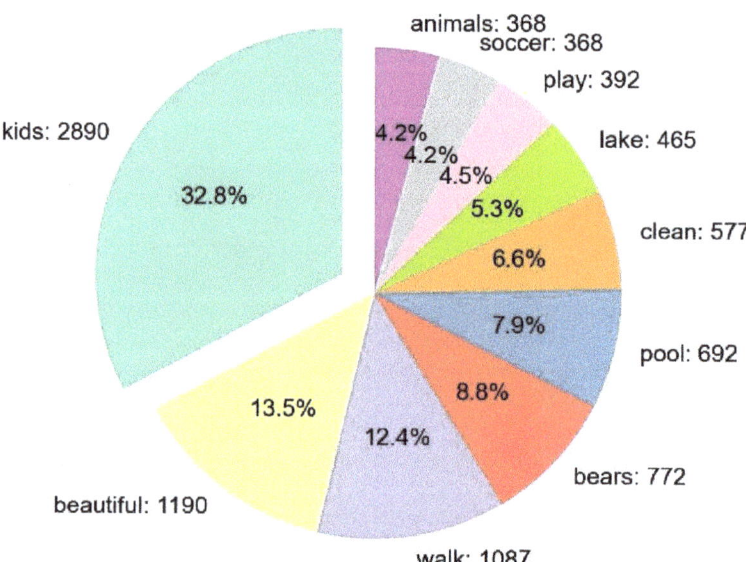

Figure 8. Pie chart of the top 10 park reviews: keywords frequency for all parks.

Results of the pie charts indicate that one park may be poorly perceived or unappreciated by users because it might be occupied by people who are unhoused and have an abundance of trash. Overall, 'homeless people' and 'bottles' are the two keywords degrading the rating of a park while 'kids', 'beautiful', and 'walk' are the three most-received keywords for high-rating parks. People might rate a park as average due to a mix of positive reasons, such as children's play equipment, sport fields and courts, and negative reasons, such as gang occupancy and the presence of people who are unhoused. We also identified more sports-related keywords such as ball game, courts, pool, and gym for the medium-rating parks than the high-rating parks. As for high-rating reviews, people used words such as beautiful and clean to describe the characteristics of the park, and used keywords such as animals they see, soccer, play, animals, beach, walk to indicate their favorite activities in the park. Many of the high rating key words were related to nature and the park amenities it provides. Park users might view these qualities as health-promoting environments.

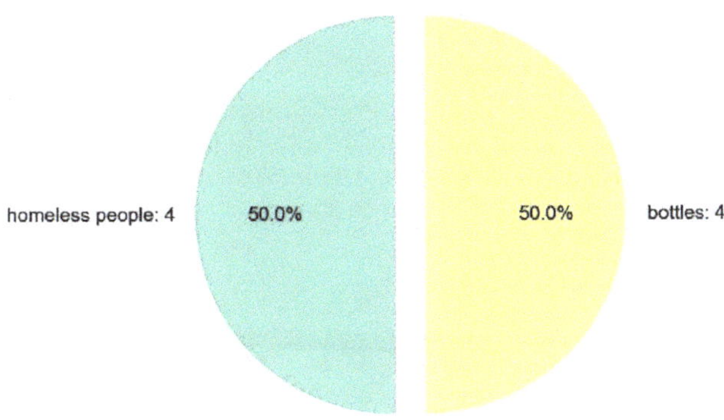

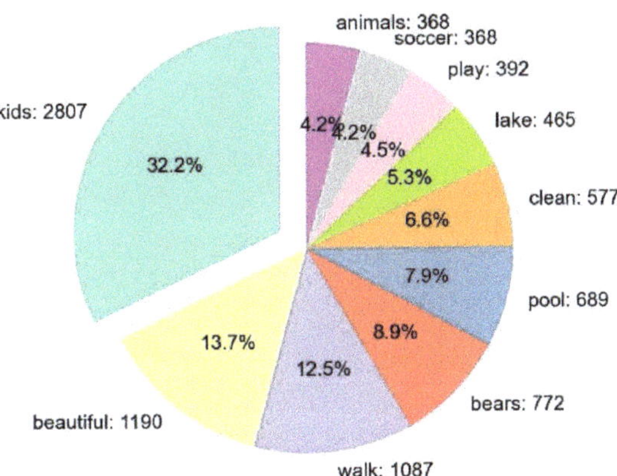

Figure 9. Pie charts of the park review keywords frequency by different ratings. From top to bottom: rating 1 to 2, rating 3 to 4, and rating 4 to 5.

In addition to the overall keyword distribution of the entire research area, six evenly distributed parks through Chicago (Burnham Park, Grant Park, Horner (Henry) Park, Marquette (Jacques) Park, McKinley (William) Park, and South Shore Cultural Center) were manually selected to explore the park features and characteristics for a more detailed analysis and individual park comparison. As shown in Figure 10, three parks are along Michigan Lake (Burnham Park, Grant Park, and South Shore Cultural Center), and three parks (Horner (Henry) Park, Marquette (Jacques) Park, and McKinley (William) Park) are located inland. We generated 6 pie charts of the top 5 mentioned keywords respectively for each park in Figure 11 to show the frequency and proportion of different types of feedbacks from visitors.

Figure 10. Distribution of 6 manually selected parks for keywords analysis.

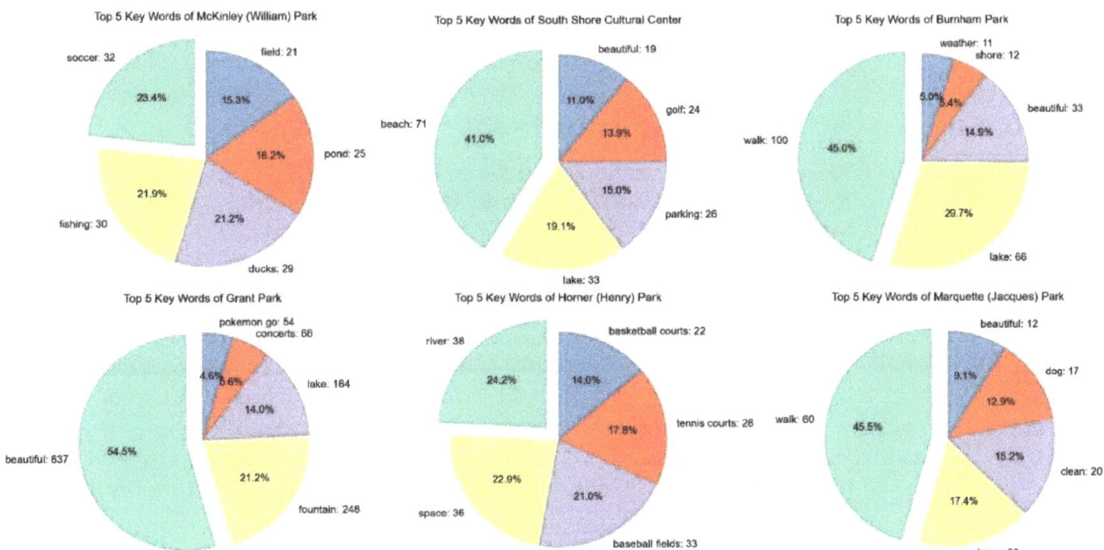

Figure 11. Pie charts of 6 selected parks for keywords analysis.

These pie charts demonstrate that each park has provided dramatically different features that attract visitors. For example, Grant parks have six features that share similar weights, including soccer (activities), fishing, field, pond, and ducks. People prize their visits to Burnham Park mostly because of its walking experience, its lake, and its beautiful view. Users like Horner (Henry) Park due to its river, space, and its basketball court, tennis court, and baseball fields. This keywords analysis would demonstrate a rough but bold picture of different parks, and almost provide a short summary of the characteristics of each park.

4. Discussion and Conclusions

As shown by the Chicago public parks Google Maps reviews, if well-utilized and effectively monitored, they can be a valuable tool to be integrated into the current city park system management. To respond to the previously mentioned challenges that parks and recreation departments are facing, Google Map reviews have several characteristics that are complementary to existing evaluation frameworks and strategies mentioned in the Introduction:

(a) The evaluation and commenting are continuously live; hence, the information is always up to date.

According to Google Product Director Russell, various channels are available for people, business owners and consumers, and others to update map data and leave comments. Google reviews of public parks are updated instantly, every time a visitor submits a response and Google Maps is also updated constantly [56]. Therefore, parks and recreation administrators have the capability to monitor users' perception of the parks by simply reviewing comments and monitoring the most recent scores of all city parks. What is invaluable is that no additional effort is required to distribute surveys and analyze the results, the feedback portal is always open, and the information is always current.

(b) Social media, especially Google Maps, is far-reaching, allowing any community members to contribute.

Social media is widely used worldwide; hence, in terms of accessibility, social media has the potential to become the most far-reaching and participatory tool in research. According to the Pew Research Center's report on social media usage from 2005 to 2015, 65% of adults use social networking sites [57]. In terms of the social disparities aforementioned in discussion of survey participation, low-income families are consistently under- sampled in traditional methods [14]. Individuals with a higher level of education and higher household income still lead the way, but more than half (56%) of the lowest income household residents use social media. Race and ethnicity are another impacting factor when public hearings and design charettes are the methods used. Yet, according to the Pew Center's research, there is no notable difference between racial or ethnic groups who used social media, with whites, Hispanics, and African Americans having 65%, 65%, and 56% use respectively [57]. We are not saying that all people who use social media will contribute to Google reviews; however, in terms of accessibility, it may be easier to leave comments on social media than physically participating in a public workshop, or submitting another online survey. People, regardless of their social status, post their opinions on Google Reviews, when they have positive and negative feelings towards parks, if they have a cellphone and Internet access. If the technology part can be bridged, Google Review has potential to become a more far-reaching opinion gathering tool than any other applications or digital survey tools.

(c)   Social media data reveals and support discoveries of environmental justice issues.

The research findings shown in Figures 6 and 7 illustrate the relationships between highly rated parks, poorly rated parks, household incomes, population densities, and level of urban development. Though only one example and one aspect, this indicates the possibility of integrating social media data to show systematic environmental injustice issues in more spectra. Echoing with previous research that the analyses of park quality could inform planning decisions [17], UGC offers an alternative way to get an overarching visitors' perception of park qualities. The method utilized in Figure 9 demonstrates the qualitative potential to roughly exhibit issue keywords for researchers and data collecting staff to start with. However, due to UGC's incomplete nature, the comprehensive factors behind environmental justice issues and limited park usage from particular groups require other types of research, such as focus groups, interviews, and participatory action research.

(d)   With thorough and professional analysis, social media records and results may even guide a city's finance and renovation priorities.

Figures 6 and 7 gave park managing staff a quick sketch of the general public's perceptions on what they like and where the parks need to be renovated. Currently, Google Review has not been widely used in the field, and the present results are not close to comprehensive or detailed. However, suppose a park district utilizes a similar UGC component in the future to solicit and encourage residents and visitors to actively offer feedback on the maintenance and status quo of all city parks. In that case, it may reveal which parks currently attract social problems, what problems they are, and how the city can improve those conditions. With the help of social media, even for cities having more than 600 parks to maintain, the findings in Figure 6 offers clear guidance on where poorly rated parks are located and the neighborhoods that need more financial investment to improve people's well-being and daily recreational opportunities. Park and recreation administrators can then use the information to identify concentrated poorly rated parks and invest money and social capital to improve them.

It must be acknowledged that, like other UGC, Google Maps review data has its limitations and cannot be a sole source of information for park and recreation management. That said, some interesting characteristics we examined are:

(1)   Reviews, and sometimes rating scores, tend to be polarized.

Research has found that social media is related to political polarization and disinformation [58]. In this project, we found the reviews to be polarized. People tended to leave comments that are either extremely positive or extremely negative. For example, in one of the parks with 2649 valid comments, 1647 comments are associated with a rating of 5/5 scores, 564 comments rated 4/5, 249 comments rated 3/5, 71 comments rated 2/5, and 112 comments rated 1 out of 5. In this case, the scores are not polarized. However, the associated comment lengths can be greater when there are scores of 5 or 1. People tend to share more when they have more to comment about, which is also one of the characteristics of participating in social media.

(2) Data mining and scaping can be tedious and require professional skills and training.

Social media has characteristics like unstructured data, is subjective, and must be solicited from a massive database. To use the data correctly and effectively, more than one research method is needed, such that extraction, coding, content analysis, and identifying relationships and statistical cluster analysis are essential [59]. This requires that those who handle the data have enough training and experience to make the data useful. It also requires other research team members to have additional eyes on data to make sure that the analytical procedures are ethical, and that the data responds to the clear objectives of the research. These criteria might be challenging for park and recreation departments; however, this encourages forging a collaboration of park and recreation departments with research institutions or local universities.

(3) Review quantities are key to convincing results. For parks with less comments, there is greater potential for biased results.

Social media data and big data also has "noise," which include advertisements, marketing messages, robot-produced content, and/or non-relevant conversations. Some social media have over 70% non-relevant or noise message content [60]. Hence, there needs to be a large enough sample of reviews to reduce bias and increase objectivity. In this research, the noise was between 20–40%, leading to a park with 600 reviews generating less than 400 reviews to work with. To expand the effectiveness of social media research about park usage, extra effort is needed to ensure widespread, effective participation of park users. Moreover, reviewers' identity is difficult to identify, which may lead to false calculations or false results of the research.

(4) Reporting bias still exists.

Though social media is far reaching and include more information and reach a wider community than those who typically could participate in a community workshop, UGC still has its natural reporting bias. Online reviews are inherently incomplete since they would not capture the opinions of those who do not have access to internet, or who do not leave their comments or write a review [61]. However, in some cases, these biases can be eliminated by different strategies. As an example, research finds that people tend to leave comments when they are satisfied compared to unsatisfied, suggesting that the bias can be rectified by an inverse probability weighting approach [61]. Moreover, UGC can be combined with other research methods, such as traditional sociological methods including interviews and focus groups, and participatory methods, to retrieve additional data from particular populations who are underrepresented in the world of social media.

Author Contributions: Conceptualization, Y.H. (Yiwei Huang) methodology, Y.H. (Yiwei Huang), Y.H. (Yuhan Huang) and Z.L.; software, Z.L.; validation, Y.H. (Yiwei Huang) Z.L. and Y.H. (Yuhan Huang); formal analysis, Z.L. and Y.H. (Yuhan Huang); investigation, Y.H. (Yiwei Huang), Z.L. and Y.H. (Yuhan Huang); resources, Y.H. (Yiwei Huang); data curation, Y.H. (Yiwei Huang), Z.L. and Y.H. (Yuhan Huang); writing—original draft preparation, Y.H. (Yiwei Huang), Z.L. and Y.H. (Yuhan Huang); writing—review and editing, Y.H. (Yiwei Huang), Z.L. and Y.H. (Yuhan Huang); visualization, Y.H. (Yiwei Huang), Z.L. and Y.H. (Yuhan Huang); supervision, Y.H. (Yiwei Huang); project administration, Y.H. (Yiwei Huang); funding acquisition, Y.H. (Yiwei Huang). All authors have read and agreed to the published version of the manuscript.

Funding: This research received no external funding.

Conflicts of Interest: The authors declare no conflict of interest.

References

1. Konijnendijk, C.C.; Annerstedt, M.; Nielsen, A.B.; Maruthaveeran, S. Benefits of urban parks. In *A systematic Review. A Report for IFPRA*; Alnarp International Federation of Parks and Recreation Administration: Copenhagen, Denmark, 2013.
2. Hurd, A.R.; Anderson, D.M. *The Park and Recreation Professional's Handbook*; Human Kinetics: Champaign, IL, USA, 2011.
3. Chiesura, A. The role of urban parks for the sustainable city. *Landsc. Urban Plan.* **2004**, *68*, 129–138. [CrossRef]
4. Ulrich, R.S. Natural versus urban scenes: Some psychophysiological effects. *Environ. Behav.* **1981**, *13*, 523–556. [CrossRef]
5. Kaplan, S.; Talbot, J.F. Psychological benefits of a wilderness experience. In *Behavior and the Natural Environment*; Springer: Boston, MA, USA, 1983; pp. 163–203.
6. Kuo, F.E.; Bacaicoa, M.; Sullivan, W.C. Transforming inner-city landscapes: Trees, sense of safety, and preference. *Environ. Behav.* **1998**, *30*, 28–59. [CrossRef]
7. Luttik, J. The value of trees, water and open space as reflected by house prices in the Netherlands. *Landsc. Urban Plan.* **2000**, *48*, 161–167. [CrossRef]
8. Błaszczyk, M.; Suchocka, M.; Wojnowska-Heciak, M.; Muszyńska, M. Quality of urban parks in the perception of city residents with mobility difficulties. *PeerJ* **2020**, *8*, e10570. [CrossRef] [PubMed]
9. Boone, C.G.; Buckley, G.L.; Grove, J.M.; Sister, C. Parks and people: An environmental justice inquiry in Baltimore, Maryland. *Ann. Assoc. Am. Geogr.* **2009**, *99*, 767–787. [CrossRef]
10. Cohen, D.A.; Marsh, T.; Williamson, S.; Derose, K.P.; Martinez, H.; Setodji, C.; McKenzie, T.L. Parks and physical activity: Why are some parks used more than others? *Prev. Med.* **2010**, *50*, S9–S12. [CrossRef]
11. Rung, A.L.; Mowen, A.J.; Broyles, S.T.; Gustat, J. The role of park conditions and features on park visitation and physical activity. *J. Phys. Act. Health* **2011**, *8*, S178–S187. [CrossRef]
12. Donahue, M.L.; Keeler, B.L.; Wood, S.A.; Fisher, D.M.; Hamstead, Z.A.; McPhearson, T. Using social media to understand drivers of urban park visitation in the Twin Cities, MN. *Landsc. Urban Plan.* **2018**, *175*, 1–10. [CrossRef]
13. Zhang, S.; Zhou, W. Recreational visits to urban parks and factors affecting park visits: Evidence from geotagged social media data. *Landsc. Urban Plan.* **2018**, *180*, 27–35. [CrossRef]
14. Scott, D.; Munson, W. Perceived constraints to park usage. *J. Park Recreat. Adm.* **1994**, *12*, 79–96.
15. City of Seattle, Seattle Legislative Department. Seattle Parks and Recreation. Recreation Division Evaluation. 2018. Available online: https://www.seattle.gov/documents/Departments/ParksAndRecreation/PoliciesPlanning/Seattle_Legislative_SPR%20Recreation%20Evaluation%208-23-18.pdf (accessed on 23 August 2018).
16. Pesavento, L.C. Reorganizing the Chicago Park District: From patronage to professional status. *J. Leis. Res.* **2000**, *32*, 116–120. [CrossRef]
17. Byrne, J.; Wolch, J.; Zhang, J. Planning for environmental justice in an urban national park. *J. Environ. Plan. Manag.* **2009**, *52*, 365–392. [CrossRef]
18. Rigolon, A. A complex landscape of inequity in access to urban parks: A literature review. *Landsc. Urban Plan.* **2016**, *153*, 160–169. [CrossRef]
19. Rigolon, A. Parks and young people: An environmental justice study of park proximity, acreage, and quality in Denver, Colorado. *Landsc. Urban Plan.* **2017**, *165*, 73–83. [CrossRef]
20. Takyi, S.A.; Seidel, A.D. Adaptive management in sustainable park planning and management: Case study of the city of Vancouver Parks. *J. Urban Ecol.* **2017**, *3*, juw009. [CrossRef]
21. Sessoms, H.D. *Eight Decades of Leadership Development: A History of Programs of Professional Preparation in Parks and Recreation: 1909–1989*; National Recreation and Park Association: Ashburn, VA, USA, 1993.
22. Wyrwoll, C. *Social Media: Fundamentals, Models, and Ranking of User-Generated Content*, 1st ed.; Springer: Vieweg, Wiesbaden, 2014; pp. 11–45.
23. Timoshenko, A.; Hauser, J.R. Identifying customer needs from user-generated content. *Mark. Sci.* **2019**, *38*, 1–20. [CrossRef]
24. O'connor, P. User-generated content and travel: A case study on Tripadvisor. com. In Proceedings of the ENTER, Innsbruck, Austria, 9–12 August 2008; Volume 2008, pp. 47–58.
25. Purves, R.; Edwardes, A.; Wood, J. Describing place through user generated content. *First Monday* **2011**, *16*, 9. [CrossRef]
26. Hamstead, Z.A.; Fisher, D.; Ilieva, R.T.; Wood, S.A.; McPhearson, T.; Kremer, P. Geolocated social media as a rapid indicator of park visitation and equitable park access. *Comput. Environ. Urban Syst.* **2018**, *72*, 38–50. [CrossRef]
27. Barry, S.J. Using social media to discover public values, interests, and perceptions about cattle grazing on park lands. *Environ. Manag.* **2014**, *53*, 454–464. [CrossRef]
28. Song, Y.; Zhang, B. Using social media data in understanding site-scale landscape architecture design: Taking Seattle Freeway Park as an example. *Landsc. Res.* **2020**, *45*, 627–648. [CrossRef]
29. Lee, K.; Yu, C. Assessment of airport service quality: A complementary approach to measure perceived service quality based on Google reviews. *J. Air Transp. Manag.* **2018**, *71*, 28–44. [CrossRef]
30. Mathayomchan, B.; Taecharungroj, V. "How was your meal?" Examining customer experience using Google maps reviews. *Int. J. Hosp. Manag.* **2020**, *90*, 102641. [CrossRef]
31. Shah, M.; Pabel, A.; Martin-Sardesai, A. Assessing Google reviews to monitor student experience. *Int. J. Educ. Manag.* **2020**, *34*, 610–625. [CrossRef]

32. Koerniawan, M.D.; Dewancker, B.J. Visitor Perceptions and Effectiveness of Place Branding Strategies in Thematic Parks in Bandung City Using Text Mining Based on Google Maps User Reviews. *Sustainability* **2019**, *11*, 2123. [CrossRef]

33. Cronon, W. *Nature's Metropolis: Chicago and the Great West*; WW Norton & Company: New York, NY, USA, 2009.

34. Orsi, J.M.; Margellos-Anast, H.; Whitman, S. Black–white health disparities in the United States and Chicago: A 15-year progress analysis. *Am. J. Public Health* **2010**, *100*, 349–356. [CrossRef]

35. Kolak, M.; Bradley, M.; Block, D.R.; Pool, L.; Garg, G.; Toman, C.K.; Wolf, M. Urban foodscape trends: Disparities in healthy food access in Chicago, 2007–2014. *Health Place* **2018**, *52*, 231–239. [CrossRef]

36. Allen, D.J. *Lost in the Transit Desert: Race, Transit Access, and Suburban Form*; Routledge: London, UK, 2017.

37. Chicago Park District. History of Chicago's Parks. 2021. Available online: https://www.chicagoparkdistrict.com/about-us/history-chicagos-parks (accessed on 6 July 2021).

38. Chicago Park District Official Website. Available online: https://www.chicagoparkdistrict.com/ (accessed on 6 July 2021).

39. Borrego, Á.; Navarra, M.C. What users say about public libraries: An analysis of Google Maps reviews. *Online Inf. Rev.* **2020**, *45*, 84–98. [CrossRef]

40. Kim, R.I.; Bernard, S. Comparing online reviews of hyper-local restaurants using deductive content analysis. *Int. J. Hosp. Manag.* **2020**, *86*, 102445. [CrossRef]

41. Lestari, T. The language used by Indonesian local guides in Google Maps reviews. In Proceedings of the International Seminar on Language Maintenance and Shift (LAMAS) 7, The Vitality of Local Languages in Global Community, Semarang, Indonesia, 19–20 July 2017.

42. Pesaresi, M.; Ehrlich, D.; Ferri, S.; Florczyk, A.; Freire, S.; Halkia, M.; Julea, A.; Kemper, T.; Soille, P.; Syrris, V. GHS Built-Up grid, Derived from Landsat, Multitemporal (1975, 1990, 2000, 2014). European Commission, Joint Research Centre (JRC). 2015. Available online: https://data.europa.eu/89h/jrc-ghsl-ghs_built_ldsmt_globe_r2015b (accessed on 16 November 2021).

43. Kennedy, C.M.; Oakleaf, J.R.; Theobald, D.M.; Baruch-Mordo, S.; Kiesecker, J. Managing the middle: A shift in conservation priorities based on the global human modification gradient. *Glob. Chang. Biol.* **2019**, *25*, 811–826. [CrossRef]

44. Scott, S. The TkWWW Robot: Beyond Browsing. In Proceedings of the Second World Wide Web Conference '94: Mosaic and the Web, Chicago, IL, USA,, 17–20 October 1994.

45. Kobayashi, M.; Takeda, K. Information retrieval on the web. *ACM Comput. Surv.* **2000**, *32*, 144–173. [CrossRef]

46. Goyvaerts, J. Regular Expression Tutorial—Learn How to Use Regular Expressions. Available online: www.regular-expressions.info (accessed on 31 October 2021).

47. Yin, P. Kernels and density estimation. In *The Geographic Information Science & Technology Body of Knowledge*, 1st ed.; Wilson, J.P., Ed.; University Consortium Geographic Information Science: Washington, DC, USA, 2020. [CrossRef]

48. Taverniers, S.; Bosma, S.B.; Tartakovsky, D.M. Accelerated multilevel Monte Carlo with kernel-based smoothing and latinized stratification. *Water Resour. Res.* **2020**, *56*, e2019WR026984. [CrossRef]

49. Kalinic, M.; Krisp, J.M. Kernel density estimation (KDE) vs. hot-spot analysis–detecting criminal hot spots in the City of San Francisco. In Proceedings of the 21st Conference on Geo-Information Science, Lund, Sweden, 12–15 June 2018.

50. Hart, T.; Zandbergen, P. Kernel density estimation and hotspot mapping: Examining the influence of interpolation method, grid cell size, and bandwidth on crime forecasting. *Polic. Int. J. Police Strateg. Manag.* **2014**, *37*, 305–323. [CrossRef]

51. Tiefelsdorf, M. Some practical applications of Moran's I's exact conditional distribution. *Pap. Reg. Sci.* **1998**, *77*, 101–129. [CrossRef]

52. Manepalli, U.R.; Bham, G.H.; Kandada, S. Evaluation of hotspots identification using kernel density estimation (K) and Getis-Ord (Gi*) on I-630. In Proceedings of the 3rd International Conference on Road Safety and Simulation, Indianapolis, IN, USA, 14–16 September 2011; pp. 14–16.

53. Boots, B.N.; Getis, A. *Point Pattern Analysis*; Sage Publications: Thousand Oaks, CA, USA, 1998.

54. Getis, A.; Ord, J.K. The analysis of spatial association by use of distance statistics. *Geogr. Anal.* **1992**, *24*, 189–206. [CrossRef]

55. Anselin, L. The local indicators of spatial association—LISA. *Geogr. Anal.* **1995**, *27*, 93–115. [CrossRef]

56. Russell, E. Google Maps Platform: 9 Things to Know about Google's Maps Data: Beyond the Map. 2019. Available online: https://cloud.google.com/blog/products/maps-platform/9-things-know-about-googles-maps-data-beyond-map (accessed on 19 November 2021).

57. Perrin, A. Social media usage. *Pew Res. Cent.* **2015**, *125*, 52–68.

58. Tucker, J.A.; Guess, A.; Barberá, P.; Vaccari, C.; Siegel, A.; Sanovich, S.; Nyhan, B. Social media, political polarization, and political disinformation: A review of the scientific literature. 2018. Available online: https://ssrn.com/abstract=3144139 (accessed on 19 March 2018).

59. Chan, H.K.; Wang, X.; Lacka, E.; Zhang, M. A mixed-method approach to extracting the value of social media data. *Prod. Oper. Manag.* **2016**, *25*, 568–583. [CrossRef]

60. Tsou, M.H. Research challenges and opportunities in mapping social media and Big Data. *Cartogr. Geogr. Inf. Sci.* **2015**, *42* (Suppl. 1), 70–74. [CrossRef]

61. Chen, H.; Zheng, Z.; Ceran, Y. De-biasing the reporting bias in social media analytics. *Prod. Oper. Manag.* **2016**, *25*, 849–865. [CrossRef]

 *land*

Article

The Central Arizona Conservation Alliance Programs: Use of Social Media and App-Supported Community Science for Landscape-Scale Habitat Restoration, Governance Support, and Community Resilience-Building

Aireona B. Raschke *, Jeny Davis and Annia Quiroz

Desert Botanical Garden, Desert Horticulture and Conservation, Phoenix, AZ 85008, USA; jdavis@dbg.org (J.D.); aquiroz@dbg.org (A.Q.)
* Correspondence: araschke@dbg.org

Abstract: Land managers are currently faced with a nexus of challenges, both ecological and social, when trying to govern natural open spaces. While social media has led to many challenges for effective land management and governance, the technology has the potential to support key activities related to habitat restoration, awareness-raising for policy changes, and increased community resilience as the impacts of increased use and climate change become more apparent. Through the use of a case study examining the work of the Central Arizona Conservation Alliance's social media ambassadorship and its app-supported community science projects, we examine the potential and realized positive impact that technology such as social media and smartphone apps can create for land managers and surrounding communities.

Keywords: social media; land governance; community resilience; online technology; community science; biodiversity conservation

Citation: Raschke, A.B.; Davis, J.; Quiroz, A. The Central Arizona Conservation Alliance Programs: Use of Social Media and App-Supported Community Science for Landscape-Scale Habitat Restoration, Governance Support, and Community Resilience-Building. *Land* **2022**, *11*, 137. https://doi.org/10.3390/land11010137

Academic Editors: Cecilia Arnaiz Schmitz, Nicolas Marine and María Fe Schmitz

Received: 15 December 2021
Accepted: 11 January 2022
Published: 15 January 2022

Publisher's Note: MDPI stays neutral with regard to jurisdictional claims in published maps and institutional affiliations.

1. Introduction

1.1. Social and Ecological Challenges Abound for Land Managers

Land managers are faced with a myriad of social and ecological challenges, which are often compounded by a lack of sufficient resources and complicated social and ecological objectives [1,2]. Among these challenges are climate change and the impact of invasive species on the land, rapid urban development, habitat loss, and increased recreational use, which can bring problematic behavior on trails, campgrounds, and backcountry areas [3].

There is a noted acceleration of global habitat loss, fragmentation, and degradation due to a variety of human activities [4]. Rapid urban development and the growth of tourism and its associated infrastructure has led to the increase of urban–wildland interfaces, which are challenging to manage and present increased wildfire risk [5,6]. As needs for resources have thus far increased in connection to land, changing patterns of consumerism and a growing human population have also increased pressures on natural resources. Human activities, even removed from growing urban areas, can cause detrimental changes to habitats. These include the introduction of invasive species, increasing areas of edge habitat, and the changing fire and weather regimes. All of these present land governance challenges and can result in the loss of biodiversity and key ecosystem services [7].

These human-caused ecological challenges are compounded and linked with climate change, and are impacted by generally high levels of uncertainty regarding best practices for adaptable and effective management [8]. Climate change is already known to cause ecological changes, from shifting the ranges of individual species to complete ecological transformations. These changes may be incremental over the span of many years, or caused by the increased intensity and number of disasters such as wildfires, floods, and storms [9,10]. Climate change also impacts local communities, often in detrimental ways [11,12].

In conjunction with natural forces, land managers must also plan for and mitigate changes to habitats due to human usage, including activities considered consumptive (such as resource extraction) and nonconsumptive (such as outdoor recreation) [2,7]. Of particular interest to this study are those "nonconsumptive" activities that have increased in their intensity consistently over time, for example, hiking, camping, OHV, shooting, mountain biking, etc. This is taxing the natural infrastructure, impacting the health of the landscape, and testing the innovation of land managers who are operating with limited resources while trying to ensure the safety of users [13]. The growing intensity of use is further complicated by changes in user behavior, influenced by traditional marketing, social media, tourism trends, and global conditions (e.g., the COVID-19 pandemic) [14,15].

1.2. Demonstrated Negative Impacts of Social Media and Apps on Land Management and Related Conservation

As with many aspects of modern life, social media and the use of phone applications (henceforth, "apps") has shaped the opportunities and challenges that land managers face worldwide. Each may influence user behavior such that there is an increase in the intensity of use in previously low-impact areas, degradation of sensitive habitats and archeological sites, handling of wildlife and artifacts, and dangerous behaviors driven by online clout culture and trends [16].

Among social media platforms, Instagram (iOS version 217.0; Android version 216.1.0.21.137; Menlo Park, CA, USA) has received considerable attention in this regard, although all platforms have the potential to cause similar issues depending on their level of popularity at any one time [16,17]. Using Instagram as an example, social media can influence behavior at scale by introducing large groups of people to landscapes previously unknown to them. The patterns (both algorithmic and human) of popularity among posts and images of these landscapes may drive some users to behave in ways perceived by them to receive more attention on the platform [18].

The rapid increase in the popularity of places such as Horseshoe Bend, AZ, USA, which has become an Instagram staple, has required infrastructural hardening of the site in order to manage impacts of increased traffic [19]. Further, images with animals, showing the account owner deeply immersed in the environment or depicting risky behavior, can result in more popular posts. Some notorious results of this include the death of a dolphin calf who was removed from the ocean by beachgoers and held for photographs for a prolonged time [20] and the many images of recreational users approaching charismatic wildlife such as bison [21]. Desire for depicted immersion also led to the degradation of the California poppy super bloom in 2019 [22]. There is a growing rise in recreational users who have damaged resources, and/or suffered injury and even death while posing for photos in the hope of attracting interest on Instagram and other platforms [23,24].

Among apps not otherwise considered as social media platforms, AllTrails (iOS version 14.3.0; Android version 14.2.0; San Francisco, CA, USA) serves as an example of the impacts these tools can have for land management. AllTrails is a popular crowd-sourced trail guide and tracker with information on hiking, biking, and OHV trails all over the world. While this is a powerful tool for users looking to explore and navigate via their phones, guides and maps are created by users, which means that they may represent non-established trails and can lack important safety, permitting, and access information. Users may be accessing sensitive areas and/or circumventing planned trail networks, and/or continue to use areas designated for rehabilitation or trail closure, due to these guides. Land managers may also find an increase in the need for rescues due to misrepresentation of trail conditions [25]. These guides can be edited and removed by land managers, but AllTrails only represents one of many such apps and the scale of the user-base-produced guides far outstrips that of land management staff [26].

1.3. Potential for Support of Land Management and Associated Community Well-Being by Social Media and Other Apps

Although the challenges presented by social media and other apps require the attention of land managers, there are also positive potential applications of these online tools including support for conservation activities via data collection, awareness raising for policy changes and increased community support, and enhanced community resilience to natural disasters related to climate change and biodiversity loss.

Land management and conservation have considerable and immediate data demands for informed decision-making. However, there is a lack of time or resources to collect sufficient information, and often actions must move forward with the available data [27,28]. Community science is one method to supplement and address data needs, as apps can serve as a means for collecting, collating, and even analyzing data [29]. Some common examples of apps that can support data gathering and community science include iNaturalist (iOS Version 3.2.4; Android Version 1.25.12; San Francisco, CA, USA), ESRI's (Redlands, CA, USA) Field Maps (iOS Version 21.4.0; Android version 21.4.0), and EDDMapS (iOS Version 1.0.9; Android version 1.1.8; Tifton, GA, USA), among others. Some of these will be explored in the case studies below.

Social media platforms can be used to elevate land manager messaging and create a sense of community between the public and land managers. This facilitated communication and increased transparency has been found to ease tension around changing policy and increasing cooperation [30]. The sense of community and avenues for free discussion via social media can also support inclusion, outreach and education, can increase support for ongoing land management activities, and increase public engagement surrounding planning processes [31,32].

Finally, research on community resilience would suggest that when properly applied, social media and other online technologies may be effective tools for supporting communities through natural disasters, such as those linked to climate change and biodiversity loss [33,34]. Community resilience may be defined as (1) a system's ability to return to a particular state after perturbations (such as natural disasters) and/or (2) an individual's ability to recover from disturbances (such as natural disasters) [35]. Social capital, or the sense of community and goodwill among members of a group, is understood to be closely linked with community resilience such that best practice would dictate the consideration of both infrastructural and social mitigation of disaster impacts [36]. Of the nine elements commonly associated with community resilience across the literature, social media and other online apps are likely to support five, including: local knowledge, community networks and relationships, communication, preparedness, and mental outlook [37].

It will be key for land managers to utilize digital tools to their benefit in this age of rapid change and social-media-driven trends. Ambassadorships and community science programs are two potential methods for harnessing the potential power of these digital tools to support land governance.

1.4. Introduction to the Central Arizona Conservation Alliance

The Central Arizona Conservation Alliance (CAZCA) is an initiative of Desert Botanical Garden (DBG), founded in 2012 out of Phoenix, Arizona, USA. The Alliance, as of 2021, consists of more than 60 partner organizations including parks and recreation departments at city and county levels, land managing federal and state agencies, and local nonprofits working on biodiversity conservation and community well-being, with DBG as the backbone organization [38]. The objective of CAZCA is to facilitate collaborative conservation efforts across the Central Arizona region, with a focus on Maricopa County and its associated HUC 10 watersheds, in order to create a network of natural open spaces. This network includes habitat blocks across the urban–rural gradient, as well as habitat corridors of varying sizes, and integrates already existing parks and preserves with habitats that are yet to be protected. These lands would serve as recreational areas for local residents and visitors, while maintaining thriving Sonoran Desert ecosystems.

CAZCA is designed to accomplish this by leveraging expertise and resources from organizations across the region and working towards the vision outlined above, as guided by the collaboratively developed Regional Open Space Strategy for Maricopa County (ROSS). The ROSS integrated leadership and feedback from more than 50 partner organizations to outline objectives for four primary goals includes: (1) protect and connect ecosystems across the region, (2) sustain and restore habitats protected in the present and future, (3) create spaces for local champions to love and support the network of open spaces, and (4) continue this work via the coordination and elevation of partner organizations and their work [39].

Functionally, as a fairly young collaborative conservation initiative as per the collective impact framework [38], CAZCA has and continues to accomplish many of its goals via collaborative programs and projects. Many of these focus on collective strategy development and decision-making, as well as on-the-ground conservation efforts such as invasive plant management and native plant material development. However, CAZCA has also innovated on land management issues through the use of social media (Sonoran Insiders) and community science using smartphone apps (Desert Defenders and Metro Phoenix EcoFlora, henceforth, EcoFlora).

It is our objective to utilize three case studies from among CAZCA's programs, where social media and smartphone apps have been successfully applied to land management challenges and community resilience building, to illustrate the realized potential of digital technologies and explore lessons learned and challenges faced.

2. Materials and Methods

We cover three different programs here to explore the role of social media and online technologies in supporting land governance in the arid southwestern USA (Table 1). All of these programs are collaborative in nature and focus on regions of varying sizes in Central Arizona. This general spatial focus is determined by the study area of CAZCA, which is defined by the location of our institution, Desert Botanical Garden in Phoenix, AZ, USA, and the Regional Open Space Strategy for Maricopa County.

Table 1. CAZCA programs explored in this case study in order of examination.

Program Name	Focal Area	General Objectives	Technology Utilized
Sonoran Insiders	Central Arizona, USA	Public education on land management, policy changes, and responsible use.	Instagram
Desert Defenders	Maricopa County, Arizona, USA	Map and monitor invasive plant species; share information on management strategies.	ESRI Field Maps, ESRI Collector, ArcGIS Online
Metro Phoenix EcoFlora	Phoenix Metropolitan Area, Arizona, USA	Collect, analyze, and share urban biodiversity data and information; increase the understanding and appreciation of plant life.	iNaturalist, SEINet, Instagram, Facebook, Twitter, Zoom

2.1. Sonoran Insiders' Programmatic Methodology

The Sonoran Insiders program seeks to create a community with local social media influencers to elevate messages about responsible use and raise awareness of the work performed by organizations to protect and maintain natural areas. The successes and lessons learned from this work illustrate the power and potential for social media to support land governance. It is a collaborative effort that includes leadership from CAZCA, the National Forest Foundation, and the Tonto National Forest as of 2019–2021.

The primary social media platform of focus for this program was Instagram. This platform was selected due to its historic impact on public lands and natural open spaces globally [14,17]. Furthermore, it remains one of the most popular social media platforms, with roughly one billion users monthly in 2021 [40]. Instagram also strikes an impactful balance between visual and written storytelling in a way that facilitates strong messaging opportunities more suited to the needs of awareness-raising than Twitter (which previous to 2021, edited image display sizes automatically and limited the characters of a single post) or Facebook (which has severely cut its users' organic reach in recent years).

The Sonoran Insider social media ambassadors, or influencers that participate in the program, were initially identified based on their location, follower counts, and the quality of content in their Instagram feed. Local ambassadors self-selected, but no spatial boundary was drawn for the project. The collaborative agreement with the ambassadors included requirements for participation in in-person events, such that the participants themselves could determine if they were close enough to attend regularly.

In terms of follower numbers, we determined that influencers with 800+ followers would be ideal for our project needs and the resources available for the project. Our team's expert experience in the field indicated that this number would provide a robust minimum follower count and would avoid excluding most passionate ambassadors with smaller audiences. Furthermore, larger influencers often require payment for partnerships with them and we did not have such funds. We also prioritized the quality of influencer content in regards to responsible recreation practices and stewardship. We initially recruited local influencers who already demonstrated some interest in the outdoors or the environment through the use of local outdoor recreation hashtags such as #hikearizona and #explorearizona.

Ambassadors were then recruited through targeted digital outreach, either via direct messages on Instagram or via email. Upon joining, they signed a non-formal agreement to attend at least five events per year, create two to four posts on the events and another two to four posts about related themes such as Leave No Trace, wildfire prevention, invasive plant species management, etc. To assist the ambassadors with the sharing of key messages, we developed briefs that succinctly summarized the theme, messages, and calls to action. These include relevant hashtags, links, and sample text.

Some examples of responsible-use messages elevated by the program include: (1) methods for preventing human-caused wildfire and (2) exploration of the damage that off-trail recreation can have on microbiomes and non-charismatic species such as desert biocrusts. With the large increase in usage of public lands throughout 2020 due to COVID-19 impacts [41], ensuring these messages are reaching both new and long-term users is a key to lessening the impact of recreation on habitats and infrastructure. This is being accomplished at a scale that would not be within the budgets of many land managers [42].

2.2. Desert Defenders' Programmatic Methodology

There are a variety of invasive plant species (including buffelgrass (*Cenchrus ciliaris*), fountain grass (*Cenchrus setaceus*), and stinknet (*Oncosiphon piluliferum*)) that are known to threaten both local community well-being and the habitats of the Sonoran Desert. Some outcompete native species for habitat and resources and impact the ecosystem services of Central Arizona, while many vastly increase wildfire fuel loads which has led to larger, more commonly occurring fires [43–45]. As these species are present across land jurisdictions and at a large and rapidly changing scale, effective management requires cooperation and sufficient data on where these species are, how their ranges change over time, and the success of treatment efforts.

Except in a few cases, land managers in this region lack the resources to accomplish all of this. CAZCA, with McDowell Sonoran Conservancy, Maricopa County Parks and Rec, White Tank Mountains Conservancy, Friends of the Tonto, the City of Phoenix Parks and Rec, the City of Tempe Parks and Rec, and Arizona Sustainability Alliance, set out to address this challenge via technology-supported community science in a project called Desert Defenders (DD).

The collaborative team has been utilizing ESRI's Collector and Field Maps smartphone apps to gather data annually on the presence of eight focal invasive plant species. With the support of these apps, the professional team trains volunteer, community scientists how to identify the focal species and how to add data points on the regional map demarcating the location of the plants to best support local land managers each year. The process of adding data to the map is simple; community scientists download the ESRI Collector or Field Maps app onto their smartphone after they are trained to identify the focal invasive plant species. They then travel into the field, and use their smartphones and FieldMaps/Collector to mark GIS points in real time with data about the plant species found, what stage of life those plants are at, how many there are, and whether that area has been treated to manage invasive plants previously. Mapping activities can be initiated quickly and adapted to changing conditions, e.g., wildfires, early or particularly strong monsoons prompting invasive plant growth. The same apps are then utilized to track and monitor treatments by drawing polygons around those areas and can help determine best re-treatment approaches.

2.3. EcoFlora Programmatic Methodology

The EcoFlora project leverages community science to collect data on urban biodiversity, specifically plant life, and enhances community relationships with nature through technology, engagement, and communication. The project focuses on understanding the impacts of urbanization, accessible biodiversity data, and increasing the understanding and appreciation of plant life. By connecting with the community, contributing to local conservation efforts, and studying urban biodiversity, the project contributes to building community resilience.

EcoFlora is a collaborative project initiated by the New York Botanical Garden in 2016. In 2019, a National Leadership Grant from the Institute of Museum and Library Services was awarded to the New York Botanical Garden and partner gardens to expand the EcoFlora model across the United States. The partner garden recipients include the Chicago Botanic Garden, Denver Botanic Gardens, Desert Botanical Garden, and Marie Selby Botanical Gardens. Within the Desert Botanical Garden, the CAZCA team operates EcoFlora in the Phoenix metropolitan area (Figure 1).

The project collects data on urban biodiversity through iNaturalist (https://www.inaturalist.org/ (accessed on 11 January 2022)), a free web-based platform and app that is a joint initiative of the California Academy of Sciences and the National Geographic Society. Project members create observations with the mobile phone app, or upload images taken with a camera to the website. Photos can be coupled with suggested identifications, various annotations, and observation fields such as life stage, phenology, and associated organisms. iNaturalist neatly houses projects and observation data on their servers for free. This data is downloadable and open source, making it more accessible and useful for conservation efforts, land managers, and organizations. EcoFlora compares data from iNaturalist with the legacy data from the Southwestern Environmental Information Network (SEINet) (https://swbiodiversity.org/ (accessed on 11 January 2022)), a digital portal for floristic information that provides open access to floras, herbarium data, and collections. Species lists have been created with this comparison data and will provide preliminary information about plant biodiversity changes in the Phoenix metropolitan area over time.

EcoQuests are monthly challenges in addition to the overall project that ask members to observe specific species or ecological interactions. This is a venue for collecting specific data and information, and for collaboration. Working with community scientists, EcoFlora can provide supplemental data and support for other projects, organizations, and municipalities. One example is the project collaborated with the Great Milkweed Grow Out (GMGO) program at the Desert Botanical Garden to supplement data concerning western monarch butterfly decline. The EcoFlora model can also be implemented and utilized by other organizations; the Maricopa County Parks and Recreation Department modeled their EcoBlitz program after EcoFlora.

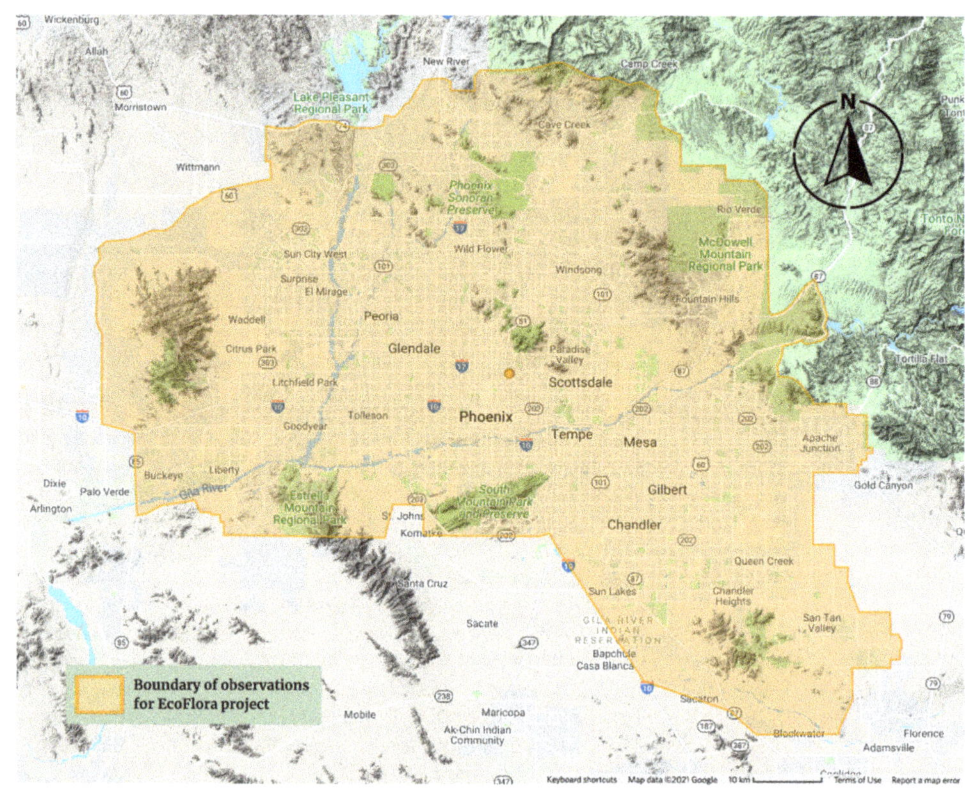

Figure 1. EcoFlora project boundary created with iNaturalist, using Google Maps data.

Engagement is the gateway to involvement and success in the project. Without an engaged group of community scientists, the project would not be able to gather sufficient data or increase community knowledge of biodiversity and plant life. The iNaturalist website includes a direct messaging feature, and this was extremely helpful in the early stages of the project to reach users already invested in using iNaturalist in the Phoenix metropolitan area. Further engagement has been garnered with video conferencing software and apps, such as Zoom (San Jose, CA, USA) and Google Hangouts, developed by Google (Mountain View, CA, USA), which allow the project to host virtual events and information and training sessions. This was essential when gathering in person was not feasible due to the COVID-19 pandemic. In-person engagement has included bird watching, moth lighting, and various botany-themed events. Through a monthly e-newsletter and social media and associated apps, the project has been able to maintain engagement and communication with members, inform them of project developments, and provide opportunities to increase their environmental literacy and plant appreciation. Social media specifically has provided the opportunity to communicate and engage with a wider audience.

3. Results

The quantitative social media and data-production results for each project will be represented in this section, along with qualitative outcomes for the congruity of our case study explorations.

3.1. Sonoran Insiders

Over the course of Y1, the number of Insider posts increased as the community of ambassadors grew and the program engaged more with the participants through events, online discussions, and social media collaborations (e.g., Instagram take-over and Insider-led hiking webinar). This trend culminated in the largest number of posts in December 2020 ($n = 52$; Figure 2).

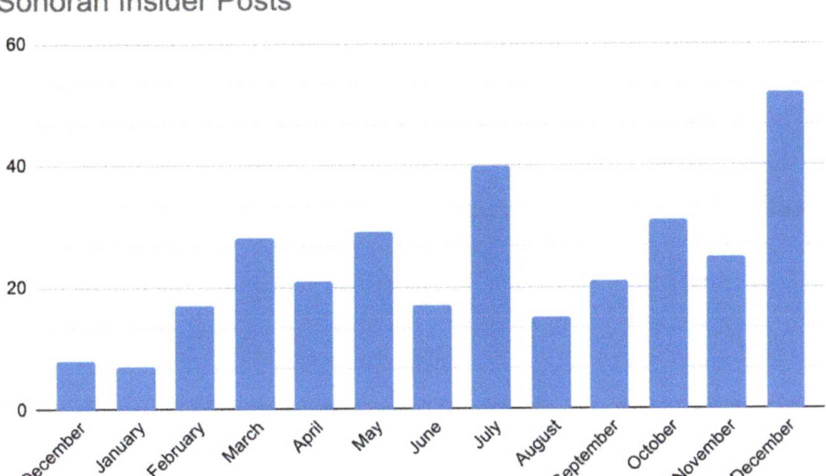

Figure 2. Number of posts on Instagram with the hashtag #sonoraninsiders by month, from December 2019–December 2020.

The impacts of ambassador posts were measured via their reach on Instagram. The term "reach" here refers to the number of unique users that saw the Instagram post using the #sonoraninsiders hashtag on any given day. This number was calculated by adding all of the followers of each account that engaged with the post using the hashtag. The Sonoran Insiders hashtag was used 311 times during year one, accumulating 57,059 likes. We had an annual reach of 5,996,329 people and an average monthly reach of 935,641 people. For reference, the estimated cost of reaching 6 million people with an online ad campaign could range from $18,000 to $60,000, with an average cost per thousand impressions of $3–$10 across the main advertisement platforms [46]. Due to the focused nature of our ambassador audiences, it is possible that the value of this outreach is underestimated, as the posts produced are organically targeting relevant audiences to land managers.

Support for conservation is produced in two ways through this program: (1) by elevating responsible-use behaviors to help protect sensitive environmental and cultural elements of the landscape and (2) raising awareness for volunteer opportunities and charitable support of conservation programs and events. Themes communicated through this social media outreach are shaped by the leading collaborators of the Sonoran Insiders program, in conjunction with other land-managing partners involved in the monthly events on which messaging is based. For example, in early 2020, the program partnered with the McDowell Sonoran Conservancy (MSC) for a behind-the-scenes tour of ecological and restoration research run in the McDowell Mountains Preserve. MSC then led the crafting of the brief, as well as the narrative and calls-to-action presented at the event itself.

3.2. Desert Defenders

The Desert Defenders community science mapping activities have been active from December 2018 to July 2021, with increasing participation from local land managers over time. This has produced a regional map (Figure 3) of the focal invasive species from lands managed by Maricopa County Parks and Recreation, and the City of Scottsdale, City of Tempe, and City of Phoenix parks and recreation departments. This map includes

6824 observations from more than 100 community scientists. Table 2 illustrates the number of data points for each of the focal invasive plant species.

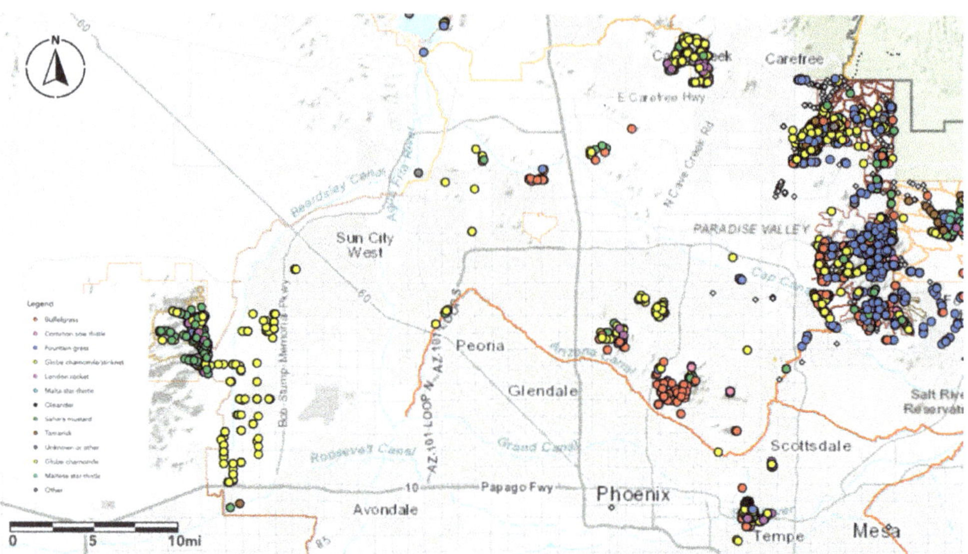

Figure 3. Example map for Desert Defenders; colored points represent different species of focal invasive plants identified across the region as of Spring 2021.

Table 2. Focal invasive plant species and number of data points for each produced by community scientists from 2018–2021.

Species Name (Common)	Number of Data Points
Buffelgrass	659
Common sow thistle	66
Fountain grass	1788
Globe chamomile	1704
London rocket	600
Maltese star-thistle	159
Oleander	38
Sahara mustard	1130
Tamarisk	137
Unknown or other	251

There are several key impacts from this project. First, the data gathered by community scientists through the apps has given land managers access to annual spatial data on invasive plant species across the landscape (Table 2). In 2021, thanks to these efforts, a regional map was created, with minimal monetary investment from land managers outside of the modest staff time and the ESRI licenses necessary for volunteers to use the app. This cost could be further limited through the use of free apps such as EDDMapS, launched and maintained by the University of Georgia. In any case, this data allows for annual planning for treatments that make the most efficient use of limited resources. It has also provided information for the movement of invasive plant species across the region, facilitating collaboration and management among land managers across jurisdictions.

3.3. EcoFlora

Key impacts from the project include open-access biodiversity data, alleviating plant invisibility and increasing support for conservation, and community resilience through involvement in community science. As mentioned above, iNaturalist and SEINet are both open-access platforms, making the data and collections gathered through EcoFlora available to anyone interested in viewing or using them. The project began in February 2020 and as of December 2021, 361 people have become project members and 37,319 observations (21,175 plants) of 2577 species (1158 plant species) have been made by 252 observers, with 1802 identifiers (https://www.inaturalist.org/projects/metro-phoenix-ecoflora, accessed on 14 December 2021) (Figure 4). iNaturalist observations have provided a wealth of open-source data, which would otherwise be extremely difficult to achieve without an abundance of community scientists.

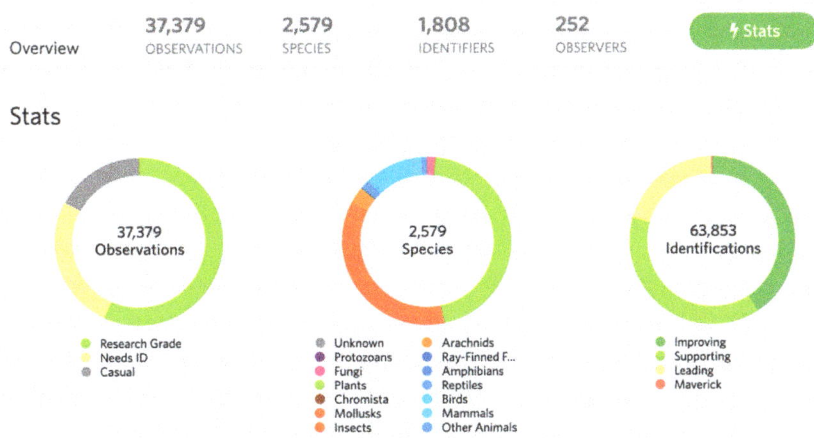

Figure 4. Screenshot from EcoFlora project page on iNaturalist website showing current project stats.

People and communities can directly participate in EcoFlora in a way that is not intimidating or technologically overwhelming. Project members can make observations in their own neighborhoods, in a place they have local knowledge of, and can connect their lived experience with. They understand their needs, wants, and challenges, and feel the ramifications of policy and daily changes better than those outside do. They are more likely to notice a new plant species or understand where street trees that provide shade would be most valuable. Through iNaturalist, project members can use the species information, observation images, and maps to support conservation and intersectional environmental efforts in their community. For the EcoQuest in October 2020, project members observed 624 ocotillo plants (*Fouquieria splendens*), bringing the total number of observations in the Phoenix metropolitan area to 1056. With this observation data, community members can see possible corridors for pollinators and wildlife that connect to open space (Figure 5). This can be used by the community to advocate for connectivity corridors, in turn contributing to landscape management efforts and community well-being through increased nearby nature in urban areas [47]. Project members can communicate and build relationships with one another through interaction on the iNaturalist platform and the project's social media accounts. These venues allow people to actively take part in community conversations about urban biodiversity.

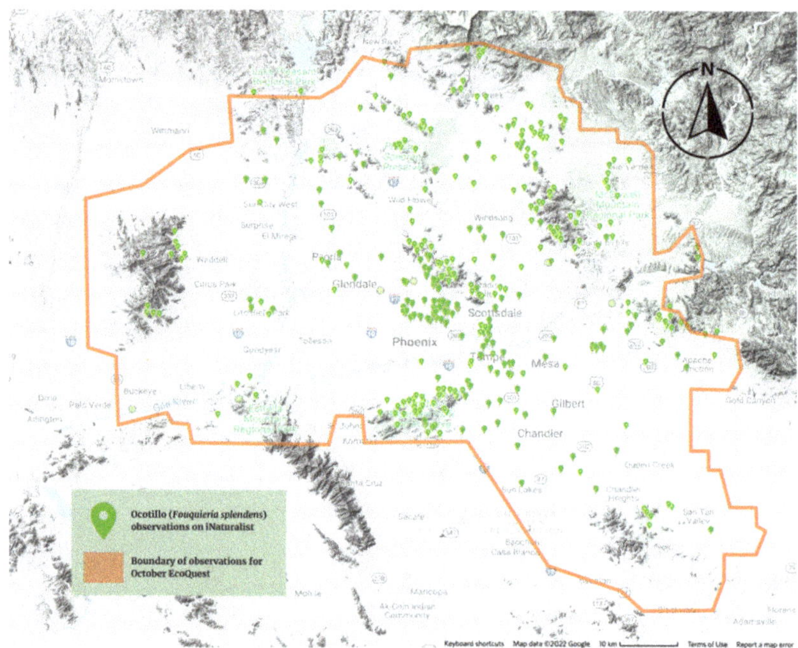

Figure 5. iNaturalist map showing observations of ocotillo (*Fouquieria splendens*) and possible connectivity corridors in the Phoenix metropolitan area, accessed on 5 January 2021.

Events hosted by EcoFlora give people the opportunity to connect with scientists, local leaders, organizations, and professionals. EcoQuestions, for example, are virtual question and answer sessions that provide the opportunity for project members and the community at large to learn more about urban biodiversity and plant science. In total, project events and training sessions have been attended approximately 270 times, with the project having 1081 followers collectively on Facebook, Instagram, and Twitter. Increased support for conservation ideally follows increased awareness and understanding of plant life and biodiversity. Collaborations, events, and social media presence equate to the public learning more about plants and biodiversity, alleviating plant invisibility (previously plant blindness), or the tendency for people to overlook plants and view them with lower regard than other life forms [48].

EcoFlora has a positive impact on the mental and physical health of the community through encouraging physical activity and providing mental stimulation. Project members have repeatedly stated that EcoFlora has provided them with an outlet, specifically throughout the COVID-19 pandemic. Making observations can be done alone or when socially distanced, and is an activity that can safely be done outdoors. EcoQuest challenges also give project members something new to look forward to every month, contributing to well-being through anticipation of future positive events [49].

4. Discussion

In regards to land governance, social media and other online applications are often posed as challenges. For example, clout-chasing may cause users to engage in behaviors that risk their safety or the natural resources themselves [23,24]. These tools can cause rapid increases in area use over a relatively short period of time, making management and infrastructure development a game of catch-up [19]. With limited resources, both in terms of monetary support and staff capacity, land governance may be faced with resulting changes that damage resources, degrade trust, and create new use-norms that undermine conservation efforts.

These are all serious concerns. However, much of the evidence for such impacts is rhetorical, with increasing calls for closer studies of these impacts, to understand when they manifest and how [32]. At the same time, there are large potential benefits for both users and land managers that include increased equity of access and much-needed social support for land governance activities. As we will demonstrate below, our models specifically demonstrate the positive benefits of social media and other online applications (such as iNaturalist and ESRI Field Maps) in regards to support for (1) biodiversity conservation, (2) land governance and policy changes, and (3) building community resilience.

4.1. Sonoran Insiders

The Sonoran Insiders program contributes to community resilience in a variety of ways in the face of disasters such as those caused by climate change and biodiversity loss. In particular, when considering the common elements of community resilience as identified by Patel et.al. in their 2017 systematic review [37] of the concept, four common elements are positively impacted by this application of social media to land governance. (1) Social media through the Insiders program has increased local knowledge around land management, related disasters, and modes for community involvement in protecting natural resources and local infrastructure from disasters. (2) The Insiders program builds community networks and relationships between land managers and the public. (3) It increases communication between and among land managers, program ambassadors, nonprofit partners, and the public. (4) Through all of this, messaging and connections formed through the project have the potential to increase hope among community members and, thus, improve mental outlooks.

There is considerable need for volunteer support for land management activities ranging from infrastructure maintenance to ecological research. Volunteers can be difficult to recruit, and the costs for finding a sufficient volunteer force can include marketing costs, as well as the time and capacity of volunteer coordinators (where available) [50]. Thus, alleviating those needs through effective social media messaging can represent another economic and social support for land managers.

Social media ambassadors and related social media communications have also provided support for land managers in cases of contested policy changes. For example, in 2021, the City of Phoenix city council voted to institute trail closures during extreme heat days. This decision was made following an increase in the number of rescues on the popular and difficult Echo Canyon and Piestewa Summit trails, with the associated health impacts on rescue personnel. While the rationale for these trail closures was logical and promoted safety among both trail users and city staff, there was considerable pushback from some sectors of the public to these closures [51]. A lack of cooperation with safety measures and other land management policies can undermine safety and create resource sinks. Thus, effective communication with trail-use communities is essential, in this case, among both local users and tourists. Social media platforms, particularly those run by local influencers, have the potential to reach both groups. Furthermore, communication from other users, as opposed to perceived authority figures, can enhance the acceptance of new rules and use patterns [52].

While this program has seen a variety of high-impact successes, even while at a pilot stage, there are challenges to be accounted for when considering the social media ambassadorship model. First, while monetary investments are low compared to equivalent marketing expenses, staff time may be considerable. Consistent engagement among ambassadors and between ambassadors and program staff is also essential for trust building. This also requires a significant time investment by staff. For CAZCA's program, this was mitigated through collaboration among several organizations, which allowed for sharing of the planning, organizational, and community-building duties.

4.2. Desert Defenders

In conjunction with the raw data provided, the DD program bolsters land managers in a variety of social ways. The training modules designed by the collaborative group include extensive knowledge on the role of invasive plants in ecosystem shifts and increases in wildfire risk for the community scientists. Thus, through involvement in the program, members of the public gain hands-on conservation experience and have become advocates and supporters for management efforts. In some cases, these community scientists have led the mapping projects themselves in support of small municipalities, reached out to their home owners associations (HOAs) about invasive plants in local landscapes, and raised money to purchase equipment for invasive plant management efforts [29].

The data produced by the project also supports community resilience, particularly in regards to pressing wildfire threats in the southwest. In particular, this data is essential for the effective management of these invasive plant species such that the community and the surrounding habitats can be protected from heightened wildfire risk and biodiversity loss. Furthermore, this data supports preparedness in the community and can be used to identify areas with high wildfire fuel loads. The process used to develop this data, via its reliance on more than 100 community scientists across the region, also increases local knowledge on the focal invasive plant species and their impacts on wildfire risk. Community scientists are also provided with increased community networks and relationships with land managers and scientists.

There are challenges to applying these ESRI apps to land management, however, particularly in cases of limited expert capacity and volunteer coordination. While ESRI's GIS tools have powerful analytical capabilities, they are complex and require professional levels of expertise to take full advantage of their functionality. In the case of this collaborative community science project, the lack of a dedicated GIS professional has hampered comprehensive use of the data gathered. Land managers utilizing similar community science models would likely see more returns if a GIS manager can assist with the program. Similarly, skilled volunteer coordination is an essential element of a successful community science program, even with the support of powerful, user-friendly apps. Collaborators with the Desert Defenders project have seen the most success mapping in cases of: (1) a dedicated and highly trained force of stewards, (2) community scientists led by rangers on joint mapping and removal efforts, and (3) single "champion" volunteers or interns who take the lead in smaller areas.

4.3. EcoFlora

Of the aforementioned nine core elements most often associated with community resilience, community science innately contributes to community resilience most strongly through local knowledge, community networks and relationships, communication, health, governance and leadership, resources, and mental outlook [37]. Community science projects can become community networks, providing the opportunity for people to develop relationships with one another that they may not have had otherwise, promoting social capital, enhancing social cohesion and community trust, and contributing to community resilience [36]. The data, information, and workload contributed by community scientists can be used to shape policy or empower community involvement in local government and leadership, and is at a wider range and frequency than could be accomplished by scientists or land managers alone [53].

Community science can contribute to conservation by directly affecting the conservation science influencing decision-making [54]. As mentioned previously, the project collects a wealth of open-source data and information that can be used by the community, organizations, or land managers. Educational resources, training sessions, and workshops through the project give the community access to and experience with technological and physical tools that exist indefinitely. Empowered with and given access to science knowledge, tools, and experience, communities can be more involved in decision-making and political advocacy, and feel more confident in conservation participation and activities surrounding urban biodiversity. This contributes to collective and self-efficacy, as well as mental outlook, by providing people with a way to be directly involved in science and be better prepared to respond to disasters such as wildfires. By making contributions that help scientists understand more about urban biodiversity, gaining science knowledge and experience, and contributing to local conservation efforts, EcoFlora project members are building community resilience.

The EcoFlora project has not been without challenges, namely data quality, study area considerations, and sustaining and growing engagement. Data quality can be lacking on the iNaturalist platform. Observations by community scientists can be of poor quality, for example, uploading out-of-focus images or cut flowers in a bouquet. The identification feature of the iNaturalist platform does not require any kind of credentials. This can lead to users making identifications at an exponential rate, simply to increase the number of observations they have identified or to appear more knowledgeable. Other users can make incorrect identifications simply because they are trying to learn and become better at identifying organisms. To alleviate the data-quality challenge, the EcoFlora project has created collateral to guide project members in making quality observations and has asked local experts to assist in monitoring identification accuracy.

When deciding on the study area for EcoFlora, there were multiple factors to consider. The Phoenix metropolitan area is a large and rapidly increasing region that does not have a defined boundary. It consists of numerous cities and towns and has a variety of land uses. Distinguishing urban, suburban, and exurban areas for the boundary was not feasible, while previously created Keyhole Markup Language (KML) files did not exist. Highways and roads can provide clean lines for a boundary, but they can inequitably divide communities and create border vacuums [55]. In creating a study area, EcoFlora sought to include the multiple parks that surround the Valley, be considerate of tribal lands, and not exclude under-resourced communities. A hand-drawn boundary in iNaturalist was created, using the CAZCA study area as a base and keeping these elements and objectives in mind.

Initiating and sustaining engagement is a struggle that many smaller conservation organizations face. EcoFlora is fortunate to have the backing and support of CAZCA, DBG, and the EcoFlora partner gardens. Without this existing support, it would have been substantially more difficult to make the public aware of the project. Even with this support, engaging the community and drawing participants to the project has been challenging, specifically considering the COVID-19 pandemic. It was not possible to host in-person workshops and events that could increase interest and participation in the project with large groups of people. Engagement has been largely virtual, which has presented its own learning curve. When feasible, in-person events with limited attendance (10 or less people) while following pandemic guidelines have resulted in fluctuating attendance. Without the digital distribution of EcoFlora information through iNaturalist, DBG, and CAZCA, along with online technologies such as video conferencing software, social media, and e-newsletter, the project would not have been able to initiate or sustain the engagement that it has.

5. Conclusions

Social media has often been considered a source of increasing challenges to land management of public lands and natural open spaces worldwide, but these tools also have supportive potential which is of considerable importance as natural disasters increase with climate change and resource shortfalls for land managers become more common. Our three case study programs illustrate a variety of ways that land managers can use social media and other online applications (iNaturalist and ESRI FieldMaps) to enhance governance and local well-being. We have demonstrated that each model provides support for (1) conservation activities, (2) land governance and policy changes, and (3) community resilience. In conjunction with this, two of the projects also provided spatial data relevant for land management via community science efforts that utilized online applications as user-friendly data-gathering interfaces.

While capacity and expertise are necessary to launch ambassadorships and community science, programs that make the best use of social media and online technologies ensure resources can be kept to a minimum. Innovation and the potential for such efforts to alleviate landscape governance and land management resource challenges is considerable. The case studies outlined lend practical support for the application of different kinds of online technologies to the challenges faced by land managers, and also outline models for applying these tools in effective ways. They are shown to provide marketing and outreach that might otherwise be cost prohibitive, while also building relationships and trust between land managers, partner organizations, and the public [16,42,46]. Data on changing conditions is also essential to effective landscape-scale governance and biodiversity conservation efforts, but the temporal and spatial extent of that need can also be prohibitive without community support. Online technologies, particularly phone apps, can facilitate the creation of community science projects that can supply data, while also developing public support and understanding for land managers [28,29].

In conjunction with the various ways in which programs linked with social media and online technologies support land managers directly, these efforts can enhance community resilience to unprecedented changes in the environment and disasters such as wildfires, hurricanes, and pandemics, among others. In particular, of the elements commonly identified across fields as elements of community resilience [37], those that have been found to be positively influenced by these efforts include local knowledge, community networks and relationships, communication, resources, preparedness, and mental outlook. The impacts of social media and app-supported efforts by land managers could be further improved through design and planning that explicitly targets community resilience as a primary objective.

Finally, the challenges outlined in these case studies, as well as the moderate resource needs of these models, can be sourced via collaborative efforts among organizations with common goals. While complex, collaborative efforts facilitate the leveraging of limited but shared resources for the greater good and also enhance the effectiveness of land governance among varied assemblages of stakeholders and constituents [56,57].

Author Contributions: Conceptualization, A.B.R., A.Q. and J.D.; Methodology, A.B.R., A.Q. and J.D.; Software, A.B.R., A.Q. and J.D.; Validation, A.B.R., A.Q. and J.D.; Formal Analysis, A.B.R. and J.D.; Investigation, A.B.R. and J.D.; Resources, A.B.R.; Data Curation, A.B.R., A.Q. and J.D.; Writing—Original Draft Preparation, A.B.R. and J.D.; Writing—Review and Editing, A.B.R., A.Q. and J.D.; Visualization, A.B.R., A.Q. and J.D.; Supervision, A.B.R.; Project Administration, A.B.R.; Funding Acquisition, A.B.R. All authors have read and agreed to the published version of the manuscript.

Funding: This research was funded by the Nina Mason Pulliam Charitable Trust and Institute of Museum and Library Services. The APC was funded by the Nina Mason Pulliam Charitable Trust.

Institutional Review Board Statement: Not applicable.

Informed Consent Statement: Not applicable.

Data Availability Statement: Not applicable.

Acknowledgments: We would like to acknowledge the work of the many partner organizations and individuals that made all of these projects possible. For the Sonoran Insiders, we thank our leading partners, the National Forest Foundation and the U.S. Forest Service. For Desert Defenders, we thank the leadership of McDowell Mountain Conservancy, Maricopa County Parks and Recreation, City of Phoenix, City of Tempe, White Tank Mountains Conservancy, and Friends of the Tonto. For the EcoFlora, we thank New York Botanical Garden as the founding EcoFlora organization; Chicago Botanic Garden, Denver Botanic Gardens, and Marie Selby Botanical Gardens as supporting partner organizations; the many contributing project members and collaborators; Volunteers in the Garden (VIGs); and the Marketing Communications team at Desert Botanical Garden.

Conflicts of Interest: The authors declare no conflict of interest. The funders had no role in the design of the study; in the collection, analyses, or interpretation of data; in the writing of the manuscript, or in the decision to publish the results.

References

1. World Bank. *Sustainable Land Management: Challenges, Opportunities, and Trade-Offs*; The World Bank: Washington, DC, USA, 2006.
2. Thomas, S.L.; Reed, S.E. Entrenched ties between outdoor recreation and conservation pose challenges for sustainable land management. *Environ. Res. Lett.* **2019**, *14*, 115009. [CrossRef]
3. Buckley, R.C. Grand challenges in conservation research. *Front. Ecol. Evol.* **2015**, *3*, 128. [CrossRef]
4. Braje, T.J.; Erlandson, J.M. Human acceleration of animal and plant extinctions: A Late Pleistocene, Holocene, and Anthropocene continuum. *Anthropocene* **2013**, *4*, 14–23. [CrossRef]
5. Radeloff, V.C.; Helmers, D.P.; Kramer, H.A.; Mockrin, M.H.; Alexandre, P.M.; Bar-Massada, A. Rapid growth of the US wildland-urban interface raises wildfire risk. *Proc. Natl. Acad. Sci. USA* **2018**, *115*, 3314–3319. [CrossRef]
6. Theobald, D.M.; Romme, W.H. Expansion of the US wildland–urban interface. *Landsc. Urban Plan.* **2007**, *83*, 340–354. [CrossRef]
7. Scanes, C.G. *Animals and Human Society*; Academic Press: Cambridge, MA, USA, 2018; pp. 451–482. [CrossRef]
8. Cooney, R.; Dickson, B. (Eds.) *Biodiversity and The Precautionary Principle: Risk, Uncertainty and Practice in Conservation and Sustainable Use*; Routledge: New York, NY, USA, 2012.
9. Bellard, C.; Bertelsmeier, C.; Leadley, P.; Thuiller, W.; Courchamp, F. Impacts of climate change on the future of biodiversity. *Ecol. Lett.* **2012**, *15*, 365–377. [CrossRef]
10. Hallegatte, S. *Natural Disasters and Climate Change*; Springer Int.: New York, NY, USA, 2016.
11. Carleton, T.A.; Hsiang, S.M. Social and economic impacts of climate. *Science* **2016**, *353*, 1112. [CrossRef]
12. Kaganzi, K.R.; Cuni-Sanchez, A.; Mcharazo, F.; Martin, E.H.; Marchant, R.A.; Thorn, J.P.R. Local Perceptions of Climate Change and Adaptation Responses from Two Mountain Regions in Tanzania. *Land* **2021**, *10*, 999. [CrossRef]
13. Winter, P.L.; Selin, S.; Cerveny, L.; Bricker, K. Outdoor recreation, nature-based tourism, and sustainability. *Sustainability* **2020**, *12*, 81. [CrossRef]
14. Leung, D.; Law, R.; Van Hoof, H.; Buhalis, D. Social media in tourism and hospitality: A literature review. *J. Travel Tour. Mark.* **2013**, *30*, 3–22. [CrossRef]
15. Morrison, A.M. *Marketing and Managing Tourism Destinations*; Routledge: London, UK, 2013.
16. Martin, S. Real and potential influences of information technology on outdoor recreation and wilderness experiences and management. *J. Park Recreat. Admin.* **2017**, *35*, 98–101.
17. Nosowitz, D. Why "Instagram hikers' are national parks' saviors—And scourges. *NY Mag.* 2015. Available online: https://nymag.com/intelligencer/2015/12/why-outdoors-nerds-hate-instagram-hikers.html (accessed on 13 November 2021).
18. Florenthal, B. Young consumers' motivational drivers of brand engagement behavior on social media sites: A synthesized U&G and TAM framework. *J. Res. Interact. Mark.* **2019**, *13*, 351–391. [CrossRef]
19. Carlton, J. Instagram turns obscure sites into social media destinations. *Wall Str. J.* 2019. Available online: https://www.wsj.com/articles/instagram-turns-obscure-u-s-sights-into-social-media-destinations-11562410801 (accessed on 13 November 2021).
20. Bale, R. Another baby dolphin killed by selfie-taking tourists. *Nat.Geo.* 2017. Available online: https://www.nationalgeographic.com/animals/article/wildlife-watch-baby-dolphin-killed-tourists-argentina (accessed on 14 November 2021).
21. Miller, M.E. Bison selfies are a bad idea: Tourist gored in Yellowstone as another photo goes awry. *The Washington Post* 23 July 2015.
22. Stone, Z. Under the influence of a 'Super Bloom'. *The New York Times*. 23 March 2019. Available online: https://www.nytimes.com/2019/03/23/style/super-bloom-california-instagram-influencer.html (accessed on 14 November 2021).
23. Bansal, A.; Garg, C.; Pakhare, A.; Gupta, S. Selfies: A boon or bane? *J. Family Med. Prim. Care* **2018**, *7*, 828–831. [CrossRef] [PubMed]
24. Bertaglia, T.; Dubois, A.; Goanta, C. Clout Chasing for the Sake of Content Monetization: Gaming Algorithmic Architectures with Self-moderation Strategies. *Morals Machines* **2021**. [CrossRef]
25. Francovich, E. As more people hike, trail applications offer a bounty of sometimes false information. *The Billings Gazette*. 2021. Available online: https://billingsgazette.com/lifestyles/recreation/as-more-people-hike-trail-applications-offer-a-bounty-of-sometimes-false-information/article_d42b92a3-4b35-54a4-897f-02824300cb56.html#:~:text=topical-,As%20more%20people%20hike%2C%20trail%20applications%20offer,bounty%20of%20sometimes%20false%20information&text=Hiking%20trail%20apps%20have%20been,can%20lead%20to%20private%20property.&text=Knowles%2C%20Spokane%20County'%20park%20planner,to%20fielding%20calls%20from%20hikers (accessed on 14 November 2021).
26. Schweikhard, A. Go take a hike: Online hiking resources. *C RL News* **2019**, *80*, 112–115. [CrossRef]
27. McClenachan, L.; Ferretti, F.; Baum, J.K. From archives to conservation: Why historical data are needed to set baselines for marine animals and ecosystems. *Conserv. Lett.* **2012**, *5*, 349–359. [CrossRef]
28. Costello, M.J.; Vanhoorne, B.; Appeltans, W. Conservation of biodiversity through taxonomy, data publication, and collaborative infrastructures. *Conserv. Biol.* **2015**, *29*, 1094–1099. [CrossRef] [PubMed]
29. Chandler, M.; See, L.; Copas, K.; Bonde, A.M.; López, B.C.; Danielsen, F.; Turak, E. Contribution of citizen science towards international biodiversity monitoring. *Biol. Conserv.* **2017**, *213*, 280–294. [CrossRef]

30. Lodge, M. Accountability and transparency in regulation: Critiques, doctrines and instruments. In *The Politics of Regulation: Institutions and Regulatory Reforms for the Age of Governance*; Jordana, J., Levi-Faur, D., Eds.; Edward Elgar Publishing: Cheltenham, UK, 2004; pp. 124–144.

31. Flores, D.; Kuhn, K. Latino Outdoors: Using storytelling and social media to increase diversity on public lands. *J. Park Recreat. Admi.* **2018**, *36*, 47–62. [CrossRef]

32. Doyon, T.M. From Screen to Summit: An Investigation of Claims about Social Media Use for Outdoor Recreation Purposes. Master's Thesis, Humboldt State University, Arcata, CA, USA, 2020.

33. Feng, S.; Hossain, L.; Paton, D. Harnessing informal education for community resilience. *Disaster Prev. Manag.* **2017**, *27*, 43–59. [CrossRef]

34. Dharmasena, M.G.I.; Toledano, M.; Weaver, C.K. The role of public relations in building community resilience to natural disasters: Perspectives from Sri Lanka and New Zealand. *J. Commun. Manag.* **2020**, *24*, 301–317. [CrossRef]

35. Berkes, F.; Ross, H. Community resilience: Toward an integrated approach. *Soc. Nat. Resour.* **2013**, *26*, 5–20. [CrossRef]

36. Aldrich, D.P.; Meyer, M.A. Social capital and community resilience. *Am. Behav. Sci.* **2015**, *59*, 254–269. [CrossRef]

37. Patel, S.S.; Rogers, M.B.; Amlôt, R.; Rubin, G.J. What do we mean by 'community resilience'? A systematic literature review of how it is defined in the literature. *PLoS Curr.* **2017**, *9*. [CrossRef]

38. Kania, J.; Kramer, M. Collective Impact. Stanf. *Soc. Innov. Rev.* **2011**, *9*, 36–41. [CrossRef]

39. Beute, S.; Arndt, L. *The Regional Open Space Strategy for Maricopa County*; Central Arizona Conservation Alliance: Phoenix, AZ, USA, 2018.

40. Omnicore. Instagram by the Numbers. 2021. Available online: https://www.omnicoreagency.com/instagram-statistics/ (accessed on 12 November 2021).

41. Beery, T.; Olsson, M.R.; Vitestam, M. COVID-19 and outdoor recreation management: Increased participation, connection to nature, and a look to climate adaptation. *J. Outdoor Recreat. Tour.* **2021**, *36*, 100457. [CrossRef]

42. NRPA. 2021 NRPA Agency Performance Review. 2021. Available online: https://www.nrpa.org/siteassets/2021-agency-performance-review_final.pdf (accessed on 12 November 2021).

43. Lyons, K.G.; Maldonado-Leal, B.G.; Owen, G. Community and ecosystem effects of buffelgrass (Pennisetum ciliare) and nitrogen deposition in the Sonoran Desert. *Invasive Plant Sci. Manag.* **2013**, *6*, 65–78. [CrossRef]

44. Hedrick, P.W.; McDonald, C.J. Stinknet, A New Invasive, Non-native Plant in the Southwestern United States. *Desert Plants* **2020**, *36*, 5–16.

45. Tellman, B. Is Your Landscape Threatening the Desert? *Desert Plants.* 2002, pp. 21–25. Available online: https://repository.arizona.edu/handle/10150/555903 (accessed on 15 December 2021).

46. Topdraw. Online Advertising Costs in 2021. 2020. Available online: https://www.topdraw.com/insights/is-online-advertising-expensive/ (accessed on 20 November 2021).

47. Sandifer, P.; Sutton-Grier, A.; Ward, B. Exploring connections among nature, biodiversity, ecosystem services, and human health and well-being: Opportunities to enhance health and biodiversity conservation. *Ecosyst. Serv.* **2015**, *12*, 1–15. [CrossRef]

48. Knapp, S. Are humans really blind to plants? *Plants People Planet* **2019**, *1*, 164–168. [CrossRef]

49. Luo, Y.; Chen, X.; Qi, S.; You, X.; Huang, X. Well-being and Anticipation for Future Positive Events: Evidences from an fMRI Study. *Front. Psychol.* **2018**, *8*, 2199. [CrossRef] [PubMed]

50. Russell, R.; Kirkbride, E. Conservation of an Overlooked Resource–Volunteers. *Rangelands* **2004**, *26*, 20–23. [CrossRef]

51. Miller, B. Pilot Program to Close Phoenix Hiking Trails on Extreme Heat Days Now Permanent. Fox 10 News. 2021. Available online: https://www.fox10phoenix.com/news/pilot-program-to-close-phoenix-hiking-trails-on-extreme-heat-days-may-become-permanent (accessed on 15 November 2021).

52. Tyler, T.R. Psychological perspectives on legitimacy and legitimation. *Annu. Rev. Psychol.* **2006**, *57*, 375–400. [CrossRef] [PubMed]

53. Pudifoot, B.; Cárdenas, M.L.; Buytaert, W.; Paul, J.D.; Narraway, C.L.; Loiselle, S. When It Rains, It Pours: Integrating Citizen Science Methods to Understand Resilience of Urban Green Spaces. *Front. Water* **2021**, *3*, 654493. [CrossRef]

54. Newman, G.; Chandler, M.; Clyde, M.; McGreavy, B.; Haklay, M.; Ballard, H.; Gray, S.; Scarpino, R.; Hauptfeld, R.; Mellor, D.; et al. Leveraging the power of place in citizen science for effective conservation decision making. *Biol. Conserv.* **2017**, *208*, 55–64. [CrossRef]

55. Jacobs, J. *The Death and Life of Great American Cities*; Random House Inc.: New York, NY, USA, 1961; pp. 257–259.

56. Koontz, T.M.; Jager, N.W.; Newig, J. Assessing collaborative conservation: A case survey of output, outcome, and impact measures used in the empirical literature. *Soc. Nat. Resour.* **2020**, *33*, 442–461. [CrossRef]

57. Gray, M.; Micheli, E.; Comendant, T.; Merenlender, A. Climate-Wise Habitat Connectivity Takes Sustained Stakeholder Engagement. *Land* **2020**, *9*, 413. [CrossRef]

 *land*

Article

Analysis of the Image of Global Glacier Tourism Destinations from the Perspective of Tourists

Fan Tang [1,2], Jianping Yang [1,*], Yanxia Wang [1,2] and Qiuling Ge [1,2]

[1]   State Key Laboratory of Cryospheric Science, Northwest Institute of Eco-Environment and Resources, Chinese Academy of Sciences, Lanzhou 730000, China
[2]   The School of Resources and Environment, University of Chinese Academy of Sciences, Beijing 100049, China
*   Correspondence: jianping@lzb.ac.cn

Abstract: Glaciers are attracting increasing attention in the context of climate change, and glacier tourism has also become a popular tourist product. However, few studies have been conducted concerning the image of glacier tourism destinations. To address this gap in the literature, in this study, we extracted destination images from 138,709 visitor reviews of 107 glacier tourism destinations on TripAdvisor using latent Dirichlet allocation (LDA) topic modeling, identified destination image characteristics using salience−valence analysis (SVA), and analyzed the differences in glacier tourism destination image characteristics across seasons and regions. According to the findings, the image of a glacier tourism destination consists of 14 dimensions and 53 attributes, with landscapes and specific activities representing the core image and viewing location and necessity representing the unique image. We identified significant seasonal and regional differences in the image of glacier tourism destinations. Finally, we discussed the unique image of glacier tourism destinations, the reasons for differences in the images, and the characteristics of different glacier tourism regions. This research could assist in the scientific management of their core images by glacier tourism destinations, as well as in the rational selection of destinations and travel timing by glacier tourists.

Keywords: glacier tourism; perceived destination image; destination image uniqueness; user-generated content; online reviews; TripAdvisor; latent Dirichlet allocation; salience−valence analysis; destination management

Citation: Tang, F.; Yang, J.; Wang, Y.; Ge, Q. Analysis of the Image of Global Glacier Tourism Destinations from the Perspective of Tourists. *Land* **2022**, *11*, 1853. https://doi.org/10.3390/land11101853

Academic Editors: Cecilia Arnaiz Schmitz, Nicolas Marine and María Fe Schmitz

Received: 27 September 2022
Accepted: 18 October 2022
Published: 20 October 2022

Publisher's Note: MDPI stays neutral with regard to jurisdictional claims in published maps and institutional affiliations.

1. Introduction

Mountains are among the most popular destinations for tourists, with their spectacular landscapes, majestic views, and unique and comfortable valleys [1]. Although nature, wilderness, topography, remoteness, and climate limit the development of mountain areas, these features also represent the strengths of mountain tourism [2]. Mountain tourism is growing at an unprecedented rate, playing an important role in the global tourism landscape as an obvious means of achieving sustainable development in mountain areas [3] and is considered an important tool for local economic development and environmental management [4].

Glacier tourism in general is a subcategory of mountain tourism and plays an important role in creating mountain landscapes and enhancing the connotation and visibility of mountain tourism [5]. Glacier tourism has become one of the most popular tourism projects worldwide, creating considerable value for tourists and local communities alike. The distinctive landscape and artistic features of glaciers are perceived to provide aesthetic value to tourists [6], and the evidence provided by glaciers with regard to climate change makes glacier tourism extremely valuable with respect environmental education and popular science [7]. The value of glacier tourism has drawn a sizable influx of visitors. The world's most famous glacier tourism destinations include the Alps [8], New Zealand's west coast [9], Canada's Columbia Icefields [10], China's Greater Shangri−La [11], and others, with more than one million people visiting these locations each year. The arrival of tourists brings enormous economic value to local communities. More than USD 81 million each year is directly contributed to the economy by tourism associated with glaciers in New Zealand [12], and benefits with a value of more than USD 71 million were generated by tourism to China's Jade Dragon Snow Mountain in 2016 [13]. Furthermore, glacier tourism provides employment for local residents [14], and the building of related facilities may provide some indirect economic advantages to local communities [15].

However, today's mountain tourism is being hit by climate change and the COVID pandemic. Because mountain tourist infrastructure and activities rely on alpine temperature, topography, beauty, and seasonal cycles, climate change is having and will continue to have an impact on both current and future tourism growth in mountain regions, with consequences for residents in tourism−dependent mountain communities [16]. The booming glacier tourism industry has been negatively affected by the ongoing retreat of glaciers as the climate warms [17]. Melting glaciers will degrade the quality of the glacial scenery [11], increase the risk of rockfall during tourism activities [18], impair the tourist experience, and decrease the number of tourists visiting glaciers. Between 2003 and 2009, the number of visitors to Norway's Jostedalsbreen National Park decreased by 38%. The primary causes of this reduction were changes in glacier morphology and accessibility [19]. Another group of academics believes that although climate change has accelerated the melting of glaciers, it has also increased new glacier tourism opportunities and visitor motivation to engage in "last chance tourism (LCT)" [20–22], which encourages tourists to experience this type of tourism before it is endangered [23], increasing the number of tourists. In addition, despite the considerable negative impact of the COVID pandemic on tourism, the therapeutic effects of natural landscapes could bring more opportunities for tourism in a post−COVID era [24]. Therefore, glacier tourism may remain popular over the coming decades or even reach a new peak of development.

In this context, glacier tourism destination management is particularly significant. The destination is the core of tourism [25], and the management of the destination is an important factor affecting the development of glacier tourism [26]. Scientific management of tourism destinations can assist with adaptation to the negative effects of climate change [27], increase visitor satisfaction, and promote regional economic development. However, the majority of existing research on glacier tourist destination management has focused on climate change adaptation [8,28,29], suggesting that glacier tourism destinations should adopt adaptive measures, such as management changes, developing new activities, enhancing educational activities, and changing the seasonality and spatiality of activities [5,27]. Such adaptive management measures represent a long−term strategy oriented toward climate change, with the goal of achieving sustainable development of glacier tourism. In contrast, the core image represents the main attraction of a glacier tourism destination and is the main factor influencing the glacier tourism experience of tourists. Targeting the needs of tourists and improving their satisfaction by improving the core image of the destination is perhaps more effective in the short term. Destination image is critical for destination management decisions and positioning [30]. Destination image creates brand value, is a crucial competitive asset [31], and is a powerful management tool for tourism [32]. Understanding the image of a destination helps tourism operators to attract more visitors and predict their behavioral intentions [33]. Therefore, identifying the image of glacier tourism destinations from the perspective of tourists is important for the management of glacier tourism destinations. However, few studies have been conducted concerning the experience of tourists at glacier tourist destinations and their perception of the image of the destination.

Currently, there are two main paradigms for tourism destination image research: structured and unstructured [34]. The structured paradigm refers to researchers' attempt to construct a framework for a destination image based on relevant theories, under which subimages can be divided, mainly by means of structured questionnaires. The unstructured paradigm involves distilling and summarizing the respondents' free descriptions of destination image to capture the destination image. Early destination image studies relied mostly on structured questionnaire and interview data, but with the development of the Internet, user-generated content (UGC) has proliferated, causing not only a paradigm shift from structured to unstructured destination image studies, but also in traveler-generated content (TGC), the data sources for destination image studies experienced a shift from travel blogs to online travel reviews [35]. Thus, online travel reviews (OTR) based on social media has become an important data source for destination image research.

Therefore, in this study we extracted images of glacier tourism destination and analyzed the differences in their characteristics based on tourist reviews of glacier tourism locations on TripAdvisor. Specifically, we wanted to achieve the following research objectives (ROs):

1. The creation of an overall image of global glacier tourism destinations. Reviews of glacier tourism destinations were aggregated by country, and the potential image themes of glacier tourism locations in these countries were extracted separately and finally combined to form the overall image of global glacier tourism locations.
2. Core image recognition of tourist destinations with glaciers. Features of the glacier tourist destination's image were analyzed using the significance and positivity of the destination image as indicators.
3. An analysis of the image characteristics (indicating the degree of importance and positivity of the image) of glacier tourism destinations in various seasons. Reviews of glacier tourism destinations were gathered by season, and the destination images in different seasons were extracted.
4. An analysis of the image characteristics of glacier tourism destinations in various regions. The glacier tourist destinations were divided into six regions—North America, South America, Nordic, the Alps, the Qinghai–Tibet Plateau, and New Zealand—and the destination images were extracted for each region.

This study, on the one hand, closes a research gap concerning the image of glacier tourism destinations and, on the other hand, may assist agencies managing glacier tourism destinations to better understand the perceptions and experiences of visitors and therefore make better management decisions. In addition, the reported results may assist travelers in making trip decisions according to how closely their preferences match the image of glacial tourism destinations.

2. Key Concepts and Definitions

2.1. Glacier Tourism and the Glacier Tourism Destination

Glacier tourism arose with pilgrimage, expedition, and mountaineering in the 18th century; developed in the 20th century with mass tourism; and has been popular since the 1980s with leisure and experiential tourism activities. There is currently no accepted definition of glacier tourism due to the differing disciplinary backgrounds and research objectives of scholars engaged in research in this field. Pralong and Reynard describe glacier tourism as a synthesis of several types of tourism in glacier areas, such as geology tourism, mountain tourism, and adventure tourism [36]. Liu et al. define glacier tourism simply as tourism activities such as sightseeing, scientific research, exploration, and popular science education that take place in a glacier area [37]. According to Wang et al., glacier tourism refers to alpine tourist experiences or activities for which glacier resources or glacial relics represent the primary attraction [38]. Purdie expands the scope of glacier tourism by stating that in addition to activities that take place on a glacier, glacier tourism also includes activities that take place in adjacent areas, such as on glacial lakes, glacier inlets, or fjords [9]. Although a uniform understanding of glacier tourism has not yet been established, it is evident that it consists of glacier tourism resources and glacier tourism activities. A tourist destination is a location with unique natural or manmade features that attract non−native tourists to experience a variety of activities [39]. Therefore, unlike the concept of glacier tourism, the glacier tourism destination consists of glacier tourism resources, tourism infrastructure, service communities, transportation access, etc.

It is located close to the natural body of the glacier and within its surrounding extended area [40]. Glacier tourism destinations can be regarded as market−oriented multifunctional carriers based on glacier resources.

2.2. Tourism Destination Image

A tourism destination image is described as the totality of the impressions, feelings, and beliefs of tourists about a destination [41]. It starts to take shape before tourists arrive, and the visitor's experience will cause their image of the destination to evolve dynamically [35]. Echtner and Ritchie construct the destination image using three axes: functional–psychological, common–unique, and attribute–holistic, and propose a combination of standardized measures and open−ended questions to generate the destination image [34]. Gartner suggests that a destination image comprises three parts: cognitive, affective, and conative [42]; this definition has been widely accepted by tourism researchers [43,44]. The cognitive image is constructed in the tourist's mind based on facts about the destination and is the sum of what the individual knows or believes about the destination [45]. The affective image refers to the individual's emotional responses or appraisals, which reflect their feelings about the destination [46], and the identification of an emotional image helps tourists to pursue benefits that match the emotions associated with the destination, thus creating a more positive image of the destination [47]. Conative image is the motivation, preference, or behavioral intention of the visitor after being influenced by cognitive and emotional images [48]. Therefore, destination image theory proposes that cognitive and affective images represent an individual's subjective associations or impressions about the attributes of a destination [42], and the conative image depicts the individual's own idealized and desired future condition [49].

In summary, in this paper, we define the destination image of glacier tourism as the impressions, feelings, and behavioral intentions of tourists toward glacier tourism resources and activities, tourism infrastructure, and other elements. It includes cognitive images consisting of the glacier landscape, glacier activities, tourism transportation, tourism services, etc.; affective images consisting of excitement, enjoyment, worthiness, etc.; and conative images consisting of behaviors, such as willingness to recommend or revisit glacier tourism locations.

3. Data and Methodology

3.1. Data Sources

The data used in this study were derived from glacier visitor online review data from TripAdvisor. Depending on the source of information, the medium of destination image formation can be divided into induced (emanating from the destination promoters), organic (transmitted between individuals), and autonomous (produced independently of the previous categories) [42]. Organic agents include, along with the experience itself, the opinion of users and consumers that spreads through word−of−mouth marketing (WoM) in conversations with relatives, friends, colleagues, or acquaintances [35]. As a form of electronic word−of−mouth, online reviews contain a mixture of facts, opinions, impressions, emotions, etc., published and disseminated to others by travelers [50], which can influence the decision making of consumers and managers [51,52]. Consumers are more likely to trust online consumer evaluations than information provided by the operator because the former may contain important details that the latter is reluctant to make public [53]. Furthermore, TripAdvisor, as one of the world's top travel service platforms, is involved in travel marketplaces in a variety of nations and languages. In February 2022, the platform's official website reported that it had amassed one billion online comments and opinions. In addition, TripAdvisor protects its reputation by preventing bogus reviews via a regulatory system, has established a certain level of credibility and user trust in the sector, and has become a crucial database for researchers [54,55].

3.2. Data Collection

Some research has offered pertinent listings for locations worldwide that are popular for glacier tourism. We combined these listings and searched on the TripAdvisor website using the keywords "glacier" and "snow mountain." Following the consolidation of some of the retrieved projects, such as "Mendenhall Glacier" and "Mendenhall Glacier Visitor Center," a total of 107 glacier tourism destinations (Figure 1) from 16 countries (Table 1) were ultimately discovered. Using a Python crawler that we created, in June 2022, we automatically crawled 138,709 glacier visitor reviews. The fields that were crawled contained "username", "hometown", "comments", "date", and "score". The earliest review in this set was posted in July 2003, and the most recent was posted in June 2022. Our dataset includes English, Spanish, Portuguese, Chinese, German, and other languages and was uniformly converted to English by calling the Google Translate API.

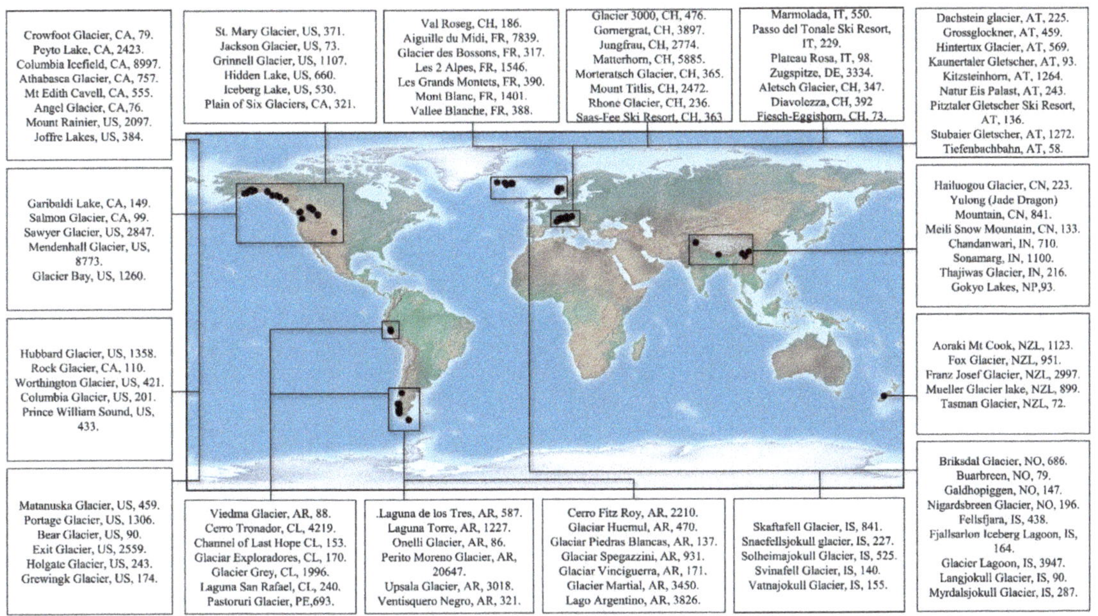

Figure 1. Location of the glacier tourist attractions. (The boxes indicate the name of the glacier, the abbreviation of the country, and the number of comments from visitors. Data from https://www.tripadvisor.com/ (accessed on 12 June 2022)).

Table 1. Descriptive statistics of collected glacier travel reviews.

Country	No. of Attractions	Reviews	Percentage	Country	No. of Attractions	Reviews	Percentage
Argentina	14	37,169	26.80%	India	3	2026	1.46%
Austria	9	4319	3.11%	Italy	3	877	0.63%
Canada	10	13,566	9.78%	Nepal	1	93	0.07%
Chile	5	6778	4.89%	New Zealand	5	6042	4.36%
China	3	1197	0.86%	Norway	4	1108	0.80%
France	6	11,881	8.57%	Peru	1	693	0.50%
Germany	1	3334	2.40%	Switzerland	12	17,466	12.59%
Iceland	10	6814	4.91%	United States	20	25,346	18.27%

3.3. Data Preprocessing

The text data produced by user comments are unstructured and contain a lot of noisy information, which may seriously interfere with the results when used directly [56]. Therefore, data preprocessing has become a fundamental step in text data analysis [44,57,58]. In this study, the NLTK package in Python was selected for data preprocessing, comprising a set of text processing packages for classification, tokenization, stemming, parsing, and semantic reasoning. First, all text is converted to lowercase, and special punctuation is removed. Secondly, word splitting and loading of stop words were performed. Then, the text stop words were set, in addition to generic words such as "is", "it", "that", etc. In addition, the names of some glacier tourist places, such as "Zugspitze", "Mendenhall", etc., were deactivated. Finally, POS filtering and word stemming were performed. Figure 2 shows the process for achieving the research objectives.

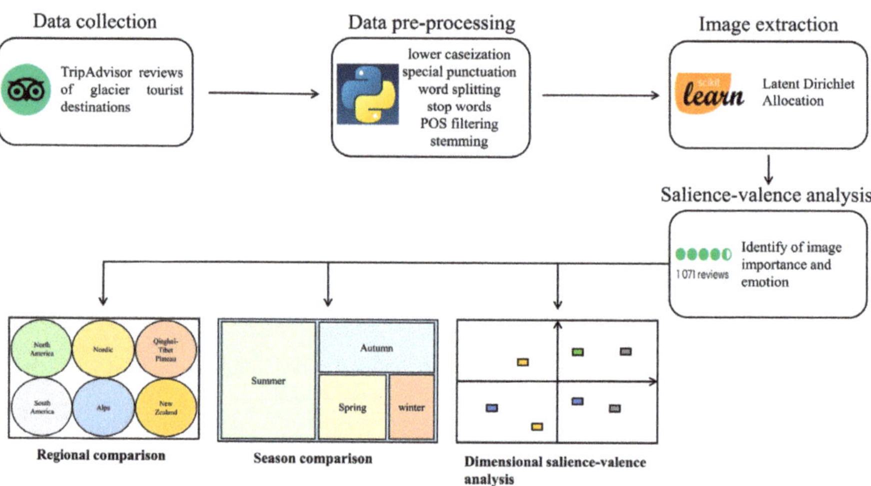

Figure 2. Research process.

3.4. LDA Topic Model

Latent Dirichlet allocation (LDA) is a generative probabilistic model used to process text data, representing text as a random mixture of potential topics in which each topic is characterized by a distribution of words [59]. The use of the LDA topic model can allow potential topics to be discovered from a large amount of unstructured textual big data [60], which helped us to construct an image of glacier tourism destinations in the world. These potential topics need to be named, usually with reference to the few words with the highest probability within the topic, as performed by one researcher and determined by another researcher. Figure 3 presents a graphical representation of LDA adapted from Blei et al. [59]. In the figure, α denotes the Dirichlet distribution of the first document topic, from which the topic distribution of document θ (polynomial distribution) is obtained, and from θ, a series of topics (z) can be derived. β denotes the topic–word Dirichlet distribution, and the word distribution (φ) (polynomial distribution) corresponding to topic z is generated by sampling from β. Finally, the words (w) are generated by combining z and φ. One word is extracted from each topic, and these words are connected to obtain a document. This is repeated several times, generating a large number of documents in the corpus. Finally, it is compared with the original document to determine the best way to distribute the points of the Dirichlet distribution.

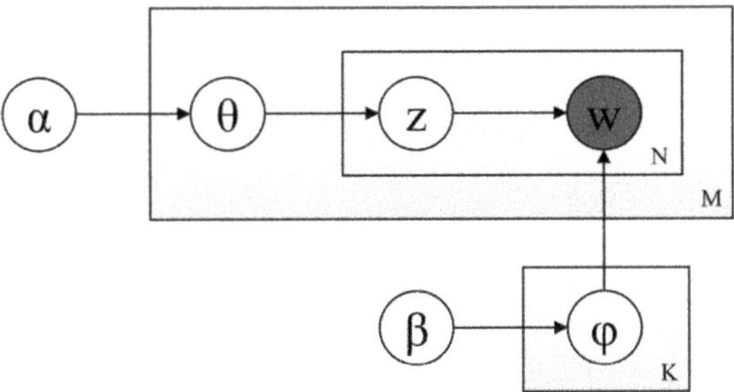

Figure 3. Graphical model representation of LDA.

The scikit−learn library in Python makes it easy to implement LDA modeling. However, it is first necessary to determine the values of the four main parameters: alpha, beta, n_topics, and n_iter (number of iterations). The corpus−level hyperparameters alpha (α) and beta (β) directly affect the LDA results, with smaller alpha values implying fewer dimensions per comment and smaller beta values resulting in fewer words per dimension [58]. Perplexity represents the uncertainty of the trained model with respect to which topic a document (d) belongs; the lower the perplexity, the more effective the model [52]. Therefore, in this study, perplexity was chosen as the basis for determining the values of the alpha, beta, and n_topic parameters. With reference to the experience of the Taecharungroj study, the alpha was set between 1 and 0.1, and the beta was set between 0.1, 0.01, and 0.001 [58]. When constructing the overall image of global glacier tourist locations (RO1) and determining the topic number, on the one hand, we required more theme numbers to ensure a more comprehensive extracted image, but on the other hand, we required a lower perplexity level. We ultimately set the number of topics to 5 when the visitor had posted fewer than 1000 comments. When the visitor had posted more than 1000 comments, the number of topics was set to 11. For the number of topics in RO3 and RO4, the number of topics with the lowest confusion was selected. In addition, alpha was set to 1, and beta was set to 0.1 for the lowest perplexity, and n_iter was taken to be 2000 to ensure convergence of results.

3.5. Salience–Valence Analysis

To achieve RO2, in this study, we used the diagnostic tool salience−valence analysis (SVA) developed by Taecharungroj et al. to identify the importance and emotional color (positive or negative) of each image [57], where salience is expressed as the total number of visitor reviews for the image, and valence is expressed as:

$$\text{Image Valence} = (5 \text{ bubble reviews} - \text{Others bubble reviews})/(\text{Total reviews of the image}) \tag{1}$$

Although TripAdvisor officially states that 5−bubble reviews from visitors indicate excellence and 3−bubble reviews are the average, the average score for all reviews in this study was 4.68. We observed that when tourists give a rating of fewer than 5−bubbles, it means that the tourist destination has at least some factors that make tourists feel dissatisfied. Therefore, in this study, we set reviews with 5−bubble ratings as "above average" reviews and all the other levels as "below average" reviews.

4. Findings

4.1. Overall Image of the Glacier Tourist Destination

Images of glacier tourism destinations (Appendix A, Table A1) in 16 countries were extracted using the LDA model and aggregated to obtain a global glacier tourism destination image system consisting of 14 dimensions and 53 attributes (Table 2). Among the 14 dimensions, the landscape dimension was dominated by attributes such as *mountain, landscape, glacier, ice,* and *glacial lake.* The specific activities of glacier tourism included *hiking, skiing, cruising,* and *helicopter sightseeing.* The most important means of transportation for glacier tourism are *cable cars, boats,* and *trains. Whale, bear,* and *seal* were the main animals encountered during glacier tours. The most important infrastructure items were *roads* and *visitor centers.* Landscape features included *color* and *magnificence.* The price image was mainly described in terms of *tickets.* The tourism environment included *accessibility* and *weather/climate.* Viewing locations were specifically represented as *viewpoints* and *viewing platforms* where visitors could take photographs. *People* and *food* were used as separate dimensions and attributes, with *people* mainly indicating the atmosphere among visitors and the type of companions. The visitor experience was mainly expressed in terms of worthwhile, *enjoyable,* and *kindness.* Glacier tourism necessities included *water* and an *oxygen tank.* The *other* attributes were results that were difficult to interpret, had little relevance, or were modeled randomly or incorrectly. Appendix A Table A2 shows the image of glacier tourism destinations in the attribute dimension and their main topic words.

Table 2. Results of overall destination images.

Dimension	Percentage	Attribute	Percentage	Dimension	Percentage	Attribute	Percentage
Landscapes	30.83%	Ice	3.70%	Local Infrastructure	13.99%	Road	6.26%
		Ice Field	1.19%			Piste	0.13%
		Glacier	4.41%			Skywalk	1.08%
		Snow	1.99%			Visitor Center	5.02%
		Mountain	7.74%			Restaurant	1.46%
		Glacier Lake	2.26%			Parking	0.04%
		Valley	0.33%	Landscape Features	4.37%	Color	1.98%
		Waterfall	2.40%			Sound	0.36%
		Landscape	6.81%			Altitude	0.17%
Specific Activity	18.40%	Hike	8.81%			Slope	0.53%
		Ski	2.84%			Magnificent	1.33%
		Self−Driving Tour	0.91%	Price	0.93%	Price	0.09%
		Climb	0.83%			Ticket	0.84%
		Ride Horse	0.15%	Travel Environment	8.00%	Accessibility	4.59%
		Cruise	2.17%			Weather/Climate	2.64%
		Helicopter Sightseeing	1.90%			Season	0.55%
		Adventure	0.79%			Crowdedness	0.22%
Transportation	10.05%	Cable Car	3.64%	Viewing Location	2.93%	Viewpoint	2.29%
		Bus	0.84%			Viewing Platform	0.64%
		Troll Car	0.08%	People	2.23%	People	2.23%
		Train	1.84%	Experience	0.71%	Worthwhile	0.45%
		Boat	3.54%			Kindness	0.10%
		Sledge	0.11%			Enjoyable	0.16%
Animals	3.37%	Seal	0.35%	Necessity	1.30%	Water	1.23%
		Whale	2.01%			Oxygen Tank	0.07%
		Bear	1.01%	Other	0.30%	Other	0.30%
Food	2.59%	Food	2.59%				

Figure 4 intuitively shows the salience–valence analysis at the dimension level of the glacier tourist destination image. Figure 4 shows that the landscape is the most salient (30.83%) and valent (0.63) image associated with glacier tourism locations, vastly outnumbering other images, showing that the landscape of glacier tourism destinations is the most popular factor for tourists. The salience of specific activities (18.4%), infrastructure (13.99%), transportation (10.05%), and tourism environment (8%) were also prominent and contained many types of secondary attributes, which have a significant impact on glacier tourism. Although the infrastructure salience was high, its valence (0.45) was lower than the overall average, and building and maintaining improved infrastructure is a common challenge for glacier tourism destinations. Among the other images in the dimension of glacier tourism, animal salience (3.37%) was not high, although its valence (0.73) was the highest, showing that it is a major highlight of glacier tourism. Conversely, price had a low salience (0.93%) and the lowest valence (0.31). As one Australian tourist commented on Zugspitze in March 2018: "The view from the gondola is fantastic, but unfortunately the price−performance ratio is not right at all". Therefore, although the gorgeous landscape of glacier tourism can provide a high−efficiency valence to tourists, the negative feelings induced by expensive pricing must also be considered.

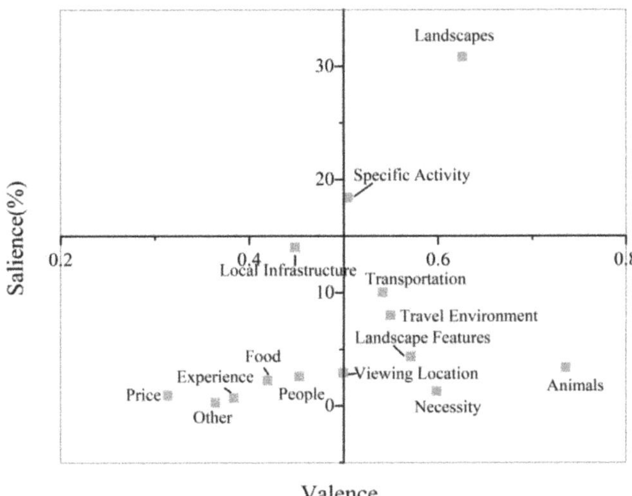

Figure 4. Dimensional salience–valence analysis.

Figure 5 shows the attribute−level valence of the image of glacier tourism destinations, with high attribute valence including *seal* (0.88), *ice* (0.77), *whale* (0.75), *glacier* (0.67), *cruising* (0.65), *color* (0.65), and *water* (0.62). On the other hand, the attributes of *price* (−0.38), *sledding* (−0.22), *skywalk* (−0.21), and *enjoyed* (0.02) had lower valence. Among the attributes, although the infrastructure valence at the dimensional level was low, the valence of *visitor center* (0.53) and *parking lot* (0.59) was high. Visitor centers are important in glacier tourism for provision of information on glacier mechanisms and glacier recession [61], thus deepening the experience of glacier tourism for tourists. For example, a visitor from Texas, USA, commented in August 2018 on the Exit Glacier visitor center: "The park visitor center is great and it is important to learn about the history and how our planet is changing". Another visitor commented: "If you are in the area, spare a day to drive to the glacier and hike to the terminus. The visitor center at the beginning of the hike is small and throws light on how the glacier has been retreating over the past few decades. The initial hike path is paved and has signs that show where the glacier terminus was and how it moved back". As a result, visitor centers play an important role in educating visitors about science and encouraging environmentally friendly behavior.

4.2. Seasonal Difference Analysis of GTD Image

Figure 6 shows the tree heat map of the image of glacier tourism destinations containing three types of information: season, dimensional image, and attribute image. The color indicates valence, and the block area indicates salience (i.e., the number of comments accounted for). As shown in Figure 6, the average valence of the four seasons did not differ significantly, with valence scores in the range of 0.52–0.56. However, summer (58,774) had a substantially higher importance score than autumn (35,952), spring (25,038), and winter (18,945), indicating that summer is still the peak season for glacier tourism locations, although glacier scenery is more impressive in the winter. Appendix A Table A3 shows the specific values of the seasonal image of glacier tourism destinations.

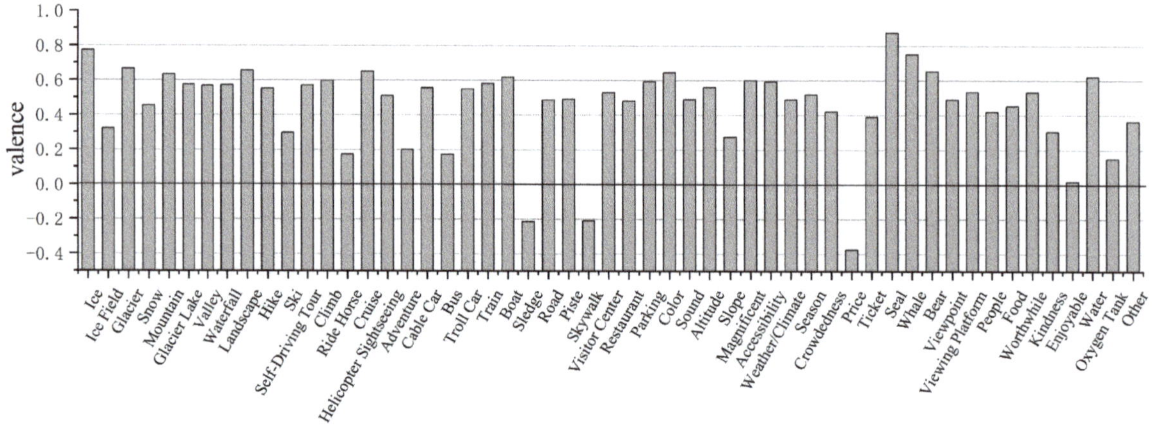

Figure 5. Attribute valence results.

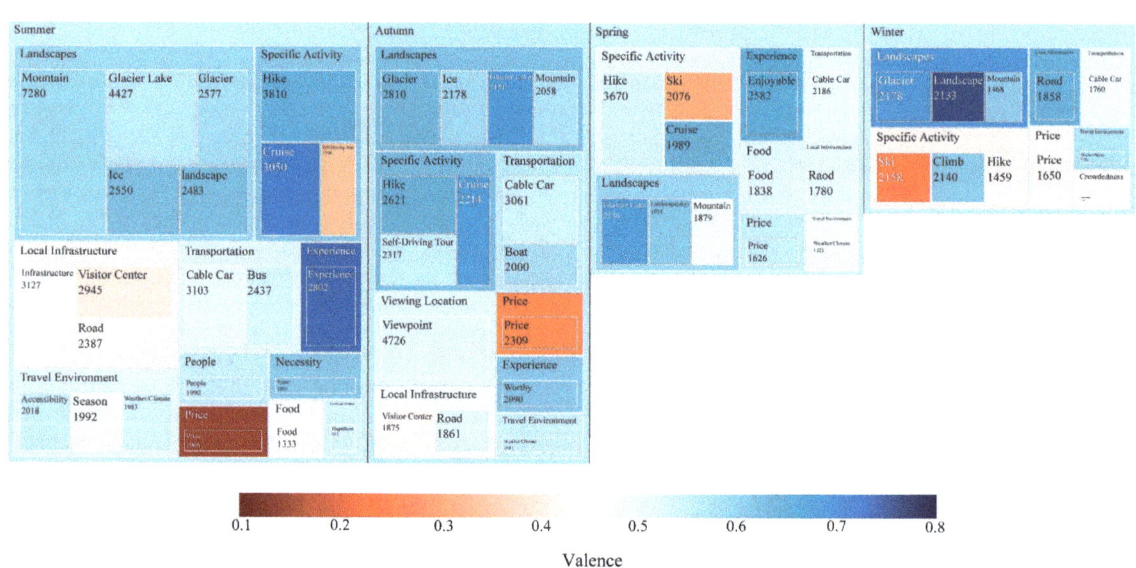

Figure 6. Seasonal differences in the image of glacier tourism destinations.

In terms of image categories, the landscape categories did not vary considerably and were mainly dominated by *mountains, glaciers,* and *glacial lakes.* Specific activities featured in each of the four seasons, with *self−driving* and *cruising* activities occurring primarily in the summer and fall, whereas *skiing* occurred only in the winter and spring, and *hiking* occurred in all four seasons as the core glacier tourism excursion. In addition, images such as *roads, cable cars, prices,* and *weather/climate* were also common to all seasons. Weather can have a direct impact on glacier tour operations, with overcast and foggy circumstances restricting visibility and accessibility to the glacier and affecting glacier tour activities, such as helicopter tours and hikes [62] and therefore valued by visitors regardless of the season. *Accessibility* is an image that was unique to summer. In the context of climate change, global glacier melting has accelerated, often forming large crevasses or producing disasters, such as ice avalanches and rockfalls, making glaciers difficult to access [63]. Rising temperatures in summer cause rapid melting of glaciers and increased tourism instability, which in turn complicates tourist access to glaciers [64]; therefore, the image of accessibility occurred mainly in summer.

In terms of image valence, although there was little difference in the average valence of the four seasons, the valence characteristics of its internal dimensions and attributes were significant. The image of the winter *landscape* (0.69) had significantly higher validity than the other seasons, and winter and spring prices were more acceptable to visitors than those in the summer and fall. Although *skiing* activities are only present in winter and spring, visitors did not seem to be satisfied with them, in contrast to hiking and cruising, for which the valence was consistently higher. The *experience* of emotional image (0.74) was the highest—valence attribute in summer, indicating that tourists were highly satisfied with the overall experience of glacier tourism in summer.

4.3. Regional Difference Analysis of GTD Image

Figure 7 shows the regional image of the glacier tourism destination consisting of three types of information: region, dimensional image, and attribute image. The color indicates the valence, and the number indicates the salience (number of comments). In terms of the inner ring, which indicates the overall valence and salience of the image of the region, the three regions of North America, South America, and the Alps had the most reviews, accounting for roughly 87.5% of total reviews, showing that these three areas are the world's top glacier tourism destinations. In terms of image valence, the Nordic (0.65) and South American (0.63) regions were the highest, followed by the North American (0.53) and Alpine (0.52) regions, whereas the New Zealand (0.31) and Tibetan Plateau (0.06) regions had the lowest valence. In terms of attribute image categories, North American (18 categories) and South American (17 categories) regions had the most attribute categories and were comprehensive glacier tourism destinations. In contrast, the New Zealand and Tibetan Plateau regions had only 11 categories of attributes, and their glacier tourism functions and elements were more singular. Appendix A Table A4 shows the specific values of the regional image of glacier tourism destinations.

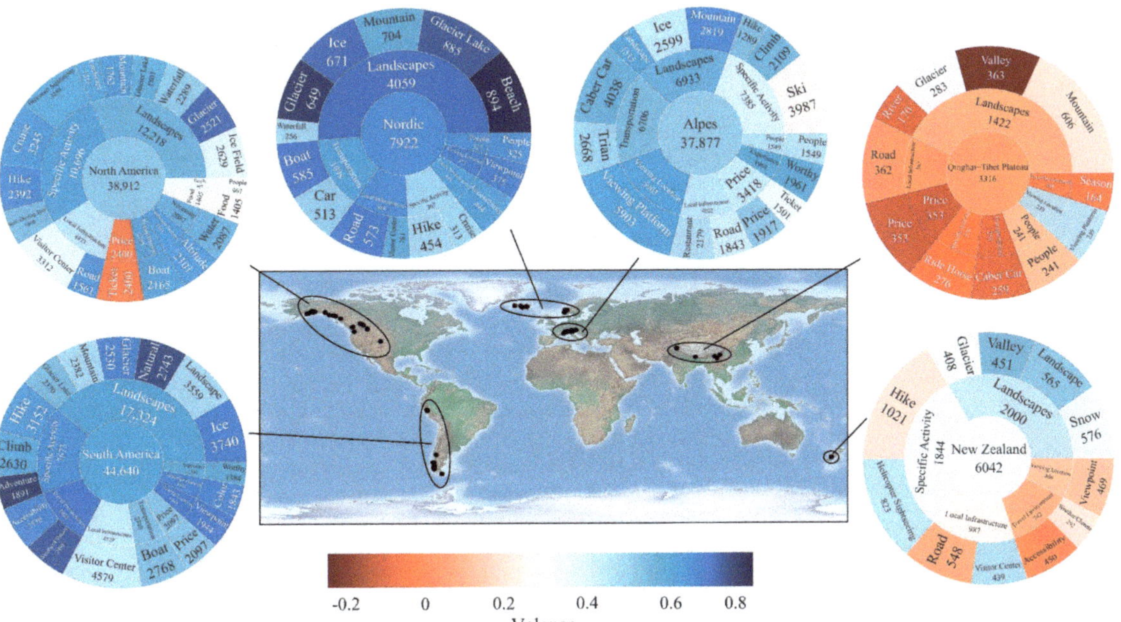

Figure 7. Regional differences in the image of glacier tourism destinations.

For the central ring, which reflects the dimension image of the region, the South American region was most popular for its travel environment, landscape features, and viewing locations, and its landscape and specific activities were most significant, although visitors had a relatively poor impression of its infrastructure. The strength of the image of the North American region lay in its specific activities and landscapes, which were the main image of its glacier tourism destinations and which were highly valued and widely recognized by tourists. Conversely, the high price of glacier tourism in North America was one of its few criticisms. The Nordic region won over visitors with its superb landscapes, and the viewing locations and infrastructure were equally popular with visitors, although they did not rate specific activities, which were relatively few and not easy to access. The Alpine region had the highest landscape, viewing location, and experience validity, as well as higher price satisfaction compared to other regions. The Qinghai–Tibet Plateau region had the lowest image valence of all the regions, with price (-0.11), tourism environment (-0.07), and transportation (-0.04) having the worst image and only the scenic location (0.47) having a slightly better image, whereas the rest of the image valence was negative or close to zero. The landscape and specific activities in the New Zealand region were relatively more popular with tourists but also had less valence than in other regions.

Finally, the outer ring reflects the image of a region's attributes. With regard to attributes under the *landscape* dimension, the naturalness of the *landscape* in the South American region (0.76) was the most prominent; North America was the most popular for *glaciers* (0.73); Northern Europe, as the region with the highest landscape valence, was highly praised by tourists for its *beaches* (0.81), *glaciers* (0.79), and *glacial lakes* (0.75); the Alpine region had the highest valence for *mountains* (0.68); the landscape of the Tibetan Plateau region seems to be unrecognized by tourists, and its highest valence was only 0.32 for *glaciers*; and although the *valley landscape* was the most criticized in the Tibetan Plateau Region, it was very different in New Zealand, reaching a valence of 0.56. In terms of specific activities, those that are unique to each region included *adventure* and *climbing* in South America, *cruising* and *hiking* in North America and Northern Europe, *climbing* and *skiing* in the Alps, *horseback* riding in the Tibetan Plateau, and *helicopter tours* in New Zealand. Although *skiing* was an image specific to the Alpine region, its satisfaction was not high, and the same applied to *horseback riding* in the Tibetan Plateau region. In terms of *price*, the valence of the regions was not low, with the exceptions of North America and the Tibetan Plateau. Notably, *accessibility* appears in the images of South America, Northern Europe, and New Zealand, suggesting that the approach and route to the glacier were important to them. The image of *viewing platforms* in the Alps and the Qinghai–Tibet Plateau was significant because glacier tourism in these two locations is mainly associated with sightseeing at the top of mountains, making the construction of viewing platforms particularly important.

5. Discussion and Conclusions
5.1. Discussion

The results of this study indicate that the image of glacier tourism locations consists of 14 dimensions. Compared to previous studies, the five most common categories were landscape, specific activities, transportation, infrastructure, and travel environment [34,43], and the image dimensions of animals, price, people, and food were also consistent with some previous research [45]. The difference is that the landscape features, viewing locations, and necessities (e.g., oxygen bottles and water) are unique images of glacier tourism. Glacial landscape features are an important reference for tourists, which produce the image of tourist destinations, and the evaluation of the glacier landscape by tourists is based on their perception and experience of glacier features, such as color, shape, texture, and sound [65]. Viewing location is also crucial to glacial landscape sightseeing. For example, tourists commented that the view of the golden mountain of sunshine (formed by sunlight hitting the top of the mountain and being reflected by the glacier) on China's Hailuogou Glacier could only be seen from specific viewpoints. The image related to necessities was also significant because glacier tourism activities mainly involve hiking, and the availability of water is indispensable; similarly, some glaciers are at high altitudes, and the air is relatively thin, so tourists participating in hiking and climbing activities may also need to carry oxygen tanks. In addition to the abovementioned cognitive images of tourism destinations, in this study, we also captured the affective image of an experience consisting of "worthwhile", "kindness", and "enjoyable", a type of image that is often difficult to capture in unstructured tourism destination image constructs [44]. The conative image is more associated with the actions of the visitor after the trip, such as revisiting and recommending the destination [48,66]. Thus, although the conative image cannot be captured, the cognitive and affective images are sufficient to represent the overall image of the destination [42,67].

Among the images of glacier tourism destinations, landscapes and specific activities have high salience and high valence and are thus the core images of glacier tourism. Additionally, there is little of the human element to attract visitors to glacier tourism sites, in contrast to the image of mixed tourism sites (e.g., city tourism), where culture, entertainment, and experience are perhaps more important [44,45,58]. Similarly, the phenomenon of lower infrastructure valence is easy to explain. Glaciers are usually located in high−latitudes or high−altitude mountains far from urban areas, with fragile natural environments that are prone to rockfall due to glacier recession [18], so the infrastructure construction and maintenance costs of glacier tourism sites are higher than those of the general tourism type, resulting in a lower valence of the image. The price image has the lowest valence, perhaps because tourists are required to pay extra for the protection of the glacier [68]. However, the high cost of travel is not unique to glacier tourism destinations, and tourist dissatisfaction with travel prices seems to be common across all types of travel [57,69,70].

The analysis of the seasonal image of glacier tourism destinations shows that glacier tourism sites have distinctive summer and winter characteristics. Visitors were significantly more satisfied with the winter landscape than the summer landscape, mainly because the summer is affected by high temperatures, leading to snow melting and glacier recession, whereas the low winter temperatures allow glaciers to be replenished by snowfall [71], and the combination of snow with mountains and forests can itself have a strong visual impact. Although tourist satisfaction with glacier skiing was not high, the number of glacier skiing visits was significantly higher in winter than in summer because of snowfall and microclimate fluctuations [72]; therefore, skiing is another feature of glacier tourism undertaken in winter. Glaciers are more stable in winter, so for glacier tourism, glacier−climbing activities are more appropriate in winter. Another advantage of winter glacier tourism is the price [68]. As it is the low season, tour operators usually attract tourists by lowering the prices of entrance fees and hotels [73], with the result that tourists are significantly more satisfied with the price in winter than in summer. On the other hand, summer glacier tourism is characterized by a diversity of landscapes, activities, and tourist service features; comfortable temperatures; and a climate more suitable for outdoor sports [74], hiking to experience the natural charm of the wilderness, or cruising the fjords to see glaciers and whales [75]. As a consequence, the experience image of the summer received the highest rating of the four seasons. The spring and fall seasons are not as distinctive as winter and summer, but they are still good times for glacier tourism.

Although the main attraction of glacier tourism is wholly based on the geographic environmental elements of the glacier destination [40], differences in regional characteristics can produce different characteristics at different glacier tourism sites. The South American and North American regions are well−developed and comprehensive glacier tourism destinations, with diverse and efficient landscape types and tourism services. However, the South American region is more prominent for adventure activities, such as hiking and climbing in glacier areas. The North American region is more diverse, and in addition to hiking, glacier tours by cruise ship, helicopter, or self−drive are also significant [76]. Northern Europe possesses the ultimate natural scenery, and cruises on glacial lakes are a key feature [77]. The Alps are a typical mountain glacier tourism destination, with world−renowned glacier ski resorts featuring skiing, climbing, and sightseeing by cable car and train [29]. The Qinghai–Tibet Plateau region is also a mountainous glacier tourism area, but there is no skiing, and mountaintop sightseeing and horseback riding in the valley are its specialties [78]. Glacier tours in the New Zealand region feature wilderness hikes and glacier tours by helicopter [79].

5.2. Conclusions

In this study, we crawled 138,709 online reviews (text data) of 107 glacier tourism destinations on TripAdvisor and extracted images of glacier tourism destinations using the LDA theme model, with some interesting results. First, we found that the world glacier tourism destination image system consists of 14 dimensions and 53 attributes. Landscape and specific activities were the core elements of glacier tourism. Landscape features, viewing locations, and necessities had unique significance for glacier tourism, whereas infrastructure and prices were influenced by the environment of glacier tourism sites, resulting in low valence. Secondly, the image of glacier tourism destinations varied significantly on a seasonal basis, with summer glacier tourism sites having a variety of image types, tourism features, and a comfortable climate. Glacier tourism in winter was regarded as better in terms of landscape and price, and skiing activities were more appropriate in winter. Finally, the regional differences in the image of glacier tourism destinations were obvious. The world glacier tourism market is mainly concentrated in South America, North America, and the Alpine region, and each regions has its own glacier tourism characteristics.

Our study has certain theoretical and practical implications. We extracted images of glacier tourism destinations based on tourist reviews of glacier tourism destinations worldwide, complementing research on glacier tourist experiences and destination image perceptions, as well as enriching unstructured research on destination images. In addition, the destination image analysis method based on the LDA theme model and SVA proposed in this study could help tourism managers to identify the key attributes of tourism destinations for scientific decision making and planning to enhance their attractiveness. Finally, the analysis of seasonal and regional variations in the image of glacier tourism destinations presented in this study could help tourists match their preferences when choosing glacier tourism destinations and the appropriate timing for their trips.

This study is subject several limitations that need to be addressed in future research. First, the study sample may be biased; we used data from TripAdvisor glacier traveler reviews, and glacier destinations and traveler reviews that did not appear on that platform were not taken into account. Secondly, the crawled visitor comments were written in multiple languages, and because of the large sample size, we converted them to English uniformly by calling the Google Translate API, a process that may lead to changes in the semantics of some words. In addition, for the image season analysis part of the destination, we used the visitor writing date for the season classification; however, this date may lag behind the visitor tour date, possibly biasing the results. Finally, traveler-generated content belongs to the organic information sources of destination image, and it is an interesting exercise to compare destination images derived from different information sources (induced, organic, and autonomous) [80]. In addition, with advances in machine learning and natural language processing capabilities, images and videos can serve as new sources of data in the study of destination images [81].

Author Contributions: Conceptualization and methodology, J.Y. and F.T.; data resources and curation, F.T., Y.W. and Q.G.; analysis, J.Y., F.T. and Y.W.; visualization, F.T.; writing—original draft preparation, F.T.; writing—review and editing, J.Y.; funding acquisition, J.Y. All authors have read and agreed to the published version of the manuscript.

Funding: This research was supported by the Strategic Priority Research Program of the Chinese Academy of Sciences (grant no. XDA23060704).

Institutional Review Board Statement: Not applicable.

Informed Consent Statement: Not applicable.

Data Availability Statement: The data presented in this study are available upon request from the corresponding author.

Acknowledgments: Many thanks to Qin Ji and Shunshun Qi for their help with this article.

Conflicts of Interest: The authors declare no conflict of interest.

Appendix A

Table A1. Results of national LDA topics.

Argentina	Austria	Canada	Chile	China	France	Germany	Iceland
Hike	Mountain	Adventure	Accessibility	Altitude	Cable Car	Snow	Accessibility
Accessibility	Slope	Ice Field	Worthy	Cable Car	Climb	Train	Glacier Lake
Ice	Crowdedness	Hike	Mountain	Kindness	Ticket	Mountain	Mountain
Glacier	Viewing Platform	Skywalk	Glacier Lake	Accessibility	People	Restaurant	Road
Boat	Restaurant	Glacier Lake	Road	Landscape	Train	Viewing platform	Seal
Mountain	Ice	Self−Driving Tour	Food	Ticket	Landscape	Cable Car	Boat
Food	Ski	Bus	Landscape	Landscape	Season	Weather	Glacier
Landscape	Accessibility	Mountain	Hike	Season	Ski	Road	Weather
Road	Other	Road	Sound	Oxygen Tank	Mountain	Ski	Hike
Visitor Center	Piste	Viewpoint	Viewpoint	Viewing Platform	Magnificent	Climb	Landscape
Color	Cable Car	Water	Boat	Weather	Weather	Ticket	Viewpoint

India	Italy	Nepal	New Zealand	Norway	Peru	Switzerland	United States
Ride Horse	Cable Car	Adventure	Hike	Waterfall	Landscape	Cable Car	Waterfall
Mountain	Slope	Mountain	Accessibility	Troll Car	Road	Road	Helicopter Sightseeing
Sledge	Ski	Hike	Landscape	Cruise	Viewpoint	Restaurant	Hike
Viewpoint	Mountain	Glacier Lake	Viewpoint	Climb	Hike	Hike	Visitor Center
Price	Altitude	Valley	Road	Accessibility	Accessibility	Mountain	Cruise
People			Glacier	Weather		Snow	Bear
Enjoyable			Weather	Parking		Ski	People
Snow			Visitor Center	People		Train	Mountain
Accessibility			Snow	Mountain		Weather	Water
Road			Helicopter Sightseeing	Hike		People	Glacier
Ice			Valley	Road		Landscape	Whale

Table A2. Attribute dimensional glacier tourism destination image and its main topic words.

Ice	Ice Field	Glacier	Snow	Mountain	Glacier Lake	Valley	Waterfall	Landscape	Hike	Ski	Self−Driving Tour	Climb	Ride Horse
ice	ice	glacier	snow	mountain	lagoon	valley	waterfall	landscape	hike	ski	drive	climb	ride
piece	field	iceberg	ice	view	lake	scenery	lake	scenery	walk	slope	trip	rock	horse
immense	explore	sight	winter	summit	water	bridge	river	mountain	trail	piste	car	summit	rider
block	wind	beauty	view	cloudy	iceberg	track	size	Lake	trek	snow	hour	glacier	valley
nature	cold	blue	icy	level	floe	cross	view	snow	pace	lift	road	rope	access

Cruise	Helicopter Sightsee-ing	Adventure	Cable Car	Bus	Troll Car	Train	Boat	Sledge	Seal	Whale	Bear	Road	Piste
cruise	helicopter	adventure	car	bus	car	train	boat	sledge	seal	whale	bear	road	piste
ship	access	wild	cable	transport	troll	journey	ride	transport	watch	saw	nature	way	ski
board	tour	walk	ride	driver	access	station	ship	snow	photograph	wildlife	away	path	variety
bay	guide	camp	queue	ride	coach	tunnel	entrance	experience	water	trace	trip	route	route
passage	ride	valley	view	ticket	site	route	guide	ride	sea	sea	point	access	slope

Skywalk	Visitor Center	Restaurant	Parking	Color	Sound	Altitude	Slope	Magnificent	Price	Ticket	Accessibility	Weather/Climate	Season
skywalk	visitor	restaurant	parking	blue	thunder	altitude	slope	magnificent	price	ticket	access	weather	season
platform	center	food	car	color	sound	level	condition	view	cost	buy	walkway	rain	summer
attract	information	service	fee	water	noise	peak	area	superb	money	price	catwalk	sun	winter
scenery	history	staff	road	ice	silence	meter	level	height	fee	ropeway	guide	cloudy	spring
watch	service	place	transport	scenery	moment	summit	variety	nature	expense	office	drive	temperature	ice

Crowdedness	Viewpoint	Viewing Platform	People	Food	Worthwhile	Kindness	Enjoyable	Water	Oxygen Tank
crowd	view	view	people	food	worth	kind	enjoy	water	oxygen
lot	viewpoint	platform	family	lunch	trip	people	experience	bring	tank
people	picture	height	couple	tea	price	help	beauty	need	breathe
queue	spot	walk	group	eat	happy	impress	scenery	bottle	altitude
wait	look	peak	children	snack	scenery	friend	comfort	drink	sense

Table A3. Four seasons image statistics of glacier tourist destinations.

Spring	Freq.	%	Summer	Freq.	%	Autumn	Freq.	%	Winter	Freq.	%
Landscapes (Glacier Lake, Mountain, Landscape)	5970	23.84%	Landscapes (Glacier Lake, Glacier, Mountain, landscape, Ice)	19,317	32.87%	Landscapes (Glacier, Ice, Glacier Lake, Mountain)	9197	25.58%	Landscapes (Glacier, Mountain, Landscape)	5779	30.50
Specific Activity (Ski, Cruise, Hike)	7735	30.89%	Specific Activity (Self−Driving Tour, Hike, Cruise)	8658	14.73%	Specific Activity (Self−Driving Tour, Cruise, Hike)	7152	19.89%	Specific Activity (Hike, Ski, Climb)	5757	30.39
Transportation (Cable Car)	2186	8.73%	Transportation (Bus, Cable Car)	5540	9.43%	Transportation (Boat, Cable Car)	5061	14.08%	Transportation (Cable Car)	1760	9.29
Local Infrastructure (Road)	1780	7.11%	Local Infrastructure (Infrastructure, Road, Visitor Center)	8459	14.39%	Local Infrastructure (Road, Visitor Center)	3736	10.39%	Local Infrastructure (Road)	1858	9.81
Landscape Features	0	0%	Landscape Features (Magnificent)	913	1.55%	Landscape Features	0	0%	Landscape Features	0	0
Travel Environment (Weather/Climate)	1321	5.28%	Travel Environment (Accessibility, Weather/Climate, Season)	5993	10.20%	Travel Environment (Weather/Climate)	1681	4.68%	Travel Environment (Weather/Climate, Crowdedness)	2141	11.30
Viewing Location	0	0%	Viewing Location	0	0%	Viewing Location (Viewpoint)	4726	13.15%	Viewing Location	0	0
Price	1626	6.49%	Price	1966	3.35%	Price	2309	6.42%	Price	1650	8.71
Experience (Enjoyable)	2582	10.31%	Experience	2802	4.77%	Experience (Worthwhile)	2090	5.81%	Experience	0	0
People	0	0%	People	1990	3.39%	People	0	0%	People	0	0
Food	1838	7.34%	Food	1333	2.27%	Food	0	0%	Food	0	0
Necessity	0	0%	Necessity (Water)	1803	3.07%	Necessity	0	0%	Necessity	0	0

Table A4. Regional image of glacier tourism destinations.

Alpes	Freq.	%	New Zealand	Freq.	%	Nordic	Freq.	%
Landscapes (Ice, Mountain, Landscape)	6933	18.30%	Landscapes (Landscape, Glacier, Snow, Valley)	2000	33.10%	Landscapes (Glacier, Waterfall, Mountain, Glacier Lake, Ice, Beach)	4059	51.24%
Specific Activity (Climb, Hike, Ski)	7385	19.50%	Specific Activity (Hike, Helicopter Sightseeing)	1844	30.52%	Specific Activity (Cruise, Hike)	767	9.68%
Transportation (Trian, Cable Car)	6706	17.70%	Transportation	0	0%	Transportation (Boat, Car)	1098	13.86%
Local Infrastructure (Restaurant, Road)	4022	10.62%	Local Infrastructure (Road, Visitor Center)	987	16.34%	Local Infrastructure (Road, Visitor Center)	854	10.78%
Travel Environment	0	0%	Travel Environment (Accessibility, Weather/Climate)	742	12.28%	Travel Environment (Accessibility)	444	5.60%
Viewing Location	5903	15.58%	Viewing Location (Viewpoint)	469	7.76%	Viewing Location (Viewpoint)	375	4.73%
Price (Price, Ticket)	3418	9.02%	Price	0	0%	Price	0	0%
Experience (Worthwhile)	1961	5.18%	Experience	0	0%	Experience	0	0%
People	1549	4.09%	People	0	0%	People	325	4.10%

North America	Freq.	%	Qinghai−Tibet Plateau	Freq.	%	South America	Freq.	%
Landscapes (Landscape, Mountain, Waterfall, Glacier Lake, Ice Field, Glacier)	12,218	31.40%	Landscapes (River, Mountain, Glacier, Valley)	1422	42.88%	Landscapes (Landscape, Glacier Lake, Glacier, Ice, Mountain, Natural)	17,324	38.81%
Specific Activity (Hike, Helicopter Sightseeing, Cruise, Self−Driving Tour)	10,696	27.49%	Specific Activity (Ride Horse)	276	8.32%	Specific Activity (Climb, Hike, Adventure)	7673	17.19%
Transportation (Boat)	2165	5.56%	Transportation (Cable Car)	259	7.81%	Transportation (Boat)	2768	6.20%
Local Infrastructure (Road, Visitor Center)	4873	12.52%	Local Infrastructure (Road)	362	10.92%	Local Infrastructure (Visitor Center)	4579	10.26%
Landscape Features (Altitude)	2101	5.40%	Landscape Features	0	0%	Landscape Features (Color)	1843	4.13%
Travel Environment (Weather/Climate)	0	0%	Travel Environment (Season)	164	4.95%	Travel Environment (Accessibility, Weather/Climate)	5028	11.26%
Viewing Location	0	0%	Viewing Location	239	7.21%	Viewing Location (Viewpoint)	1944	4.35%
Price (Ticket)	2400	6.17%	Price (Price)	353	10.65%	Price	2097	4.70%
Experience	0	0%	Experience	0	0%	Experience (Worthwhile)	1384	3.10%
People	967	2.49%	People	241	7.27%	People	0	0%
Food	1405	3.61%	Food	0	0%	Food	0	0%
Animals	0	0%	Animals	0	0%	Animals	0	0%
Necessity (Water)	2087	5.36%	Necessity (Water)	0	0%	Necessity	0	0%
Other	0	0%	Other	0	0%	Other	0	0%

References

1. Nepal, S.K.; Chipeniuk, R. Mountain Tourism: Toward a Conceptual Framework. *Tour. Geogr.* **2005**, *7*, 313–333. [CrossRef]
2. Steiger, R.; Knowles, N.; Pöll, K.; Rutty, M. Impacts of climate change on mountain tourism: A review. *J. Sustain. Tour.* **2022**, 1–34. [CrossRef]
3. Ding, K.; Yang, M.; Luo, S. Mountain Landscape Preferences of Millennials Based on Social Media Data: A Case Study on Western Sichuan. *Land* **2021**, *10*, 1246. [CrossRef]
4. Cao, Q.; Sarker, M.N.I.; Zhang, D.; Sun, J.; Xiong, T.; Ding, J. Tourism Competitiveness Evaluation: Evidence from Mountain Tourism in China [Original Research]. *Front. Psychol.* **2022**, *13*, 809314. [CrossRef]
5. Wang, S.; He, Y.; Song, X. Impacts of climate warming on alpine glacier tourism and adaptive measures: A case study of Baishui Glacier No. 1 in Yulong Snow Mountain, Southwestern China. *J. Earth Sci.* **2010**, *21*, 166–178. [CrossRef]
6. Xiao, C.; Wang, S.; Qin, D. A Preliminary Study on Cryosphere Service Function and Its Value Estimation. *Clim. Chang. Res.* **2016**, *12*, 45–52.
7. Welling, J.T.; Árnason, Þ.; Ólafsdottír, R. Glacier tourism: A scoping review. *Tourism. Geogr.* **2015**, *17*, 635–662. [CrossRef]
8. Salim, E.; Ravanel, L.; Bourdeau, P.; Deline, P. Glacier tourism and climate change: Effects, adaptations, and perspectives in the Alps. *Reg. Environ. Chang.* **2021**, *21*, 120. [CrossRef]
9. Purdie, H. Glacier Retreat and Tourism: Insights from New Zealand. *Mt. Res. Dev.* **2013**, *33*, 463–472. [CrossRef]
10. Lemieux, C.J.; Groulx, M.; Halpenny, E.; Stager, H.; Dawson, J.; Stewart, E.J.; Hvenegaard, G.T. "The End of the Ice Age?": Disappearing World Heritage and the Climate Change Communication Imperative. *Environ. Commun.* **2017**, *12*, 653–671. [CrossRef]
11. Wang, S.J.; Zhou, L.Y. Integrated impacts of climate change on glacier tourism. *Adv. Clim. Chang. Res.* **2019**, *10*, 71–79. [CrossRef]
12. Tourism Resource Consultants. Glacier Country: Issues and Options for Product Development and Growth. Report Prepared for Development West Coast by Tourism Resource Consultants in Association with Boffa Miskell. 2007. Available online: www.westcoastnz.com/content/library/glacier_country_issues_and_options_report_091107_.pdf (accessed on 17 May 2012).
13. Wang, S.J.; Xie, J.; Zhou, L.Y. China's glacier tourism: Potential evaluation and spatial planning. *J. Destin. Mark. Manag.* **2020**, *18*, 100506.

14. Wilson, J.; Stewart, E.; Espiner, S.; Purdie, H. *'Last Chance Tourism' at the Franz Josef and Fox Glaciers, Westland Tai Poutini National Park: Stakeholder Perspectives*; University of Canterbury: Christchurch, New Zealand, 2014.

15. Haimayer, P. Glacier-Skiing Areas in Austria: A socio-political perspective. *Mt. Res. Dev.* **1989**, *9*, 51–58. [CrossRef]

16. Korstanje, M.E. Tourism and climate change: Impacts, adaptation and mitigation. *J. Tour. Cult. Chang.* **2016**, *14*, 86–88. [CrossRef]

17. IPCC. *Climate Change 2013: The Physical Science Basis. Contribution of Working Group I to the Fifth Assessment Report of the Intergovernmental Panel on Climate Change*; Cambridge University Press: Cambridge, UK, 2013.

18. Purdie, H.; Gomez, C.; Espiner, S. Glacier recession and the changing rockfall hazard: Implications for glacier tourism. *N. Z. Geogr.* **2015**, *71*, 189–202. [CrossRef]

19. Furunes, T.; Mykletun, R.J. Frozen Adventure at Risk? A 7-year Follow-up Study of Norwegian Glacier Tourism. *Scand. J. Hosp. Tour.* **2012**, *12*, 324–348. [CrossRef]

20. Salim, E.; Ravanel, L. Last chance to see the ice: Visitor motivation at Montenvers-Mer-de-Glace, French Alps. *Tourism. Geogr.* **2020**, *15*, 1–23. [CrossRef]

21. Abrahams, Z.; Hoogendoorn, G.; Fitchett, J.M. Glacier tourism and tourist reviews: An experiential engagement with the concept of "Last Chance Tourism". *Scand. J. Hosp. Tour.* **2021**, *22*, 1–14. [CrossRef]

22. Salim, E.; Mayer, M.; Sacher, P.; Ravanel, L. Visitors' motivations to engage in glacier tourism in the European Alps: Comparison of six sites in France, Switzerland, and Austria. *J. Sustain. Tour.* **2022**, 1–21. [CrossRef]

23. Dawson, J.; Johnston, M.J.; Stewart, E.J.; Lemieux, C.J.; Lemelin, R.H.; Maher, P.T.; Grimwood, B.S.R. Ethical considerations of last chance tourism. *J. Ecotour.* **2011**, *10*, 250–265. [CrossRef]

24. Williams, A.; Ólafsdóttir, R. Nature-based tourism as therapeutic landscape in a COVID era: Autoethnographic learnings from a visitor's experience in Iceland. *GeoJournal* **2022**. [CrossRef] [PubMed]

25. Kozak, M.; Rimmington, M. Measuring tourist destination competitiveness: Conceptual considerations and empirical findings. *Int. J. Hosp. Manag.* **1999**, *18*, 273–283. [CrossRef]

26. Iwata, S.; Watanabe, T. A proposal on glacier tourism of Pasu Glacier Group, Gojal, Northern Pakistan. *Rikkyo Univ. Bull. Stud. Tour.* **2007**, *9*, 11–26.

27. Salim, E.; Ravanel, L.; Deline, P.; Gauchon, C. A review of melting ice adaptation strategies in the glacier tourism context. *Scand. J. Hosp. Tour.* **2021**, *21*, 229–246. [CrossRef]

28. Kaenzig, R.; Rebetez, M.; Serquet, G.E. Climate change adaptation of the tourism sector in the Bolivian Andes. *Tourism. Geogr.* **2016**, *18*, 111–128. [CrossRef]

29. Salim, E.; Mabboux, L.; Ravanel, L.; Deline, P.; Gauchon, C. A history of tourism at the Mer de Glace: Adaptations of glacier tourism to glacier fluctuations since 1741. *J. Mt. Sci.* **2021**, *18*, 1977–1994. [CrossRef]

30. Pike, S.; Ryan, C. Destination positioning analysis through a comparison of cognitive, affective, and conative perceptions. *J. Travel. Res.* **2004**, *42*, 333–342. [CrossRef]

31. Grabler, S.D.K. Applying City Perception Analysis (CPA) for Destination Positioning Decisions. *J. Travel. Tour. Mark.* **2004**, *16*, 99–111.

32. Hunt, J.D. Image: A Factor in Tourism. Unpublished. Ph.D. Dissertation, Colorado State University, Fort Collins, CO, USA, 1971.

33. Afshardoost, M.; Eshaghi, M.S. Destination image and tourist behavioural intentions: A meta-analysis. *Tour. Manag.* **2020**, *81*, 104154. [CrossRef]

34. Echtner, C.M.; Ritchie, J.R.B. The Meaning and Measurement of Destination Image. *J. Tour. Stud.* **2003**, *14*, 37–48.

35. Estela, M.R. Destination Image Analytics Through Traveller-Generated Content. *Sustainability* **2019**, *11*, 3392.

36. Pralong, J.P.; Reynard, E. A proposal for a classification of geomorphological sites depending on their tourist value. *Il Quat. Ital. J. Quat. Sci.* **2005**, *18*, 315–321.

37. Liu, X.; Yang, Z.; Xie, T. Development and conservation of glacier tourist resources—A case study of Bogda Glacier Park. *Chinese. Geogr. Sci.* **2006**, *16*, 365–370. [CrossRef]

38. Wang, S.J.; Jiao, S.T.; Niu, H.W. Patterns and strategies of glacier tourism resources development in China. *J. Nat. Resour.* **2012**, *27*, 1276–1285.

39. Framke, W. The Destination as a Concept: A Discussion of the Business-related Perspective versus the Socio-cultural Approach in Tourism Theory. *Scand. J. Hosp. Tour.* **2002**, *2*, 92–108. [CrossRef]

40. Liu, L.; Zhong, L.; Yu, H. Progress of glacier tourism research and implications. *Prog. Geogr.* **2019**, *38*, 533–545.

41. Baloglu, S. The relationship between destination images and sociodemographic and trip characteristics of international travellers. *J. Vacat. Mark.* **1997**, *3*, 221–233. [CrossRef]

42. Gartner, W.C. Image formation process. In *Communication and Channel Systems in Tourism Marketing*; Uysal, M., Fesenmaier, D., Eds.; The Haworth Press: New York, NY, USA, 1993; pp. 191–215.

43. Stylos, N.; Vassiliadis, C.A.; Bellou, V.; Andronikidis, A. Destination images, holistic images and personal normative beliefs: Predictors of intention to revisit a destination. *Tour. Manag.* **2016**, *53*, 40–60. [CrossRef]

44. Wang, R.; Hao, J.X.; Law, R.; Wang, J. Examining destination images from travel blogs: A big data analytical approach using latent Dirichlet allocation. *Asia. Pac. J. Tour. Res.* **2019**, *24*, 1092–1107. [CrossRef]

45. Bui, V.; Alaei, A.R.; Vu, H.Q.; Li, G.; Law, R. Revisiting Tourism Destination Image: A Holistic Measurement Framework Using Big Data. *J. Travel. Res.* **2021**, *61*, 1287–1307. [CrossRef]

46. Hallmann, K.; Zehrer, A.; Mueller, S. Perceived destination image: An image model for a Winter sports destination and its effect on intention to revisit. *J. Travel. Res.* **2014**, *54*, 94–106. [CrossRef]

47. Klenosky, D.B. The "Pull" of Tourism Destinations: A Means-End Investigation. *J. Travel. Res.* **2002**, *40*, 396–403. [CrossRef]

48. Tasci, A.; Gartner, W. Destination Image and Its Functional Relationships. *J. Travel. Res.* **2007**, *45*, 413–425. [CrossRef]

49. Dann, G.M.S. Tourists' Images of a Destination-An Alternative Analysis. *J. Travel. Tour. Mark.* **1996**, *5*, 41–55. [CrossRef]

50. Wilson, A.; Murphy, H.; Fierro, J.C. Hospitality and Travel: The Nature and Implications of User-Generated Content. *Cornell Hosp. Q.* **2012**, *53*, 220–228. [CrossRef]

51. Kwark, Y.; Chen, J.; Raghunathan, S. Online Product Reviews: Implications for Retailers and Competing Manufacturers. *Inform. Syst. Res.* **2014**, *25*, 93–110. [CrossRef]

52. Zhao, X.R.; Wang, L.; Guo, X.; Law, R. The influence of online reviews to online hotel booking intentions. *Int. J. Contemp. Hosp. M.* **2015**, *27*, 1343–1364. [CrossRef]

53. Lee, J.; Park, D.; Han, I. The effect of negative online consumer reviews on product attitude: An information processing view. *Electron. Commer. Res. Appl.* **2008**, *7*, 341–352. [CrossRef]

54. Banerjee, S.; Chua, A.Y.K. In search of patterns among travellers' hotel ratings in TripAdvisor. *Tour. Manag.* **2016**, *53*, 125–131. [CrossRef]

55. Ahani, A.; Nilashi, M.; Ibrahim, O.; Sanzogni, L.; Weaven, S. Market segmentation and travel choice prediction in Spa hotels through TripAdvisor's online reviews. *Int. J. Hosp. Manag.* **2019**, *80*, 52–77. [CrossRef]

56. Chang, Y.C.; Ku, C.H.; Chen, C.H. Using deep learning and visual analytics to explore hotel reviews and responses. *Tour. Manag.* **2020**, *80*, 104129. [CrossRef]

57. Taecharungroj, V.; Mathayomchan, B. Analysing TripAdvisor reviews of tourist attractions in Phuket, Thailand. *Tour. Manag.* **2019**, *75*, 550–568. [CrossRef]

58. Taecharungroj, V. An analysis of tripadvisor reviews of 127 urban rail transit networks worldwide. *Travel. Behav. Soc.* **2022**, *26*, 193–205. [CrossRef]

59. Blei, D.M.; Ng, A.Y.; Jordan, M.I. Latent Dirichlet Allocation. *J. Mach. Learn. Res.* **2003**, *3*, 993–1022.

60. Guo, Y.; Barnes, S.J.; Jia, Q. Mining meaning from online ratings and reviews: Tourist satisfaction analysis using latent dirichlet allocation. *Tour. Manag.* **2017**, *59*, 467–483. [CrossRef]

61. Aal, C.; Hoye, K.G. Tourism and climate change adaptation: The Norwegian case. In *Tourism, Recreation and Climate Change*; Hall, M.C., Higham, J., Eds.; Channel View Publications: Bristol, UK, 2005; pp. 209–221.

62. Espiner, S.; Becken, S. Tourist towns on the edge: Conceptualising vulnerability and resilience in a protected area tourism system. *J. Sustain. Tour.* **2014**, *22*, 646–665. [CrossRef]

63. Espiner, S. The Phenomenon of Risk and Its Management in Natural Resource Recreation and Tourism Settings: A Case Study of Fox and Franz Josef Glaciers, Westland National Park, New Zealand. Ph.D. Thesis, Lincoln University, Lincoln, New Zealand, 2001.

64. Manandhar, S.; Vogt, D.; Perret, S.; Kazama, F. Adapting cropping systems to climate change in Nepal: A cross-regional study of farmers' perception and practices. *Reg. Environ. Chang.* **2011**, *11*, 335–348. [CrossRef]

65. Jóhannesdóttir, G.R. Landscape and Aesthetic values: Not only in the eye of the beholder. In *Conservations with Landscape*; Benediktsson, K., Lund, K.A., Eds.; Ashgate: Farnham, UK, 2010; pp. 109–124.

66. Konecnik, M.; Gartner, W.C. Customer-based brand equity for a destination. *Ann. Tourism. Res.* **2007**, *34*, 400–421. [CrossRef]

67. Alcañiz, E.B.; García, I.S.; Blas, S.S. The functional-psychological continuum in the cognitive image of a destination: A confirmatory analysis. *Tour. Manag.* **2009**, *30*, 715–723. [CrossRef]

68. Abermann, J.; Theurl, M.; Frei, E.; Hynek, B.; Schöner, W.; Steininger, K.W. Too expensive to keep—Bidding farewell to an iconic mountain glacier? *Reg. Environ. Chang.* **2022**, *22*, 51. [CrossRef]

69. Campo, S.; Yagüe, M.J. Research Note: Effects of Price on Tourist Satisfaction. *Tourism. Econ.* **2008**, *14*, 657–661. [CrossRef]

70. Peng, Y.; Yin, P.; Matzler, K. Analysis of Destination Images in the Emerging Ski Market: The Case Study in the Host City of the 2022 Beijing Winter Olympic Games. *Sustainability* **2022**, *14*, 555. [CrossRef]

71. Dar, R.A.; Rashid, I.; Romshoo, S.A.; Marazi, A. Sustainability of winter tourism in a changing climate over Kashmir Himalaya. *Environ. Monit. Assess.* **2014**, *186*, 2549–2562. [CrossRef] [PubMed]

72. Mayer, M.; Demiroglu, O.C.; Ozcelebi, O. Microclimatic Volatility and Elasticity of Glacier Skiing Demand. *Sustainability* **2018**, *10*, 3536. [CrossRef]

73. Zhang, X.; Yan, H.; Xiong, H. A pricing optimization model of scenic spot with intertemporal demand transformation. *J. Cent. China Norm. Univ. (Nat. Sci.)* **2022**, *56*, 140–145.

74. Scott, D.; Jones, B.; Konopek, J. Implications of climate and environmental change for nature-based tourism in the Canadian Rocky Mountains: A case study of Waterton Lakes National Park. *Tour. Manag.* **2007**, *28*, 570–579. [CrossRef]

75. Mölders, N.; Gende, S.; Pirhalla, M. Assessment of cruise–ship activity influences on emissions, air quality, and visibility in Glacier Bay National Park. *Atmos. Pollut. Res.* **2013**, *4*, 435–445. [CrossRef]

76. Young, C.; Gende, S.; Harvey, J. Effects of Vessels on Harbor Seals in Glacier Bay National Park. *Tour. Mar. Environ.* **2014**, *10*, 5–20. [CrossRef]

77. Welling, J.; Abegg, B. Following the ice: Adaptation processes of glacier tour operators in Southeast Iceland. *Int. J. Biometeorol.* **2021**, *65*, 703–715. [CrossRef]

78. Yuan, L.L.; Wang, S.J. Recreational value of glacier tourism resources: A travel cost analysis for Yulong Snow Mountain. *J. Mt. Sci.* **2018**, *15*, 1446–1459. [CrossRef]
79. Department of Conservation. *Aoraki/Mount Cook National Park Management Plan (Draft)*; Department of Conservation: Christchurch, New Zealand, 2018.
80. Lojo, A.; Li, M.M.; Xu, H.G. Online tourism destination image: Components, information sources, and incongruence. *J. Travel Tour. Mark.* **2020**, *37*, 495–509. [CrossRef]
81. Xiao, X.; Fang, C.Y.; Lin, H.; Chen, J.F. A framework for quantitative analysis and differentiated marketing of tourism destination image based on visual content of photos. *Tourism. Manag.* **2022**, *93*, 104585. [CrossRef]

 *land*

Article

The Relationship between Habitat Diversity and Tourists' Visual Preference in Urban Wetland Park

Jiani Zhang [1,2]**, Xun Zhu** [1,2,*] **and Ming Gao** [1,2]

1   School of Architecture, Harbin Institute of Technology, Harbin 150006, China
2   Key Laboratory of Cold Region Urban and Rural Human Settlement Environment Science and Technology, Ministry of Industry and Information Technology, Harbin 150006, China
*   Correspondence: zhuxun@hit.edu.cn; Tel.: +86-(0451)-8628-1083

Abstract: The increasing number of visitors to wetland parks has caused varying degrees of impact on wetland life. How to reduce the damage to wetland biodiversity caused by recreational activities in parks, improve tourists' recreational experience, and balance the relationship between the two are urgent problems that need to be solved. Therefore, four urban wetland parks were selected as subjects for this study. The present study utilized social media data to study the diversity of urban wetland habitats and tourists' wetland landscape preferences from the spatial dimension and explore the relationship between the two. This is a practice different from the traditional ecological research (survey, measurement, monitoring, questionnaire survey) of wetland habitat diversity assessment. The research revealed the following findings: (1) There was a significant positive correlation between habitat saturation and positive artificial elements, such as landscape structures and aerial walkways; (2) Landscape complexity is negatively correlated with landscape instantaneity and wilderness degree; (3) Habitat diversity was negatively correlated with landscape instantaneity but positively correlated with naturalness and positive artificial elements. This study proposes wetland habitat construction as a strategy to optimize the management of habitat diversity in urban wetland parks and enhance its ecological education function.

Keywords: urban wetland park; social media; habitat diversity; tourists' visual preference; mapping of habitat units

Citation: Zhang, J.; Zhu, X.; Gao, M. The Relationship between Habitat Diversity and Tourists' Visual Preference in Urban Wetland Park. *Land* **2022**, *11*, 2284. https://doi.org/10.3390/land11122284

Academic Editors: Cecilia Arnaiz Schmitz, Nicolas Marine and María Fe Schmitz

Received: 21 November 2022
Accepted: 8 December 2022
Published: 13 December 2022

Publisher's Note: MDPI stays neutral with regard to jurisdictional claims in published maps and institutional affiliations.

1. Introduction

Urban wetland is a valuable natural resource in the city. However, the rapidly expanding city has produced a lot of human interference and pressure. In order to solve the problem of urban development and environmental destruction, people built urban wetland parks [1]. Its construction goal is not always to provide ecological services but also to undertake cultural and recreational functions. It is an integrated constructed wetland (ICW) combined with human activities [2]. Nowadays, urban wetland degrades rapidly, which is mainly caused by various human activities, such as recreational construction, urban noise, and uncivilized behavior of tourists [3]. When the loss of urban wetlands reaches a certain level, the watershed cannot provide effective water quality protection, flood retention and storage, and wildlife habitat [4]. However, with the improvement of people's leisure awareness and the diversification of leisure needs, the frequency of urban wetland park visits has been greatly increased. Running, fishing, taking a walk, and viewing wetland scenery are the daily activities of residents here. Therefore, urban wetland parks, as a part of the overall urban planning, have the most acute contradiction between the protection of wetland resources and the construction of the park [5]. How to balance the habitat diversity protection and recreation construction of wetland parks and combine the functions of wetland ecosystems with landscape recreation activities are important issues to improve the natural eco-efficiency of wetlands and the value of urban social functions.

Recreation development and natural resources protection have always been the focus of attention in related disciplines, and the theoretical applications of this research are ROS [6], LAC [7], VERP [8], and other recreation development theories. Most of the application fields are forest parks [9] and nature reserves [10], but there are few applications in the research of urban wetland parks. Human beings are born with the desire to be close to nature, a large number of studies in environmental psychology have proved that the environment with natural elements has a positive effect on the recovery of human health [11–13].

In the early stage, wetland research was focused on its biodiversity and ecological service functions. Biological data and abiotic data of wetland parks are widely used as two kinds of index systems to assess the level of biodiversity. In the study of biological data indicators, the community structure of macroinvertebrates [14], birds and amphibians [15], microbial flora [16], vascular plant community [17], and vole [18] can be used as the indicator species of biodiversity in the selected habitat unit of this organism. However, it is not clear whether it can represent the level of the whole wetland park, because the interpretation percentage is about 20%, which has a certain bias [19]. The assessment of abiotic data is mainly conducted through the sampling of environmental factors, such as hydrology [20] and soil [21], in the wetland park. However, the sampling is complicated and requires professional operation and evaluation. Then, some ecologists turned their attention to the study of habitats. A habitat is the sum of ecological factors that act on organisms in a particular area, and it is one or more spatial units of different scales and sizes. Compared with the singleness of the former research index, all biological and abiotic factors within the habitat diversity assessment are included for comprehensive consideration. In this way, the results obtained have higher explanatory power and are more suitable to be used as a substitute index to measure wetland biodiversity [22,23]. The establishment of a landscape structure index system, such as area, perimeter, edge complexity, and connectivity of habitat units, makes up for the deficiency of habitat space [24–26]. In recent years, the visualization of habitat units has been achieved with the aid of computer software [27–29]. Spatial data that can be easily measured and recorded enhances the operability of wetland habitat diversity assessment. However, the research focus is still on the ecological attributes of wetland parks, ignoring that it is also disturbed by human recreation as a park. Conservation is not isolation. Now humans are gradually becoming participants in the wetland ecological process. If the protection strategy of a wetland park is proposed simply from the perspective of ecology while its recreational function as a park is ignored, this is the result of a serious disconnection between experimental design, management, and evaluation.

Tourists' landscape preference was found to be closely related to nature at the early stage of the study [30]. Human beings perceive the natural environment through their brain consciousness, thus generating landscape "element" preference and "spatial structure" preference [31]. People prefer quantitative and measurable physical elements, such as water [32] and plants [33], etc., while in their preferred spatial environment, they prefer different habitats, such as wetlands with water surface [34], open lawns [35], and wetland tree communities [36]. In addition, human beings can perceive biodiversity, which has nothing to do with tourists' personal attributes, such as age, occupation, and gender, but has something to do with perception intensity and preference [37–40]. The wetland park is favored by tourists because of its unique biodiversity, complexity, and naturalness [35]. Based on the close connection between green space and human preference cognition, as well as the increasingly fierce conflict between "recreation construction" and "resource conservation", relevant scholars try to study biodiversity and tourists' landscape preference from the following three aspects: (1) Biodiversity can promote the ecological health of green space around cities [41,42]. (2) The level of biodiversity will also affect human aesthetic preferences. For example, Matthies studied people's perception of species diversity through experiments combined with practical studies [43,44]. (3) The relationship between species richness and happiness. Jung et al. found through experiments that people have a positive evaluation of plant planting with rich species [45,46]. However, other studies have shown that the relationship between biodiversity and recreation services is not always positive. For visitors, aesthetics and naturalness are the most important attributes of urban parks, while the attraction of plants and animals is secondary [47,48].

The rise of social media data provides a more comprehensive approach combining subjective and objective factors to the uncertainty of the relationship between the two and brings new ideas to solve the contradiction between biodiversity conservation and recreation construction. Big data, such as comments [49] and geographical location information [50] released by a large number of tourists on social media platforms, are increasingly widely used as the scientific basis for in-depth research on landscape preferences. Compared with traditional data, in the era occupied by social networks, big data is more representative and spontaneous. In particular, the framing form and shooting content of tourists' photos can reflect the interaction between humans and nature in the dimensions of time and space [51]. As the research basis of landscape preference research, photographs are now widely used in the research of the "Objective-element" paradigm and "subject-perception" paradigm [52]. In the study of objective paradigm, comments on social media are quantified, reflecting that the biophysical features of the landscape are transformed into quantifiable parameters, such as type, diversity, line of elements, area proportion, etc., which constitute the parameter indicators for landscape preference evaluation [53–55]. The subjective paradigm takes the traditional Kaplan preference matrix as the core theory [56] and reflects the human perception of landscape in combination with comments and questionnaires on social media, such as naturalness, complexity, and coherence [57,58]. The preference research based on photo analysis combines the practice of environmental psychology and green space, which provides a certain reference value for enriching the human space experience and developing the potential of natural recreation space. Therefore, using social media data to study the public's preference for wetland landscapes can promote the development of a good "biodiversity perception" experience, thereby better predicting the acceptance of biodiversity management measures, and realize the construction of a "conservation-development-education" urban wetland park. Visitors' awareness of wetland biodiversity will enhance their awareness and behavior of nature conservation. However, there is a certain disconnection between habitat protection and recreation space organization in wetland parks, which requires landscape architects to make full use of environmental factors, such as water, sunlight, and heat, that affect the growth of plants when designing recreation space in wetland, so as to create a spatial organization that shows the inner organisms.

Based on the literature review, we found that the quantitative research scale of habitat diversity was mainly concentrated in cities, nature reserves, and so on. Habitat studies on park site scale tend to focus on ecological matrix and plant species but lack overall research on green space design scale related to the ecological environment. Research on the relationship between landscape preference and habitat diversity currently focuses on the correlation of single factors, such as trees and water bodies in wetlands, while other factors affecting wetland biodiversity, such as roads and structures in parks, have not been systematically considered. Research on the relationship between landscape preference and habitat diversity currently focuses on the correlation of single factors, such as trees and water, in wetlands, while other factors affecting wetland biodiversity, such as roads and structures in parks, have not been systematically taken into account. Under the social background of big data, scholars have explored related aspects, such as recreation and tourism, from different dimensions. However, in terms of biodiversity, the involvement of pictures and location information data of social media big data has broadened the research on the spatial coupling relationship between landscape construction and resource conservation. Therefore, this study takes the urban wetland parks as the research object and studies the following objectives based on practical cases:

(1) Explore the real landscape preferences of tourists in urban wetland parks and enrich the research content related to biodiversity and tourists' preferences in urban wetland parks.

(2) Through the calculation and drawing of spatial data of an urban wetland park habitat diversity map, combined with the geographical information location of photos, explore the tourists' preferred habitat spatial pattern to provide park managers and designers with scientific and effective management and design suggestions.

(3) Deeply explore the influence mechanism of landscape preference and wetland habitat diversity, try to change relevant biological factors, intervene in the biodiversity protection of urban wetland parks, and achieve the unity of conservation and development objectives so that urban wetland parks can play an important ecological role while performing recreational sightseeing and science popularization and education functions. We also seek to provide a reasonable reference for the later management of wetland park habitat protection. This is extremely important for landscape architects and urban park managers. A convenient and quick assessment method can provide developers with scientific and effective site information, allow landscape architects to participate in the protection of urban biodiversity, enhance the cultural and entertainment value of wetland parks, and achieve a relative balance between the protection and development of urban natural resources.

2. Materials and Methods

2.1. Study Area

Harbin and Changchun were selected as the research areas in this study. These two cities are located in northeast China and are important central cities in this region, as shown in Figure 1. With rivers passing through the cities, abundant tidal plain wetlands are formed, which have a good natural foundation of ecological wetlands. By screening 15 urban wetland parks in Harbin and Changchun and considering the frequency of people's recreational trips and the number of comments on public media platforms, we selected 4 urban wetland parks that are free to the public in urban core areas, namely, Harbin Qunli Wetland Park, Harbin Cultural Center Wetland Park, Changchun North Lake Wetland Park, and Changchun Nanxi Wetland Park, see Table 1 for details.

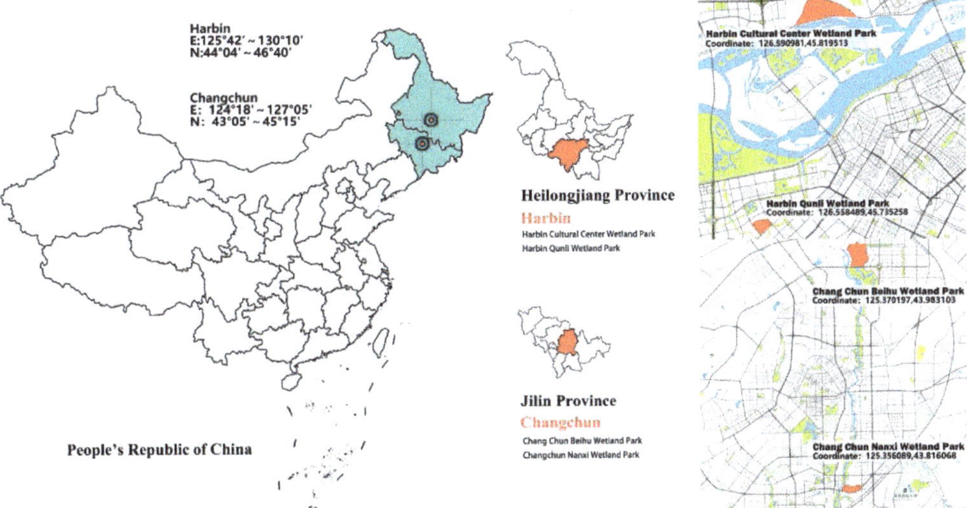

Figure 1. Location of the four urban wetland parks.

Table 1. Basic information about the four wetland parks.

Park Name	Total Area (m^2)	Wetland Area (m^2)	Wetland Type
Harbin Qunli	342,000	231,000	Natural wetlands, Artificial wetlands
Cultural Center	1,180,000	553,000	Swamp meadow, Prairie meadow
Changchun Beihu	11,970,000	3,765,000	Rivers, Lakes, Swamps, Constructed wetlands
Changchun Nanxi	310,000	112,000	Constructed wetlands, River wetlands

2.2. Data Acquisition

Basic quantitative data, such as the area and types of urban wetland parks, can be obtained from the official website of the National Forestry and Grassland Administration of China [59]. At the same time, Landsat 8 OLI_TIRS satellite data of four urban wetland parks were downloaded from the Geospatial Data Cloud [60], corrected in combination with the actual research, and the plan of the four urban wetland parks was drawn using Auto CAD 2016 in order to calculate the area, perimeter vegetation area, and quantity of wetland park [24]. Then, based on the shadow phase and multi-source remote sensing data of Landsat 8 OLI, the four wetlands were classified into multi-feature vegetation [61,62]. Then, combined with the preliminary classification results, field research, and related data records, the classification results were corrected and modified to prepare for the subsequent statistics of habitat elements.

As for the acquisition of tourists' visual preference data, the comment data of China's popular open social media platforms, such as Weibo [63], Ctrip [64], Two-step Road [65], and Qunar [66], were captured and screened. These data included the text and image content released by users during their visit to the wetland park and the gender of evaluators. The time period used was from January 2017 to May 2022, so as to avoid the problem of incorrect results due to the time difference in comment information. After screening and deleting comments unrelated to wetlands, 2880 photos with geographic information markers were obtained. According to the names of the corresponding parks, they were coded as H—Qunli Wetland Park, W—Cultural Center Wetland Park, B—North Lake Wetland Park, and N—Nanxi Wetland Park; the photo database of tourists' preferences was constructed.

2.3. Data Preprocessing

In terms of data processing, both "subjective data- visitor preference" and "objective data-habitat diversity of wetland park" were carried out at the same time, and a research framework was built to achieve the final goal, as shown in Figure 2.

In the processing of tourists' visual preference data, the photos were firstly segmented semantically [67,68] to identify the natural and artificial elements in the photos. The landscape elements identified include blue sky, water, trees, lawns, flowers, roads, people, buildings, landscape structures, facilities (garbage cans, lights, seats, publicity boards, etc.), wooden piles, cement roads, wooden walkways, etc. [69]. Then, the segmentation results of the establishment of the Java image processing library imageIO, and the use of computer programming language to obtain the photo content proportion data, the segmentation of natural elements, and artificial elements of the data for statistics and calculation, provided the visual area ratio of each element in the photo. The habitat diversity data of wetland park was calculated by the formula. We evaluated the habitat diversity of wetland parks through statistics and calculation of objective data, such as wetland area, perimeter, and number of habitat elements, and visualized the results. Finally, through correlation analysis, cluster analysis, spatial coupling, and other methods, the characteristics of wetland habitat units preferred by tourists were summarized and analyzed, and corresponding habitat protection-development strategies were proposed.

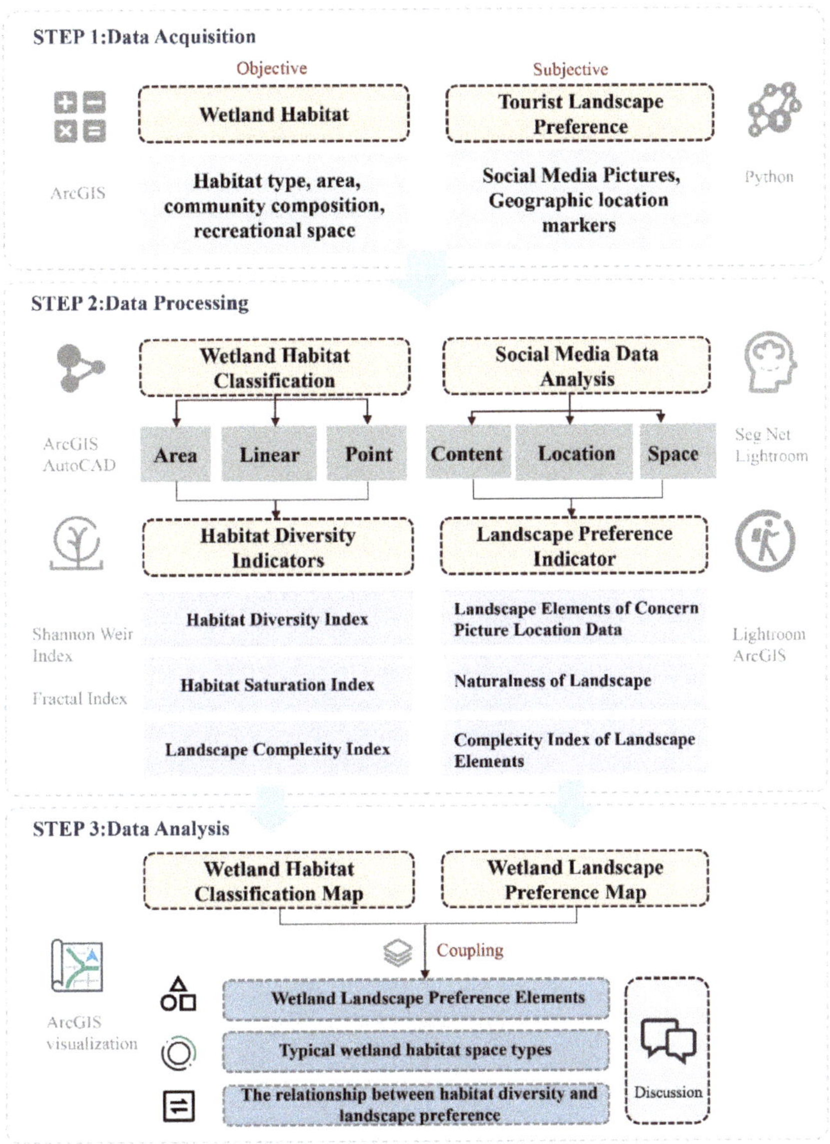

Figure 2. This shows the process for achieving the research objectives.

2.4. Visual Preference Indicator

According to the component proxy model [70], six indicators were selected as the visual preference measurement of tourists, including perceived naturalness, diversity, wildness, timeliness, and positive artificial elements. A measurement was calculated for each sample, see in Table 2 for details.

Table 2. Visual preference metrics and scoring principles [69].

Measure Value	Definition	Calculation	References
Naturalness	Naturalness of vegetation	Natural Vegetation Proportion 0–25% = 0; 25–50% = 1; 50–75% = 2; 75–100% = 3	Kaplan and Kaplan (1989) [30] Ulrich (1984) [71]
Diversity	Rich landscape elements	Number of landscape elements per view	Kellert and Wilson (1993) [72]
Wilderness	Degree of artificial construction	House + road + other = 0; few artificial independent elements = 1; wetlands with no artificial elements = 2; wild vegetation = 3	Clay and Smidt (2004) [73]
Temporality	(1) Landscape properties with seasonal changes (2) Landscape attributes (water) with weather characteristics	No time difference = 0; multi-season landscape = 1; single-season landscape = 2; meteorological landscape = 3	Kaplan and Kaplan(1989) [30] Ulrich (1984) [71]
Positive artificial element	Artificial landscape and local architecture	None = 0; 1 element = 1; 2 elements = 2; 3 elements and more = 3	Arriaza M (2004) [74]
Negative artificial element	Highways, industrial facilities, electrical wiring	None = 0; 1 element = 1; 2 elements = 2; 3 elements and more = 3	Arriaza M (2004) [74]

2.5. Habitat Diversity Indicators

2.5.1. Division of Habitat Units

Habitat unit refers to the environment related to organisms, which can be divided into spatial units in the ecosystem. Some abiotic factors together constitute the living environment [75]. Its concept is constantly updated and developed with the passage of time, and its essence reflects the characteristics of spatial or environmental patterns. In terms of cities, it is formed by urban buildings, structures, vegetation, human activities, and other factors [76]. Based on the urban park habitat classification system proposed by Belgian scholar Hermy [25], this paper divided the habitats of four wetland parks into units, see Table 3 for details. Habitat diversity was related to the area, and species richness was related to habitat diversity while the area remains unchanged [24]. Therefore, we divided the four wetland parks into 52 habitat units with an area of 50 m × 50 m, so as to facilitate the study and exploration of other variables.

Table 3. Classification standard of habitat units in urban wetland parks [25].

Habitat Unit Type	Description of Habitat Indicators		Scale Standard
Area elements/m^2	(1) (2) (3) (4) (5) (6) (7)	Natural stand (natural or semi-natural forest vegetation) Scrub (shrub) Grass (herb) Tall vegetation (including reeds) Hydrological elements (waters) Islands Parking lot (place where vehicles are parked)	Deciduous forest (deciduous forest) mixed forest Length < 10 m, area > 100 m^2
Linear elements/m	(1) (2) (3) (4)	Roads (belts for pedestrians and traffic services) Tree column (trees planted in column) Long span bridge Landscape wall, etc.	Roads with a width of ≥2 m (main road, secondary road);Roads with a width of less than 2 m (plank roads, grass walks, air corridors); Shelter forest, street tree column ≥10 m
Point element/piece	(1) (2) (3)	Single tree or shrub Shallow water and sports fields within 100 m^2 Infrastructure	Buildings, pavilions, sculptures, monuments <100 m^2

2.5.2. Habitat Diversity Calculation

Firstly, according to the classification criteria, the sample units were divided into plane, linear, and point habitat elements. Then, the area, length, and number of each cell are counted. Finally, the Shannon–Weill index and fractal index of each sample were calculated, and habitat diversity (H), habitat saturation (H′), and landscape complexity (P) were used as indicators to measure the level of habitat diversity in each sample [69]. Then, the results were analyzed visually using GIS, and a map of habitat diversity was obtained for further analysis.

(4)   Diversity of habitat

The habitat diversity index can reflect the difference between the actual biodiversity level and the ideal biodiversity level in the study area, so the managers can intervene and manage the landscape types in the site according to the index. Shannonville Diversity Index was used to calculate the specific index. This index was constructed by Merglef and Shannonville's information measurement formula in 1958. It is one of the most widely used indicators, and relevant scholars have used it to measure diversity at the species level. The specific calculation formula is as follows:

$$H' = -\sum_{i=1}^{S} (n_i/N) \ln(n_i/N) \tag{1}$$

H'—diversity index of habitat unit;
i—i-th habitat unit;
S—total number of habitat units;
n_i—area, length, or number of habitat unit i;
N—total area, total length, or the total number of habitat units in the park.

(5)   Habitat saturation index

The habitat saturation index is the ratio (%) between the actually measured diversity index and the maximum possible diversity index H'_{max}, which can be calculated, according to Formula (3), and then the habitat saturation is calculated. For the specific formula, see (2):

$$S = \frac{H'}{H'_{max}} \times 100\% \tag{2}$$

S_{max}—total number of habitat units.

$$H'_{max} = -\ln \frac{1}{S_{max}} \tag{3}$$

(6)   Complexity of landscape

The complexity of the shape of landscape elements is both an ecological function and a dimension of visual characteristics. Landscape complexity is one of the intersecting indicators of ecology and environmental aesthetics, which can not only reflect the landscape complexity of objective wetland habitats but also serve as a reference index for preference prediction. In this study, the most representative morphological index calculation model established by Krummel et al., the "girth–area" measurement index, is adopted:

$$S = 0.25P/\sqrt{A} \tag{4}$$

P—The sum of the edge perimeters possessed by individual patches in the object area (m);
A—Total area of object area (m^2);

2.5.3. Data Validation

Since the area proportion of various habitat elements in the photos can reflect the preference characteristics of tourists for habitat units, this study takes the preference database obtained after semantic analysis of photos obtained from the internet as the research variable and conducts a difference test to verify the degree of "affinity and disaffinity" between elements. Since the sample data does not obey the normal distribution, non-parametric test is chosen. In other words, SPSS 26.0 (Statistical Product Service Solutions, IBM) was used to conduct K independent sample tests on sample data. It can be seen from Table 4 that $p < 0.05$, there are significant differences among various element variables, and the data are reliable.

Table 4. Test statistics [a,b].

	Men	Buildings	Structures	Facilities	Cement Road	Wooden WALKWAY	Steps	Aerial BOARD-WALK	Sky	Water Surface	Trees	Lawns	Herbaceous Plant	Shrub	Hydrophyte
Kruskal–Wallis H	128.422	206.658	151.468	36.508	14.757	104.890	23.182	220.398	157.619	172.547	259.918	75.938	158.260	26.722	73.424
df	3	3	3	3	3	3	3	3	3	3	3	3	3	3	3
Asymp. Sig.	0.000	0.000	0.000	0.000	0.000	0.000	0.000	0.000	0.000	0.000	0.000	0.000	0.000	0.000	0.000

[a]. Kruskal–Wallis Test. [b]. Grouping variable: type

3. Visual Preferences and Habitat Diversity

3.1. Analysis of Tourists' Landscape Preference Elements

According to the semantic segmentation of 2880 photos and the element proportion, it is found that the landscape elements chosen by tourists are blue sky, trees, lawn, water, men, cement road, shrubs, buildings, hydrophyte, structures, wooden walkway, flowers, other plants, aerial boardwalk, facilities, steps, wooden walkway, and structure. Blue sky accounted for 33.21%, followed by tall trees accounting for 12.40%; herbs and trees were close to each other, accounting for 10.20%. Water area, as a popular area of wetland park, accounted for 9.00%; Cement road represents artificial traffic, including hard paved square, accounting for 6.10% of the total. Bush group and stairs are similar, accounting for 4.20% and 4.10% of the total. As one of the representatives of seasonal landscape, aquatic plants accounted for 3.00%; structures and wooden boardwalks are also popular in wetlands, accounting for 2.80% and 2.50%, respectively. Other ground cover flowers and plants as summer and autumn close-up elements accounted for 1.40%; artificial facilities, such as garbage bins and transformer boxes (0.70%), are not preferred (see Figure 3 for detail).

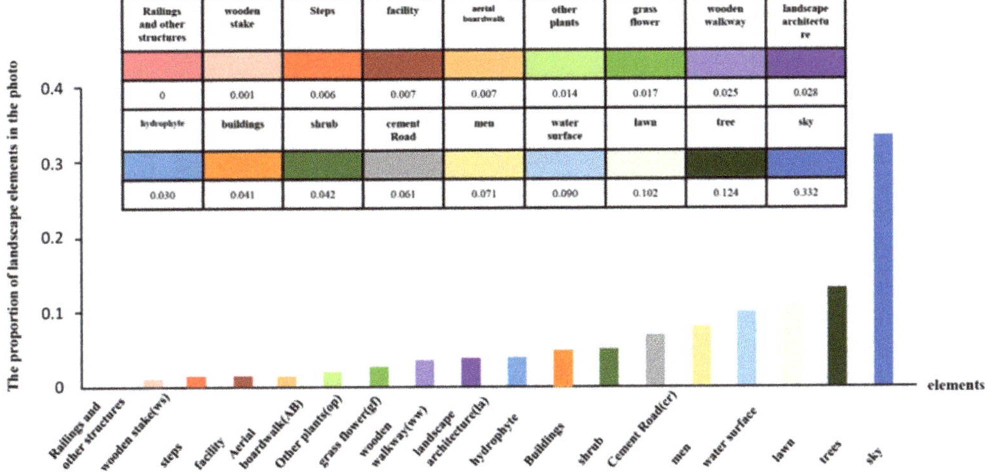

Railings and other structures	wooden stake	Steps	facility	aerial boardwalk	other plants	grass flower	wooden walkway	landscape architecture
0	0.001	0.006	0.007	0.007	0.014	0.017	0.025	0.028
hydrophyte	buildings	shrub	cement Road	men	water surface	lawn	tree	sky
0.030	0.041	0.042	0.061	0.071	0.090	0.102	0.124	0.332

Figure 3. Photo element recognition statistics.

Then, the study produced statistics on the seasons of the photos, and it was found that spring and summer were the peak seasons for tourists to visit the wetland park, while winter had the least number of visitors. In this study, landscape elements preferred by tourists in the four seasons are divided into natural elements (animals, plants, water surface, natural revetment, snow and ice, sky, and sunset) and artificial elements (park facilities, artificial landscape, road traffic) and are identified and counted according to the elements in the photos and their proportions [69]. The statistical table is as follows (see Figure 4 for detail): In terms of plant preference, in spring people prefer trees in wetland parks more. New buds and flowers, such as forsythia and elm blossom, in early spring become the preferred ornamental characteristics of trees in spring, and the preference degree of wetland landscape remains the highest among all elements. Birch, as a unique wetland tree species in northeast China, has become the plant landscape with the highest tourist preference. The preference degree of reed community landscape was the highest in summer and autumn.

In wetland parks, people prefer natural elements to artificial ones, and this preference is not affected by seasons, as shown in Figure 5. Plants were chosen as a preference by visitors in summer, followed by winter, autumn and spring; The season with the highest focus on artificial elements is summer, followed by winter, spring, and fall. This shows

Land **2022**, 11, 2284

that people pay more attention to the construction of plants and artificial landscape when visiting wetland park in winter, although the types of plants are relatively monotonous. In terms of preference for natural elements, plants still pay the most attention to natural elements in spring and summer, followed by water, sky, and animals. Insects and water birds in summer are second only to water. In autumn, the natural elements of focus are color-changing plants and reed swamps, while winter is snow, plants, ice, sky, sunset, animals.

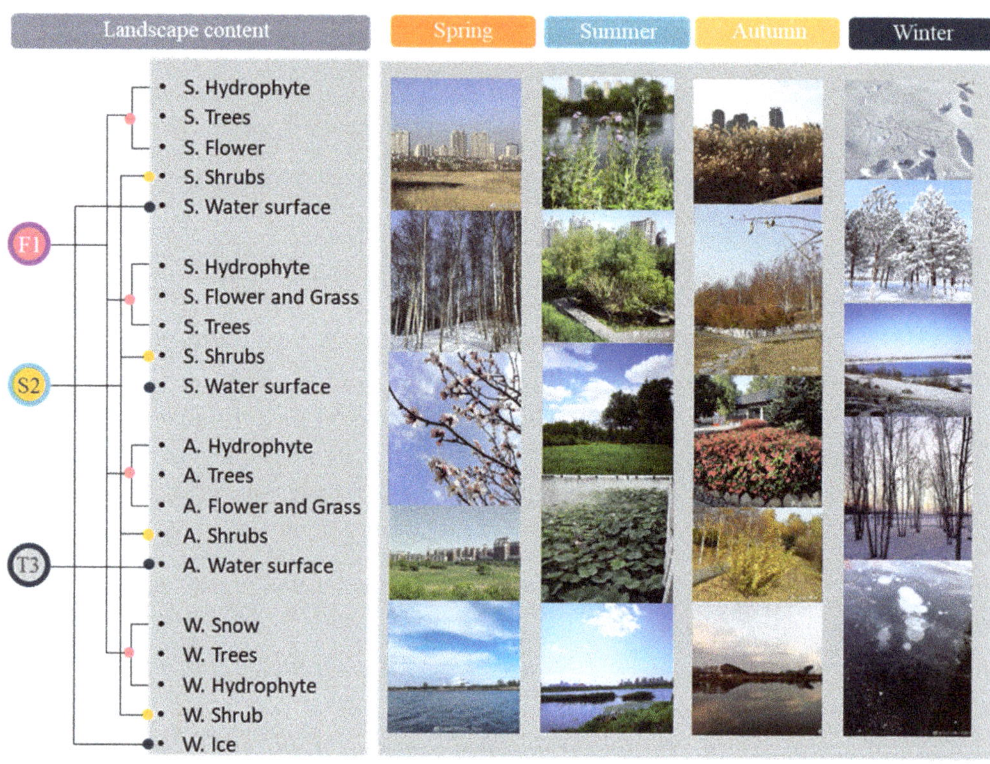

Figure 4. Focus on landscape elements in the four seasons (F1: the element with the highest attention from tourists, S2: the element with relatively little attention, T3: the element with little attention).

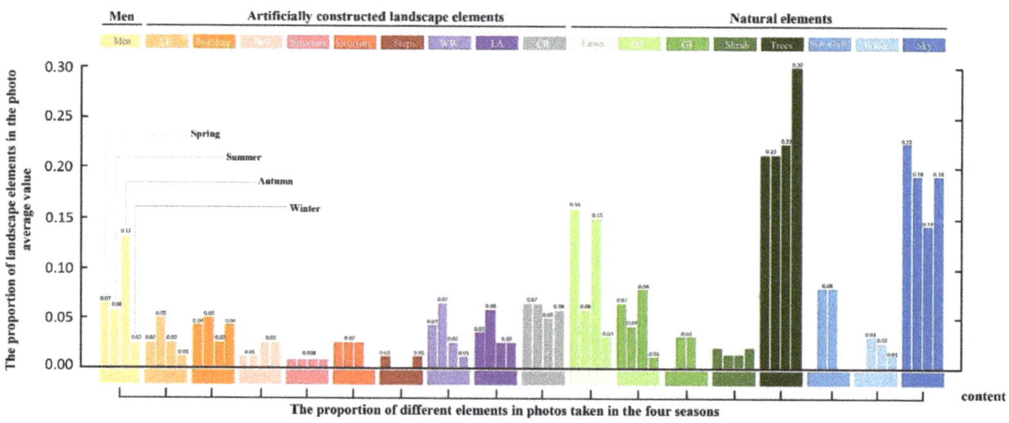

Figure 5. The four seasons focus on landscape elements.

3.2. Correlation Analysis between Visual Preference and Habitat Diversity

Based on the correlation analysis between the habitat diversity index and preference measure (Table 5), it is found that there is a significant positive correlation between habitat saturation and active artificial elements, and a significant negative correlation between landscape complexity and landscape wilderness and instantaneity. The naturalness, instantaneity of landscape, and positive artificial elements play a positive role in promoting habitat diversity. There was no correlation between habitat saturation and tourists' perception of naturalness, richness, wilderness, landscape instantaneity, and negative artificial elements. There is no correlation between landscape complexity and naturalness, richness, or negative or positive artificial elements. There was no correlation between habitat diversity and perceived richness, wilderness, or negative artificial elements.

Table 5. The correlation between the habitat diversity of the samples and the measure of visual preference [69].

	Habitat Saturation	Landscape Complexity	Habitat Diversity
Naturalness	0.239	0.014	0.355 *
Diversity	0.113	−0.023	0.186
Wilderness	0.211	−0.313 *	−0.186
Temporality	0.225	−0.514 **	−0.419 **
Positive artificial element	0.487 **	−0.218	0.304 *
Negative artificial element	0.099	−0.170	−0.129
Visual scale	0.088	−0.174	−0.011

* $p < 0.05$ ** $p < 0.01$.

(1)   Correlation analysis between saturation and active artificial elements

The aerial walkway in the positive artificial element is proportional to the habitat saturation of the wetland park. The higher the habitat saturation in the area with an aerial walkway, the closer the actual biodiversity level is to the ideal biodiversity level. The air corridor is the secondary spatial structure of the wetland park and the participation of human beings in the vertical ecological process of the wetland park. This participation method has been proven to be positive and effective in this study.

(2)   Correlation analysis of landscape complexity, temporality, and wilderness degree

There is a significant negative correlation between landscape complexity and temporality, which includes the seasonal changes of autumn and winter plants and the meteorological landscape, such as snow and ice. High temporality means that there are snow and ice and other meteorological landscapes in the field of view, so winter hampers visitors' perception of wetland habitat diversity. The seasonal changes of plants can help tourists perceive the diversity of habitats, and the landscape richness presented by autumn-colored plants is more easily perceived by tourists. Summer visitors' perception of landscape complexity is at a moderate level.

There is a negative correlation between landscape complexity and wilderness degree. The higher the landscape complexity, the better the ecology in the region. The lower the wilderness, the more man-made buildings, roads, and other artificial elements are within view and the less vegetation there is in the wilderness. This also confirms the conclusion that the aerial walkway, a positive artificial element, is positively correlated with habitat saturation. It shows that the ecology of the natural environment with artificial elements is not necessarily bad, and moderate artificial construction is beneficial to tourists' perception of landscape complexity and regional biodiversity.

(3)   Habitat diversity and temporality, naturalness, and positive artificial elements

The correlation analysis between habitat diversity and visual preference shows that there is a close correlation between habitat diversity and the naturalness of view. The area of natural vegetation in the view of high naturalness is larger, and the higher the habitat diversity index in the area is, the better the ecology is. The habitat diversity index was negatively correlated with flowers and plants but positively correlated with trees and aquatic plants. Here, we can understand that, to some extent, tourists prefer landscapes with high vegetation naturalness, and the habitat diversity index in this region is also at a high level.

There was a significant positive correlation between habitat diversity and temporality within the horizon. The more temporality the landscape had within the horizon, the lower the habitat diversity index, that is, the lower the habitat diversity in the landscape covered by snow and ice in winter, which confirms the negative correlation between the landscape complexity and the temporality and indicates the commonality of the visual aesthetic and ecological characteristic indicators [54].

There was a positive correlation between habitat diversity and positive artificial elements. Corridors, pavilions, and other landscape constructions in the park are the most preferred landscape elements for tourists in spring, and the space carrying them is also a concentrated area for tourists. Although the plant configuration in this area is rich in levels, diverse in species, and high in habitat diversity, the spring season is monotonous, which reduces tourists' cognition of plant diversity.

3.3. Coupling Results of Visual Preference and Habitat Diversity

According to the location and content of tourists' photos, we formed the tourists' view analysis diagram in the sample space. Then, we superimposed the tourist horizon analysis map and wetland habitat diversity map, analyzing and summarizing the coupling results (Figure 6). According to the visual preference of tourists, it is divided into three levels: high, medium, and low, and the habitat diversity level is described as high, medium, and low. Based on the coupling results, tourists' visual preferences for typical wetland landscapes can be divided into the following four modes: high preference–high habitat, high preference–low habitat, low preference–high habitat, and low preference–low habitat.

The high habitat–high preference pattern consisted of a diverse wetland bubble, arbor, and waterfront square habitat. The existence of this model proves that the maintenance of good wetland habitats and tourists' leisure activities can achieve a relative balance. Most of the tourist activities in this habitat space are picnics and kite flying. The habitat in the activity area is dominated by tall tree communities, so for this kind of habitat space, the original supervision of plants and organisms in the habitat area should be maintained, and corresponding supporting facilities should be added in the activity area to reduce the impact of tourists on the environment.

The types of habitat combinations with higher habitat levels but lower preference included linear traffic space, arbor habitat, wetland bubble habitat, aquatic plant habitat, and point-like structure combination. Within wetland parks, the linear upper air corridor is the main space for visitors to view and participate in the wetland habitat. This area is a space where visitors often move. Visitors can enjoy the cool, take photos, get close to the plants, and partake in other activities. Therefore, for this low-preference habitat space, we should start from two aspects. On the one hand, the connection and protection between artificial structures and the original wetland habitat should be strengthened, and multi-level plant landscape design should be carried out near the linear space to form a certain buffer protection area and reduce the damage to the internal wetland. To meet the needs of tourists, it is important to improve their natural interaction experience. On the other hand, proper ground space is added along the walkway for visitors to rest and stay, increasing the length of the natural experience.

The entrance habitat combination of the park is the most popular type for visitors to focus on, but its habitat level is not high. This type of space is dominated by the square space, which is responsible for the function of the whole evacuation park and some activity places. The functional-oriented habitat space does not affect its habitat diversity on the premise of not changing its spatial function.

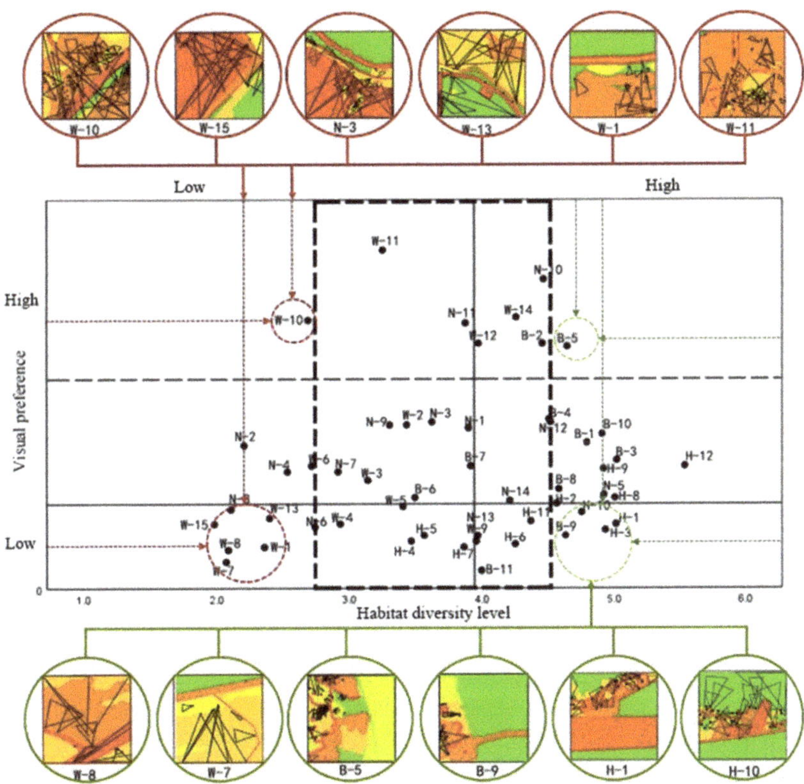

Figure 6. Analysis of views of 4 urban wetland parks.

Undesirable habitats and low-elevation habitats are combined into waterfront roads with no shoreline plants and swamps with single plant types. The existence of such habitats confirms that tourists prefer man-made vegetation spaces. In view of such habitat space problems, we should improve the vegetation design along the waterfront road according to the park space tour route. After determining the dominant plant species, artificial construction is carried out to construct the water–land ecotone formed from the water area to the waterfront, further improving the habitat diversity of the wetland on the basis of the original ecology and forming a stable ecological environment and ecological community structure.

4. Discussion

Studies have shown that tourists do not blindly pay attention to natural elements in wetland parks, instead having some unnatural elements as their first-choice recreational element in the wetland. Moreover, in areas with obvious seasonal climate changes, the season is the key to influencing tourists' preferences. There is a significant correlation between wetland habitat diversity and tourists' landscape preferences. Because of the spontaneity and statistics of social media data, it is confirmed that people are born with the perception of biodiversity [32]. Finally, cluster analysis and spatial coupling methods were used to summarize four preferred habitat types of typical urban wetland parks, and reasonable human intervention was carried out for different habitat types to improve biodiversity to a certain extent [29,77]. At the same time, with the promotion of social media, the computational dimension of the biodiversity experience is enriched, the ecological education function of urban wetland parks is strengthened, and the dynamic balance between recreation and conservation is expected to be achieved.

In the study of tourists' preference for wetland landscapes, tourists pay much more attention to the tree communities in a wetland than the characteristic aquatic plant communities in a wetland park, which is unexpected. This also proves that tourists can better capture rare values of the wetland by freely choosing the location of photos [78]. The season is an important factor affecting preferences and tourists' selection of wetland parks. People prefer plants, water surfaces, and landscapes with pruning and artificial construction, for example, people like tree-lined walking paths, rich artificial planting areas, lawns with planted trees, swamps with water surfaces, and wide water surfaces [31,32,79,80]. So we wondered what the purpose is of visiting wetland parks. Is it to experience the unique landscape of the wetland, or to use it as a traditional park green space for participation and activities? If tourists' comments on the park are combined, we can try to obtain the purpose of tourists' visit to understand the image of the wetland park in the minds of tourists, guiding people to participate more in the perception of wetland biodiversity and helping the management department to manage the core image of the wetland park [81].

Based on the mapping of habitat diversity and the location information data of tourist photos, the coupling of habitat diversity space and tourist recreation space was realized, and four typical types of habitat space were summarized. In practice, it has been proven that it is convenient and effective to take habitat diversity as the primary indicator to measure biodiversity, incorporate man-made landscape elements, such as squares and roads, into habitats, and use spatial data, such as wetland area and perimeter, to assess the biodiversity of urban wetland parks. It can help managers assess the status quo of biodiversity in target areas under limited conditions [25,78]. The diversity of urban wetland habitats is significantly related to the landscape preference of tourists. The aerial boardwalk, as a way of human participation in the vertical ecological process of a wetland park, has been proven to be positive for the protection of wetland park habitat diversity. Therefore, active artificial construction is conducive to tourists' perception of habitat diversity in wetland parks. Based on the commonality of ecological indicators and human visual indicators [54,82], more visual and ecological indicators and objective adjustable indicators can be systematically studied in future studies to determine whether there is a certain mediating effect between the three, so as to further explore the relationship between tourists' visual preference and wetland biodiversity, thereby effectively intervening in the wetland more successfully. The restoration of wetland ecological function is discussed in [83].

There is a clear lack of research on social media data in wetland parks. Although tourists' photo data are the result of tourists' freely photographing and uploading to the social media platform, the path where the photo was taken is the result of the designer's work, and the attention to the landscape is guided by the path to a certain extent. There are some undeveloped areas of wetland parks, where the tourists cannot enter. It is impossible to determine whether the undeveloped habitat types of different levels are preferred by tourists. Therefore, it is necessary to use drones and other equipment in future experiments to sample and shoot all the existing habitat types in the wetland park, enrich the experimental data, set up corresponding questionnaires, and interview a certain number of tourists. Regarding whether the social background of tourists will affect their preference for wetland parks, relevant studies have proved that professional knowledge or special hobbies will affect the preference of tourists [84]. However, since the data collection did not include the social background of the photographer, it is uncertain whether the tourists' preference for wetlands is related to a specific occupation, but it is clear that women are more enthusiastic than men when it comes to sharing landscape photos. In subsequent studies, corresponding questionnaires or survey interviews can be added to obtain tourists' attitudes towards wetland habitat through a comprehensive survey of wetland parks, which can provide certain landscape design suggestions for the planning and designers of wetland parks, so as to maintain the original landscape characteristics of wetland and enrich tourists' landscape experience.

In terms of theory, this study can be used as a basis for future research on the deep mechanism of habitat diversity and tourists' landscape preferences by using social media data at the scale of a park habitat. In the application, it provides an effective and scientific prediction method for the recreation path and space construction of the wetland park to be constructed, provides specific design strategies for the management of the existing urban wetland park in the cold region, and provides scientific construction and management basis for both the protection of urban natural resources and the recreation development.

5. Conclusions

This study found that habitat diversity was positively correlated with positive artificial elements, such as aerial boardwalk, pavilions, etc. The commonality of visual aesthetic preference and ecological characteristic index is verified, that is, there is a certain correlation between the habitat diversity of a wetland park and the preference of tourists, and tourists have a strong perception of biodiversity and preference choice. We have the opportunity to achieve a dynamic balance between habitat protection and recreation construction of wetland parks by increasing or reducing artificial elements. (1) Set the activity space of aggregation degree to improve the complex function of the entrance space; (2) increase the aerial walkway to improve wetland saturation; (3) protect the winding shoreline of the natural pond surface and improve the stability of the shoreline plant buffer zone; (4) divide "whole" into "scattered" to improve the perception of wetland complexity in winter; (5) rebuild the stagnation space of the community transition zone to avoid the disturbance of stampede activity; (6) increase positive artificial elements to reduce tourists' wilderness experience.

In future studies, the number of samples, habitat types, and coverage areas can be further expanded to supplement and improve the feedback information of tourists' preferences, such as obtaining basic information, including tourists' education background and occupation, and exploring the influencing factors of tourists' wetland preference characteristics in multiple dimensions. This study can be used as the basis of the deep mechanism between biodiversity and tourists' landscape preference at the site scale and provide a scientific basis for the construction and management of both urban natural resource protection and recreational development.

Author Contributions: Conceptualization, J.Z. and X.Z.; Methodology, J.Z.; Software, J.Z.; Validation, J.Z., M.G. and X.Z.; Formal Analysis, J.Z.; Investigation, J.Z.; Resources, J.Z.; Data Management, J.Z. and M.G.; Writing—Manuscript Preparation, J.Z.; Writing—Review and Editing, J.Z., M.G. and X.Z.; Visualization, J.Z.; Supervision, M.G.; Project Management, X.Z. All authors have read and agreed to the published version of the manuscript.

Funding: This research was funded by the National Natural Science Foundation of China—study on the global tourism space model of the Middle East Railway Heritage Corridor, grant number 51908170. Heilongjiang Forestry and Grassland Administration—study on contribution of forest and grass industry to the carbon neutrality goal of Heilongjiang Province, grant number 22252.

Data Availability Statement: Data are available on request due to privacy and ethical restrictions. The data presented in this study are available on request from the corresponding author.

Conflicts of Interest: The article is the product of the author's work during his master's degree studies. Additionally, there is no conflict of interest.

References

1. Elizabeth, B. *The Ecotourism Boom: Planning for Development and Management*; WWF: Washington, DC, USA, 1992; pp. 3–10.
2. Spieles, D.J. Wetland Construction, Restoration, and Integration: A Comparative Review. *Land* **2022**, *11*, 554. [CrossRef]
3. Saha, U.D.; Mitra, S.; Mondal, M.; Biswas, N. Assessment of the impact of anthropogenic stress on wetlands of the upper Ganga Delta, India: A multi criteria decision making approach. *Geocarto Int.* **2022**, *38*, 1–22.
4. Holland, C.C.; Honea, J.; Gwin, S.E.; Kentula, M.E. Wetland degradation and loss in the rapidly urbanizing area of Portland, Oregon. *Wetlands* **1995**, *15*, 336–345. [CrossRef]
5. Zhang, K.-L.; Zhou, X.; Gao, J.-H. Some Thoughts about Wetland, National Wetland Park and Urban Wetland Park. *Landsc. Archit.* **2012**, *6*, 108–110. [CrossRef]
6. Butler, R.W.; Waldbrook, L.A. A new planning tool: The tourism opportunity spectrum. *J. Tour. Stud.* **1991**, *2*, 2–14.

7. Stankey, G.H. *The Limits of Acceptable Change (LAC) System for Wilderness Planning*; National Agricultural Library: Beltsville, MD, USA, 1985.
8. Manning, R. Visitor experience and resource protection: A framework for managing the carrying capacity of National Parks. *J. Park Recreat. Adm.* **2001**, *19*, 1.
9. Yonghui, L.; Dan, Z. Urban Forest Park and the Reconstruction and Protection of Wildlife Habitat. *Xiandai Hortic.* **2020**, *4*, 159–160.
10. Theobald, D.M.; Hobbs, N.T.; Bearly, T.; Zack, J.A.; Shenk, T.; Riebsame, W.E. Incorporating biological information in local land-use decision making: Designing a system for conservation planning. *Landsc. Ecol.* **2000**, *15*, 35–45. [CrossRef]
11. Kaplan, S. The restorative benefits of nature: Toward an integrative framework. *J. Environ. Psychol.* **1995**, *15*, 169–182. [CrossRef]
12. Mayer, F.S.; Frantz, C.M.; Bruehlman-Senecal, E.; Dolliver, K. Why is nature beneficial? The role of connectedness to nature. *Environ. Behav.* **2009**, *41*, 607–643. [CrossRef]
13. Macaulay, R.; Lee, K.; Johnson, K.; Williams, K. Mindful engagement, psychological restoration, and connection with nature in constrained nature experiences. *Landsc. Urban Plan.* **2022**, *217*, 104263. [CrossRef]
14. Spieles, D.J.; Mitsch, W.J. Macroinvertebrate community structure in high-and low-nutrient constructed wetlands. *Wetlands* **2000**, *20*, 716–729. [CrossRef]
15. Levandowski, M.L.; Litt, A.R.; McKenna, M.F.; Burson, S.; Legg, K.L. Multi-method biodiversity assessments from wetlands in Grand Teton National Park. *Ecol. Indic.* **2021**, *131*, 108205. [CrossRef]
16. Wang, B.; Zheng, X.; Zhang, H.; Xiao, F.; Gu, H.; Zhang, K.; He, Z.; Liu, X.; Yan, Q. Bacterial community responses to tourism development in the Xixi National Wetland Park, China. *Sci. Total. Environ.* **2020**, *720*, 137570. [CrossRef] [PubMed]
17. Wang, X.; Yu, J.; Zhou, D.; Dong, H.; Li, Y.; Lin, Q.; Guan, B.; Wang, Y. Vegetative Ecological Characteristics of Restored Reed (Phragmites australis) Wetlands in the Yellow River Delta, China. *Environ. Manag.* **2012**, *49*, 325–333. [CrossRef] [PubMed]
18. Falkowska, E.; Jancewicz, E. Landscape diversity vs. population resilience of a wetland species near the limits of its range (the root vole Microtus oeconomus in Poland)—Implications for species conservation. *CATENA* **2021**, *211*, 105947. [CrossRef]
19. Bräuniger, C.; Knapp, S.; Kühn, I.; Klotz, S.; Bräuniger, C.; Knapp, S.; Kühn, I.; Klotz, S. Testing taxonomic and landscape surrogates for biodiversity in an urban setting. *Landsc. Urban Plan.* **2010**, *97*, 283–295. [CrossRef]
20. Singh, M.; Rajiv, S. Evaluating dynamic hydrological connectivity of a floodplain wetland in North Bihar, India using geostatistical methods. *Sci. Total Environ.* **2019**, *651*, 2473–2488. [CrossRef]
21. Deutschewitz, K.; Lausch, A.; Kühn, I.; Klotz, S. Native and alien plant species richness in relation to spatial hetero-geneity on a regional scale in Germany. *Glob. Ecol. Biogeogr.* **2003**, *12*, 299–311. [CrossRef]
22. Tansley A, G. The Use and Abuse of Vegetation Concepts and Terms. *Ecology* **1935**, *16*, 284–307. [CrossRef]
23. Triantis, K.A.; Mylonas, M.; Lika, K.; Vardinoyannis, K. A model for the species–area–habitat relationship. *J. Biogeogr.* **2003**, *30*, 19–27. [CrossRef]
24. Kallimanis, A.S.; Mazaris, A.D.; Tzanopoulos, J.; Halley, J.M.; Pantis, J.D.; Sgardelis, S.P. How does habitat diversity affect the species–area relationship? *Glob. Ecol. Biogeogr.* **2008**, *17*, 532–538. [CrossRef]
25. Hermy, M.; Cornelis, J. Towards a monitoring method and a number of multifaceted and hierarchical biodiversity indicators for urban and suburban parks. *Landsc. Urban Plan.* **2000**, *49*, 149–162. [CrossRef]
26. Hao, S.; Wang, C.; Lin, H. Design and assessment of biodiversity in urban wetland parks: Take Liupanshui Minghu National Wetland Park as an example. *Acta Ecol. Sin.* **2019**, *39*, 5967–5977.
27. Qiu, L.; Gao, T.; Gunnarsson, A.; Hammer, M.; von Bothmer, R. A methodological study of biotope mapping in nature conservation. *Urban For. Urban Green.* **2010**, *9*, 161–166. [CrossRef]
28. Ekici, B.; Korkut, A.; Özyavuz, M. Biotopes in the Ganos (Işıklar) Mountain and Near Surrounding Areas. *Int. J. Sci. Basic Appl. Res.* **2018**, *40*, 264–274.
29. Ekici, B. Kumul biyotoplarının haritalanması, Kurucaşile (Bartın) örneği. *Turk. J. For.* **2020**, *21*, 188–194. [CrossRef]
30. Kaplan, S.; Kaplan, R. *The Experience of Nature: A Psychological Perspective*; Cambridge University Press: Cambridge, UK, 1989.
31. Kaplan, S.; Kaplan, R. *Humanscape: Environments for People*; Duxbury Press: Belmont, CA, USA, 1978.
32. Wilson, E.O. *Biophilia*; Harvard University Press: Cambridge, UK, 1984.
33. Duan, Y.; Shuhua, L. Study of Different Vegetation Types in Green Space Landscape Preference: Compari-son of Environmental Perception in Winter and Summer. *Sustainability* **2022**, *14*, 3906. [CrossRef]
34. Dobbie, M.; Green, R. Public perceptions of freshwater wetlands in Victoria, Australia. *Landsc. Urban Plan.* **2012**, *110*, 143–154. [CrossRef]
35. Smardon, R.C. *The Future of Wetlands: Assessing Visual-Cultural Values*; Rowman & Littlefield Publishers: Lanham, MD, USA, 1983.
36. Dobbie, M.F. Public aesthetic preferences to inform sustainable wetland management in Victoria, Australia. *Landsc. Urban Plan.* **2013**, *120*, 178–189. [CrossRef]
37. Qiu, L.; Nielsen, A.B. Are Perceived Sensory Dimensions a Reliable Tool for Urban Green Space Assessment and Planning? *Landsc. Res.* **2015**, *40*, 834–854. [CrossRef]
38. Van Hecke, L.; Ghekiere, A.; Van Cauwenberg, J.; Veitch, J.; De Bourdeaudhuij, I.; Van Dyck, D.; Clarys, P.; Van De Weghe, N.; Deforche, B. Park characteristics preferred for adolescent park visitation and physical activity: A choice-based conjoint analysis using manipulated photographs. *Landsc. Urban Plan.* **2018**, *178*, 144–155. [CrossRef]
39. Wang, R.; Zhao, J.; Meitner, M.J.; Hu, Y.; Xu, X. Characteristics of urban green spaces in relation to aesthetic preference and stress recovery. *Urban For. Urban Green.* **2019**, *41*, 6–13. [CrossRef]

40. Zhang, H.; Chen, B.; Sun, Z.; Bao, Z. Landscape perception and recreation needs in urban green space in Fuyang, Hangzhou, China. *Urban For. Urban Green.* **2013**, *12*, 44–52. [CrossRef]
41. Carrus, G.; Scopelliti, M.; Lafortezza, R.; Colangelo, G.; Ferrini, F.; Salbitano, F.; Agrimi, M.; Portoghesi, L.; Semenzato, P.; Sanesi, G. Go greener, feel better? The positive effects of biodiversity on the well-being of individuals visiting urban and peri-urban green areas. *Landsc. Urban Plan.* **2015**, *134*, 221–228. [CrossRef]
42. Holy-Hasted, W.; Brendan, B. Does public space have to be green to improve well-being? An analysis of public space across Greater London and its association to subjective well-being. *Cities* **2022**, *125*, 103569. [CrossRef]
43. Lindemann-Matthies, P.; Briegel, R.; Schüpbach, B.; Junge, X. Aesthetic preference for a Swiss alpine landscape: The impact of different agricultural land-use with different biodiversity. *Landsc. Urban Plan.* **2010**, *98*, 99–109. [CrossRef]
44. Qureshi, S.; Tarashkar, M.; Matloobi, M.; Wang, Z.; Rahimi, A. Understanding the dynamics of urban horticulture by socially-oriented practices and popu-lace per-ception: Seeking future outlook through a comprehensive review. *Land Use Policy* **2022**, *122*, 106398. [CrossRef]
45. Junge, X.; Jacot, K.A.; Bosshard, A.; Lindemann-Matthies, P. Swiss people's attitudes towards field margins for biodiversity conservation. *J. Nat. Conserv.* **2009**, *17*, 150–159. [CrossRef]
46. Lindemann-Matthies, P.; Xenia, J.; Diethart, M. The influence of plant diversity on people's per-ception and aesthetic appreciation of grassland vegetation. *Biol. Conserv.* **2010**, *143*, 195–202. [CrossRef]
47. Voigt, A.; Kabisch, N.; Wurster, D.; Haase, D.; Breuste, J. Structural Diversity: A Multi-dimensional Approach to Assess Recreational Services in Urban Parks. *AMBIO* **2014**, *43*, 480–491. [CrossRef]
48. Macháč, J.; Brabec, J.; Arnberger, A. Exploring public preferences and preference heterogeneity for green and blue infrastructure in urban green spaces. *Urban For. Urban Green.* **2022**, *75*, 127695. [CrossRef]
49. Zhang, X.; Xu, D.; Zhang, N. Research on Landscape Perception and Visual Attributes Based on Social Media Data—A Case Study on Wuhan University. *Appl. Sci.* **2022**, *12*, 8346. [CrossRef]
50. Marine, N.; Arnaiz-Schmitz, C.; Santos-Cid, L.; Schmitz, M.F. Can We Foresee Landscape Interest? Maximum Entropy Applied to Social Media Photographs: A Case Study in Madrid. *Land* **2022**, *11*, 715. [CrossRef]
51. Toivonen, T.; Heikinheimo, V.; Fink, C.; Hausmann, A.; Hiippala, T.; Järv, O.; Tenkanen, H.; Di Minin, E. Social media data for conservation science: A methodological overview. *Biol. Conserv.* **2019**, *233*, 298–315. [CrossRef]
52. Daniel, T.C. Whither scenic beauty? Visual landscape quality assessment in the 21st century. *Landsc. Urban Plan.* **2001**, *54*, 267–281. [CrossRef]
53. Dronova, I. Environmental heterogeneity as a bridge between ecosystem service and visual quality objectives in man-agement, planning and design. *Landsc. Urban Plan.* **2017**, *163*, 90–106. [CrossRef]
54. Fry, G.; Tveit, M.S.; Ode, Å.; Velarde, M.D. The ecology of visual landscapes: Exploring the conceptual common ground of visual and ecologi-cal land-scape indicators. *Ecol. Indic.* **2009**, *9*, 933–947. [CrossRef]
55. Huai, S.; Chen, F.; Liu, S.; Canters, F.; Van de Voorde, T. Using social media photos and computer vision to assess cultural ecosystem services and land-scape features in urban parks. *Ecosyst. Serv.* **2022**, *57*, 101475. [CrossRef]
56. Kaplan, S.; Kaplan, R. *Cognition and Enviroment: Functioning in an Uncertain World*; Preager Press: Westport, CT, USA, 1982.
57. Tveit, M.; Ode, Å.; Fry, G. Key concepts in a framework for analysing visual landscape character. *Landsc. Res.* **2006**, *31*, 229–255. [CrossRef]
58. Zhu, X.; Chiou, S.-C. A Study on the Sustainable Development of Historic District Landscapes Based on Place Attachment among Tourists: A Case Study of Taiping Old Street, Taiwan. *Sustainability* **2022**, *14*, 11755. [CrossRef]
59. National Forestry and Grassland Administration. Available online: http://www.forestry.gov.cn/ (accessed on 10 September 2020).
60. Geospatial Data Cloud Official Website. Available online: http://www.gscloud.cn/ (accessed on 10 September 2020).
61. Laba, M.; Downs, R.; Smith, S.; Welsh, S.; Neider, C.; White, S.; Richmond, M.; Philpot, W.; Baveye, P. Mapping invasive wetland plants in the Hudson River National Estuarine Research Reserve using quickbird satellite imagery. *Remote Sens. Environ.* **2008**, *112*, 286–300. [CrossRef]
62. Zhang, C.; Chen, S.; Li, P.; Liu, Q. Spatiotemporal dynamic remote sensing monitoring of typical wetland vegetation in the Current Huanghe River Estuary Reserve. *Haiyang Xuebao* **2022**, *44*, 125–136.
63. Weibo Official Website. Available online: https://weibo.cn/ (accessed on 2 May 2022).
64. Ctrip Official Website. Available online: https://www.ctrip.com/ (accessed on 2 May 2022).
65. Liangbulu Official Website. Available online: https://www.2bulu.com/ (accessed on 2 May 2022).
66. Qunar Office Website. Available online: https://www.qunar.com (accessed on 2 May 2022).
67. Zhang, K.; Ye, C.; Chunlin, L. Discovering the tourists' behaviors and perceptions in a tourism destination by ana-lyzing photos' visual content with a computer deep learning model: The case of Beijing. *Tour. Manag.* **2019**, *75*, 595–608. [CrossRef]
68. Badrinarayanan, V.; Ankur, H.; Roberto, C. Segnet: A deep convolutional encoder-decoder archi-tecture for robust semantic pixel-wise labelling. *arXiv* **2015**, arXiv:1505.07293.
69. Jiani, Z. Study on Habitat Diversity of Wetland Park in Winter City Based on Tourists' Visual Preference. Master's Thesis, Harbin Institute of Technology, Harbin, China, 2020.
70. Dearden, P. Public participation and scenic quality analysis. *Landsc. Plan.* **1981**, *8*, 3–19. [CrossRef]
71. Ulrich, R.S. View through a window may influence recovery from surgery. *Science* **1984**, *224*, 420–421. [CrossRef]

72. Stephen R., K.; Wilson, E.O. *The Biophilia Hypothesis*; Island Press: Washington, DC, USA, 1993.
73. Clay, G.R.; Smidt, R.K. Assessing the validity and reliability of descriptor variables used in scenic highway analysis. *Landsc. Urban Plan.* **2004**, *66*, 239. [CrossRef]
74. Arriaza, M.; Cañas-Ortega, J.F.; Cañas-Madueño, J.A.; Ruiz-Aviles, P. Assessing the visual quality of rural landscapes. *Landsc. Urban Plan.* **2004**, *69*, 115–125. [CrossRef]
75. Fontana, S.; Sattler, T.; Bontadina, F.; Moretti, M. How to manage the urban green to improve bird diversity and community structure. *Landsc. Urban Plan.* **2011**, *101*, 278–285. [CrossRef]
76. Tyrväinen, L.; Kirsi, M.; Jasper, S. Tools for mapping social values of urban woodlands and other green areas. *Landsc. Urban Plan.* **2007**, *79*, 5–19. [CrossRef]
77. Hurtado-M, A.B.; Echeverry-Galvis, M.Á.; Salgado-Negret, B.; Muñoz, J.C.; Posada, J.M.; Norden, N. Little trace of floristic homogenization in peri-urban Andean secondary forests despite high anthropogenic transformation. *J. Ecol.* **2021**, *109*, 1468–1478. [CrossRef]
78. Fjellstad, W.; Eiter, S.; Puschmann, O.; Krøgli, S.O. Planning the first view: Establishing a landscape monitoring scheme based on photography. *Landsc. Urban Plan.* **2022**, *226*, 104470. [CrossRef]
79. Davidowitz, G.; Rosenzweig, M.L. The latitudinal gradient of species diversity among North Ameri-can grasshoppers (Acrididae) within a single habitat: A test of the spatial heterogeneity hypothesis. *J. Biogeogr.* **1998**, *25*, 553–560. [CrossRef]
80. Orians, G.H. Habitat selection: General theory and applications to human behavior. In *The Evolution of Human Social Behavior*; Oxford University Press: Oxford, UK, 1980.
81. Tang, F.; Yang, J.; Wang, Y.; Ge, Q. Analysis of the Image of Global Glacier Tourism Destinations from the Perspective of Tourists. *Land* **2022**, *11*, 1853. [CrossRef]
82. Karasov, O.; Heremans, S.; Külvik, M.; Domnich, A.; Burdun, I.; Kull, A.; Helm, A.; Uuemaa, E. Beyond land cover: How integrated remote sensing and social media data analysis facili-tates assessment of cultural ecosystem services. *Ecosyst. Serv.* **2022**, *53*, 101391. [CrossRef]
83. Choi, H.; Kim, H.; Nam, B.E.; Bae, Y.J.; Kim, J.G. Effect of initial planting on vegetation establishment in different depth zones of constructed farm ponds. *Restor. Ecol.* **2021**, *30*, e13488. [CrossRef]
84. Tveit, M.S. Indicators of visual scale as predictors of landscape preference; a comparison between groups. *J. Environ. Manag.* **2009**, *90*, 2882–2888. [CrossRef]

Article

Mountain Landscape Preferences of Millennials Based on Social Media Data: A Case Study on Western Sichuan

Keying Ding [1], Mian Yang [1,*] and Shixian Luo [2]

1   Art College, Sichuan Tourism University, Chengdu 610100, China; keyingding@126.com
2   Graduate School of Horticulture, Chiba University, Chiba 263-8522, Japan; 19hd4106@student.gs.chiba-u.jp
*   Correspondence: yangmiansctu@126.com; Tel.: +86-138-8028-3481

Abstract: Mountain area is one of the most important modern tourist attractions, and unique mountain landscapes are highly appealing to millennials. Millennials post their travel photos and comments on social media, and these media messages can positively influence other millennials' travel motivations. To fully understand the attraction of mountain tourist destinations to millennials, this study analyzed their landscape preferences using images posted on social media. As a case study, we analyzed the landscape resources in Western Sichuan Plateau Mountain Areas (WSPMA). We found that differences in genders, modes of transportation, and travel patterns of the millennials influenced their preferences for mountain landscapes. Our results broaden the current knowledge on mountain tourism from the perspective of millennials through social media data. Moreover, studying the landscape resources in WSPMA can facilitate the analysis of regional advantages. This will ultimately enhance tourism publicity and integrate various resources for tourism management and planning in more targeted and attractive ways.

Keywords: Western Sichuan Plateau Mountain Areas; millennials; mountain tourism; social media data; landscape preference

Citation: Ding, K.; Yang, M.; Luo, S. Mountain Landscape Preferences of Millennials Based on Social Media Data: A Case Study on Western Sichuan. *Land* **2021**, *10*, 1246. https://doi.org/10.3390/land10111246

Academic Editors: Cecilia Arnaiz Schmitz, Nicolas Marine and María Fe Schmitz

Received: 18 October 2021
Accepted: 11 November 2021
Published: 13 November 2021

Publisher's Note: MDPI stays neutral with regard to jurisdictional claims in published maps and institutional affiliations.

1. Introduction

Mountain tourism is developing at an unprecedented rate, thereby, turning to an important branch of contemporary tourism. Thus, mountain tourism nowadays plays an important role in the global tourism landscape, as it overlaps with people's desire to be close to nature and pursue health [1,2]. Mountain areas exhibit a certain degree of fragility due to their biodiversity factors and environmental/natural resource sensitivity [3]. Mountainous regions have been inhabited by aborigines for a long time, constituting a rich folklore and unique human landscape in such regions. All these characteristics make any economic activity, including tourism, which draws attention to the cultural and natural environment [4], result in the sustainable development of mountain areas being very important [5].

Tourism in mountain areas, after a long period of economic success and deep environmental transformation [6], has to be rediscovered [7]. The motivation stems from the importance of the attractiveness of tourism resources to tourists and plays a key role in tourism sustainability [8]. Also, the landscape richness is a determining factor for the development of mountain tourism [9]. From the tourism perspective, many studies have focused on the identity-related motivations in niche markets, such as backpacker tourism [10], museum visits [11], and lifestyle travelling [12]. However, further empirical research is needed to explore how identity-related motivations affect broader tourism decisions and outcomes [13]. We argue that research on landscape preferences in mountain tourism has important implications in mountain tourism development.

Millennials, defined as those born between the 1980s and 1995 [14], represent a very large proportion of the population with high purchasing power compared with the other age groups [15]. Most importantly, they represent the most important generation in the global economy, especially in tourism [16,17]. Consequently, there is a growing interest in the role of millennials in mountain tourism [18]. Mountain tourism sustainability is determined to some extent by the behavior of tourists [19], thus, studying the mountain tourism behavior of millennials can contribute to achieving sustainable development.

Cavagnaro et al. reported that millennials demonstrate interest in themes such as natural resources and prefer destinations with significant natural resources [20–25]. Giachino et al. [7] found seasonal differences in the millennials' choice of nature tourism destinations. Tieskens et al. [26,27] demonstrated that analytical studies on elements of mountain cultural landscapes are of high research value in exploring tourists' preferences for mountain landscapes. Several studies have also shown that millennials show great interest in sustainability issues in mountain tourism, and their participation is considered a necessary prerequisite for the sustainability and improvement of nature tourism [28]. Meanwhile, millennials themselves exhibit an important influence on environmental sustainability [29–31]. In particular, Sharmin et al. [32] verified how millennial tourists' awareness of environmental sustainability influences their environmentally sustainable activities when they visit nature reserves. Thus, state-of-the-art of nature tourism indicates that the preference for natural and cultural landscapes in mountainous areas is a small but important determinant of millennials' travel motivation and landscape perception. It further affects their travel behavior, and in turn affects mountain tourism sustainability. Due to this, research on millennials' natural and cultural landscape preferences in tourist destinations is significant for promoting the sustainable development of tourist destinations. In particular, tourists' preferred activities vary with age [33], gender [34], mode of transportation [35], and travel patterns [36]. However, the differences in natural and cultural landscape preferences for mountain tourism among millennials by gender, mode of travel, and trip structure is yet to be studied in detail. Thus, a study on the differences in millennials' preferences for mountain landscapes across different forms of tourism is highly desirable.

Millennials are the backbone of the population, and they extensively use online social platforms to express themselves [37]. For millennials, it is important to co-create experiences and to provide feedback on good or bad experiences during their trip [38]. Meanwhile, the information collected about millennials' travel is also utilized to promote and to develop some destinations [39]. Loda et al. [40] highlighted the importance of millennials expressing their opinions by posting reviews and photos online in promoting destinations. In addition, the potential and impact of the content they share (e.g., photos and comments) cannot be neglected. Indeed, it is important to analyze the travel behavior of millennials through social media data.

The rise of social media has opened promising prospects for landscape preference research. Platforms such as Flickr and Instagram allow users to upload photos of their environment and place them on digital maps to collectively provide a publicly available database of volunteered geographic information [41,42]. One of the main advantages of volunteered geographic information is characterized by its ability to provide insights into popular spatial choices and preferences without experimental or survey bias [43].

Social media provides a common source of publicly available user-generated data to gain insights into spatial choices and landscape preferences [27,44]. In this domain, Richards et al. [45–47] used the spatial location of photographs, combined with the actual content of the photographs, to ensure that only the relevant photographs were considered to retrieve the information about the users' landscape preferences. In the study of Hausmann et al. [48], the content of photos from social media was analyzed to understand visitors' preferences for nature-based experiences in protected areas. Furthermore, Tieskens et al. [26] analyzed the aesthetic preferences of cultural landscapes in the Dutch river landscape through social media. All these studies confirmed an intrinsic link between the content of photos posted on social media and landscape preferences at the level of natural or cultural landscapes. Although, the limited information obtained through photographs makes it difficult to conduct extensive study, it provides non-negligible research values. Using photos posted by tourists on social media, we can explore the landscape value of mountain tourism from a more subtle and specific research perspective, which provides more research references for landscape researchers and tourism managers. Therefore, we attempted to explore the landscape preferences of millennials in mountain tourism by using photos posted by tourists on social media in two dimensions: natural and human landscapes.

To sum up, mountain regions are important for tourism research. Thus, we can use social media data to understand the landscape preferences of millennials in mountain tourism. Such studies would be in line with the behavioral habits of millennials and will help scholars in studying the landscape. The produced knowledge will not only open new research directions, but will help the decision/policy-makers and tourist managers to improve the tourism attractiveness in a tailored way. This is important for promoting the sustainable development of mountain tourism. To this end, we used social media data, (1) to understand the landscape preferences of millennials in mountain tourism from both natural and human landscape perspectives; (2) to explore whether there are differences in landscape preferences in mountain tourism among millennials based on gender, transportation modes, and travel patterns; and (3) to provide reference suggestions for sustainable tourism development and planning in mountain areas for the development and utilization of landscape resources.

2. Materials and Methods

The use of big data to investigate human behavior, facilitates tourism destination planning and management especially in the field of tourism [49]. We chose Ctrip (https://www.ctrip.com, accessed on 10 July 2021), Xiaohongshu (https://www.xiaohongshu.com, accessed on 10 July 2021), and Mafengwo (https://www.mafengwo.cn, accessed on 10 July 2021), which are the popular travel social networking sites in China. We decided to use tourism social media data to replace traditional research for the following reasons. First, social media data can alleviate the limitations associated with sample size, time, location, unresponsive bias, and self-reporting errors [50]. Second, the photos provided by users capture the local environment and experience, which are based on real life rather than on research reports [51]. Third, social media is an important medium for millennials to release and receive tourism information, whose research impact is of great significance and value [40]. Our study is based on the quantitative approach of content analysis of photos posted by tourists on social media. The content analysis is a numerical research method for objective, systematic, and quantitative analysis of literature content [52] and research on communication [53]. People upload photos on the internet that they like or are interested in, unless there are special instructions [33]. To this end, studying the specific landscape elements that attract tourists is required by encoding the photo content based on landscape characteristics, and by comparing the frequency of each element. Within this context, we also analyzed the correlation between landscape elements and preferences [26].

2.1. Study Area

The mountain area in Western China is vast, with salient characteristics such as complex ecosystem, lagging economic and social development, wide distribution of ethnic minorities, and obvious landscape diversity [54]. The area is rich in natural and cultural landscape resources, where the development conditions of mountain tourism are vastly superior and are of utmost importance and significance. The Western Sichuan Plateau Mountain Areas (WSPMA)are located in minority areas as a representative of the development of mountain tourism in Western China [2], where tourism is an essential development model for ethnic minority areas in Western Sichuan [55,56]. Due to this, we considered the WSPTA, as a typical mountain tourism research destination with multicultural and rich landscape resources, for our research study.

2.2. Data Collection

Fashion or fame are the main factors affecting tourism destination selection [57,58]. In June 2021, we searched "Western Sichuan tourism destination" in the Ctrip.com (accessed on 10 July 2021) and selected the top 10 most popular tourism destinations as the specific object for this study. They include world natural heritage sites, national scenic spots, and 4A or above representative tourism destinations (Figure 1). We also searched for photos posted by users during their travel in the 10 tourism destinations which we chose through Ctrip (https://www.ctrip.com, accessed on 16 July 2021), Xiaohongshu (https://www.xiaohongshu.com, accessed on 16 July 2021), and Mafengwo (https://www.mafengwo.cn/, accessed on 16 July 2021). The analysis of shared travel photos was applied to understand the relevant outdoor recreational activities and preferences [33,42,47]. The principles for filtering data are explained here in detail. First, comments posted by visitors without any commercial behavior, were adopted to express the discriminatory criteria of their own feelings and sentiments. Then, users were required to complete the information. As mentioned, tourists' travel preferences were affected by age [33], gender [34], mode of transportation [35], travel pattern [36], and landscape preferences [42,47]. Therefore, we also collected information on the age, motivation for travel, and gender of our sample. Note that a sample must have a relatively in-depth process description and positive emotional expression of tourist behavior. Finally, the comments with salient advertising messages and copy marks were eliminated. The sample period was June 2020 to June 2021. After screening, 1230 users (450 male and 780 female) posting content were identified as the study sample, with 10,399 photos. Further, we explored millennials' landscape preferences for tourism in the WSPMA through image analysis.

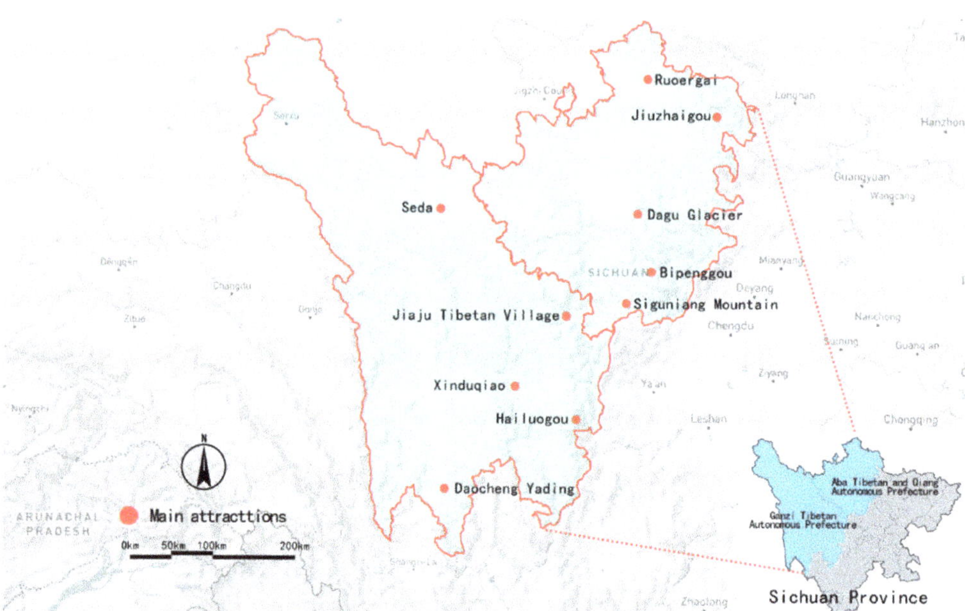

Figure 1. Study attraction locations.

2.3. Data Statistics

We identified the demographic characteristics of picture users by recording keywords in social media comments, and summarized these types of user information: gender (male and female), mode of transportation (self-driving, cycling, group following, and public transportation), and travel patterns (individuals, friends, families, and couples).

We applied the classification method based on the main themes in the picture [45], and combined China's national standard for classification, investigation, and evaluation of tourism resources (GB/t18972-2003) (https://www.mct.gov.cn/, accessed on 25 July 2021). In particular, Cao [59], and Shi [60] classified resources in the natural and cultural landscape, based on ethnic ecological areas and 11 subcategories: forest, waterscape, ice and snow, bare land, meadow, weather, heritage, customs and festivals, architecture, folk culture, and gastronomy (Table 1). Moreover, we quantified the number of pictures that belonged to multiple categories and calculated the frequency of landscape elements in each category (Table 2).

Table 1. Profile of landscape resource classification.

Main Category	Subcategory	Landscape Unit
Natural landscape	Forest	Alpine forest, Low mountain forest, Virgin forest, Slow slope forest
	Waterscape	Lake, River
	Ice and snow	Snow mountain, Snow covered land
	Bare land	River beach stone land, Alpine region
	Meadow	Sloping fields meadow, Lowland meadow, Pasture meadow, Wild grassland, Lowland meadow
	Weather	Sunrise, Sunset, Seas of clouds
Human landscape	Heritage	Building heritage
	Customs and Festivals	Ethnic customs, Ethnic art, Celebrations, Clothing
	Architecture	Residential building, Temple, Villages and towns, Diaolou structure
	Folk culture	Folk belief, Religious belief
	Gastronomy	Food, Gastronomy esthetic, Dining environment

Table 2. Profile of landscape types frequency.

Item	N	Total	Mean
natural landscape	1230	18,328	14.9
human landscape	1230	3039	2.47
forest	1230	3748	3.05
waterscape	1230	3104	2.52
ice and snow	1230	4359	3.54
bare land	1230	2869	2.33
meadow	1230	1898	1.54
weather	1230	2350	1.91
heritage	1230	475	0.39
customs and festivals	1230	480	0.39
architecture	1230	1670	1.36
folk culture	1230	255	0.21
gastronomy	1230	159	0.13

The data were analyzed using the IBM-SPSS 25 program. The results of the descriptive analysis were interpreted via the Kruskal-Wallis rank sum test by calculating samples, percentages, and frequencies based on the coding content described above. To further investigate the variability based on gender, mode of transportation, and travel patterns across the landscape types among millennials, we utilized Bonferroni adjustment for post-hoc multiple testing of non-parametric tests of consumer behavior. The confidence intervals were set at 95%, and differences in the associations were considered statistically significant at $p < 0.05$ level.

3. Result

3.1. Statistical Characteristics of Millennial Tourists

The sample of 1230 respondents consisted of 450 men (36.6%) and 780 women (63.4%) This proportion is consistent with the previous findings from Kimbrough [61] as women are keener to share on social media. From the mode of transportation perspective, 767 (62.4%) self-driving tourists accounted for a much larger proportion of trips than the 25 (2%) cycling tourists, 270 (22%) group tourists, and 168 (13.7%) public transport-using tourists combined. We suggest that the reason why most people opt self-drive rather than cycling is likely related to the high-altitude mountain environment of the WSPMA and the fact that most of the beautiful scenery can be enjoyed while on the road. Moreover, from the travel pattern standpoint, nearly 80% of tourists decide to travel in a group and only 20% travel alone. The relevant characteristics of participants are listed in Table 3.

Table 3. Basic information types of the samples.

Types	Item	N	Percentage (%)
Gender	Male	450	36.6
	Female	780	63.4
Transportation	Self-driving	767	62.4
	Cycling	25	2
	Travel agency	270	22
	Public transportation	168	13.7
Travel pattern	Individual	288	23.4
	With friends	632	51.4
	With family members	143	11.6
	Couples	167	13.6

3.2. Photo Statistics

Photos are visual elements captured by a photographer at a particular time, showing various landscape units. According to the landscape classification proposed in Table 1, we coded and analyzed the landscape characteristics of each collected photo (see Figures 2 and 3). According to the total number of landscape types in the photo, we regarded this as the landscape preference of tourists [33].

3.3. Analysis of Millennials' Landscape Preferences

At this stage of the analysis, we evaluated the differences in the aesthetics of various landscape types based on gender, mode of transportation, and travel pattern of the millennials. Table 2 shows a collation of photos posted by millennials on trips to the WSPMA. It reveals that natural landscape (14.9) appeared, on average, six times as often as human landscape (2.47). Among the landscape subcategories, ice and snow (3.54) > forest (3.05) > waterscape (2.52) > bare land (2.33) > weather (1.91) > meadow (1.54) > architecture (1.36) > heritage (0.39), customs and festivals (0.39) > folk culture (0.21), gastronomy (0.13). All landscape elements in the main category of natural landscapes appeared very frequently in the photographs. The highest frequency was identified for the snow and ice landscapes. For millennials, the appearance of meadows was the least frequent among all the natural landscapes. The main category of human landscape in the photos is less frequent than in the former, but the millennials also demonstrated an interest in architecture.

In terms of the millennials' gender differences in landscape preferences, Table 4 displays significant differences for waterscape ($p < 0.001$), weather ($p < 0.001$), heritage ($p < 0.001$), and customs and festivals ($p = 0.041$). Further, Figure 4 shows that males have a higher landscape preference for weather (665.99) and heritage (639.89), while females have higher landscape preference for waterscape (658.06) and customs and festivals (626.51). More multiple comparisons and results of gender details refer to Tables A1 and A2.

Figure 2. Landscape classification methods.

Figure 3. Landscape classification methods.

Table 4. Overall landscape preference difference.

Variable	NL	HL	F	W	IS	BL	M	Wt	H	CF	A	FC	G
Gender	0.002	0.803	0.708	**31.992**	0.385	1.708	0.336	**15.289**	**8.358**	**4.181**	0.409	0.505	1.645
Transportation	3.768	6.275	4.050	**30.09**	**18.188**	4.394	**93.808**	3.749	**13.813**	**24.736**	5.823	**26.704**	8.992
Travel pattern	**8.736**	27.933	7.977	8.583	**18.734**	4.704	**77.800**	7.233	**27.675**	**25.997**	**26.262**	**12.195**	**9.766**

Note: NL, natural landscape; HL, human landscape; F, forest; W, waterscape; IS, ice and snow; BL, bare land; M, meadow; WT, weather; H, heritage; CF, customs and festivals; A, architecture; FC, folk culture; G, gastronomy. The figures are the chi-square values, and the bold ones represent significant differences ($p < 0.05$).

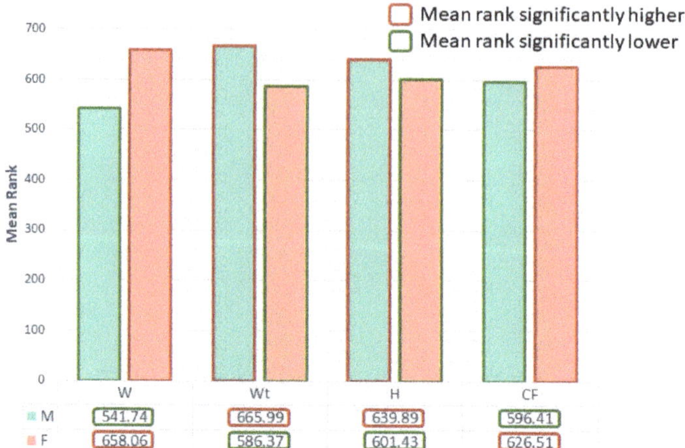

Figure 4. Kruskal-Wallis one-way ANOVA independent sample multiple comparisons of gender. Note: W: waterscape; WT: weather; H: heritage; CF: customs and festivals; A: architecture; FC: folk culture; G: gastronomy; M: male; F: female.

Differences in millennials' landscape preferences for different modes of transportation are shown in Table 4. As seen, the millennials have pronounced landscape preferences for waterscape ($p < 0.001$), ice and snow ($p < 0.001$), meadow ($p < 0.001$), and heritage ($p = 0.003$), customs and festivals ($p < 0.001$), and folk culture ($p < 0.001$). As also seen from Figure 5, millennials traveling with travel agencies exhibit higher preference for waterscape (693.58), compared with the other groups, and also exhibited higher preference for ice and snow landscape (674.11) and gastronomy (636.27) than the self-driving groups (IS = 588.19, G = 606.98). However, there was lower preference for customs and festivals (559.19) than the other groups. In addition, self-driving millennials disclosed a higher landscape preference for meadows (681.49), compared with the other groups, while the millennials that travelled by public transportation exhibited a higher landscape preference for heritage sites (665.66) than the other groups. For millennials, significant preference differences were also found between multi-landscape preferences and different modes of transportation. More multiple comparisons and results of mode of transportation details refer to Tables A1 and A2.

Differences in millennials' landscape preference were based on different travel patterns. Table 4 shows the progressive significance value in the natural landscape ($p = 0.033$), human landscape ($p < 0.001$), ice and snow ($p < 0.001$), meadow ($p < 0.001$), and heritage ($p < 0.001$), customs and festivals ($p < 0.001$), architecture ($p < 0.001$), folk culture ($p = 0.007$), and gastronomy ($p = 0.021$), is less than 0.05. From Figure 6, we found that individual millennial tourists showed higher landscape preferences for customs and festivals (676.94), but lower landscape preference for meadows (488.41) than the other groups. Millennials travelling with friends had great landscape preferences for human landscapes in heritage sites (638.9), architecture sites (645.27), and gastronomy (625.19). Millennials travelling with family showed higher preference for natural landscapes (687.26) than individual ones (587.96), but lower preference for human landscapes (478.02) on multiple sides, especially in architectural landscapes (497.4). More multiple comparisons and results of travel pattern details refer to Tables A1 and A2.

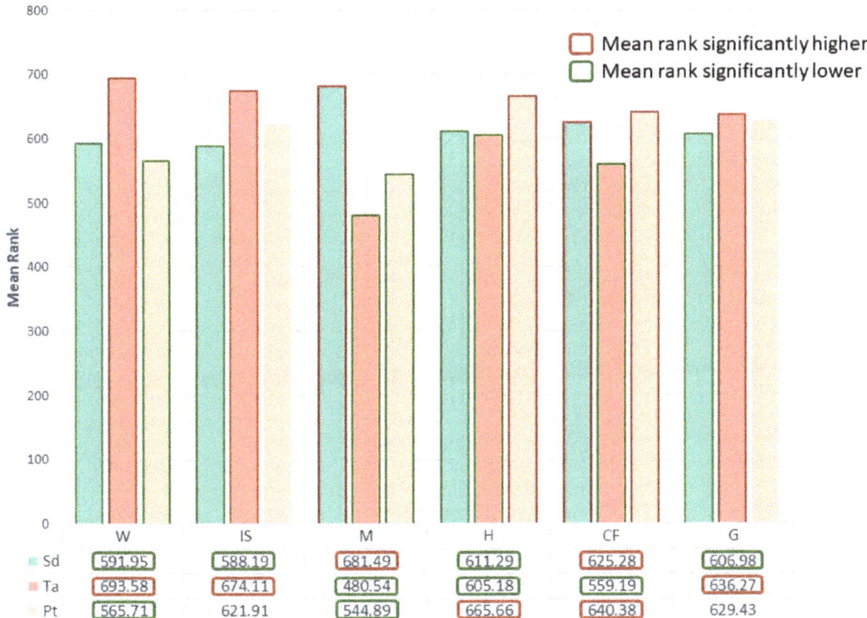

	W	IS	M	H	CF	G
Sd	591.95	588.19	681.49	611.29	625.28	606.98
Ta	693.58	674.11	480.54	605.18	559.19	636.27
Pt	565.71	621.91	544.89	665.66	640.38	629.43

Figure 5. Kruskal-Wallis one-way ANOVA independent sample multiple comparison of mode of transportation. Note: W: waterscape; IS: ice and snow; M: meadow; H: heritage; CF: customs and festivals; G: gastronomy; Sd, Self-driving; Ta, Travel Agency; Pt, Public transportation.

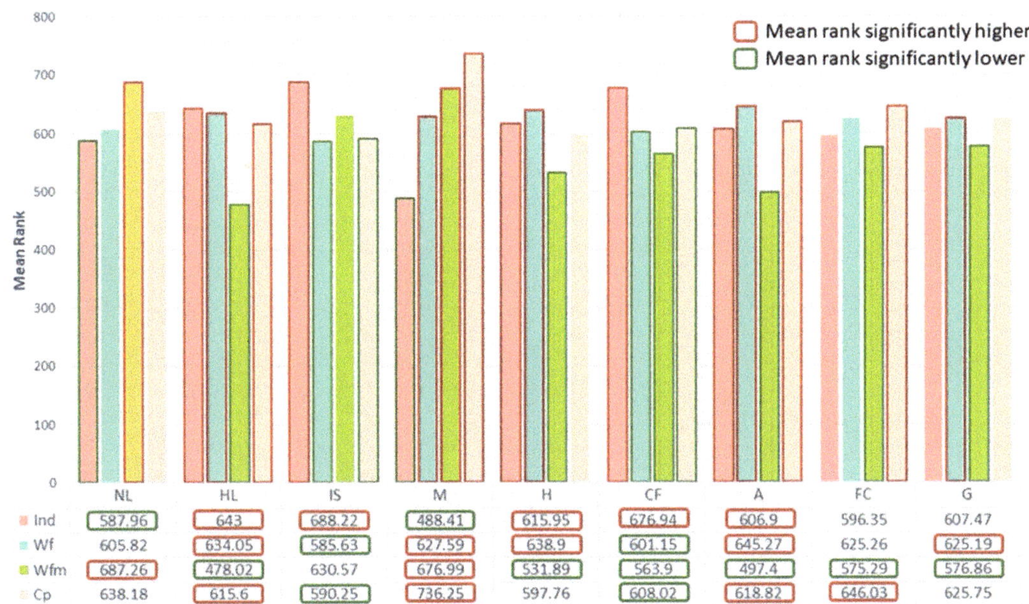

	NL	HL	IS	M	H	CF	A	FC	G
Ind	587.96	643	688.22	488.41	615.95	676.94	606.9	596.35	607.47
Wf	605.82	634.05	585.63	627.59	638.9	601.15	645.27	625.26	625.19
Wfm	687.26	478.02	630.57	676.99	531.89	563.9	497.4	575.29	576.86
Cp	638.18	615.6	590.25	736.25	597.76	608.02	618.82	646.03	625.75

Figure 6. Kruskal-Wallis one-way ANOVA independent sample multiple comparison of travel pattern. Note: NL: natural landscape; HL: human landscape; IS: ice and snow; M: meadow; H: heritage; CF: customs and festivals; A: architecture; FC: folk culture; G: gastronomy; Ind: Individual; Wf: With friends; Wfm: With family members; Cp, Couples.

4. Discussion

We selected the mountainous region of the WSPMA as the study area and examined the landscape preferences of millennials from the perspective of natural and human landscapes. This analysis was performed using social media data while considering the effects of different genders, modes of transportation, and travel patterns on tourist destination preference.

We found that, in contrast to the findings of Krueger et al. [62], self-driving travel is a prominent feature of mountain tourism among millennials. This may be on account of the plateau climate of WSPMA, and because most of the beautiful scenery is along the way, rather than in a tourist destination. Compared with public transportation, self-driving is more flexible in terms of time and travel freedom, thus, leading to better enjoyment of the unique landscape of the mountain area. We also found that architectural culture emerges most frequently in the human landscape, seemingly because the local architecture has a very strong and unique ethnic cultural style.

At the level of natural landscape, millennials demonstrated substantial interest in mountain forests, waterscape, ice and snow, and bare-land landscapes. Notably, their common denominator is the superiority and uniqueness of mountain natural landscapes. This indicates that natural factors are the primary drivers of the millennials' preferences for mountain destinations. This finding agrees with the findings of Higham's [20], who suggests that millennials show a high interest in natural landscapes. We also found that the millennials have different landscape preferences based on differences in genders, modes of transportation, and travel patterns for specific mountain landscape elements.

First, in terms of gender, both male and female groups expressed different landscape preferences. In particular, women have a higher aesthetic preference for waterscape than men, while men show a higher landscape preference for weather landscapes than women. Among the tourists with different modes of transportation, those travelling in a group cared more about the natural scenery of the mountainous areas of the WSPMA, especially waterscape and ice and snow landscape. In contrast, tourists travelling by car preferred meadow landscapes. Moreover, individual tourists were more inclined to visit ice and snow landscapes compared with the tourists using other modes of travel. There was a striking contrast in the preference results for meadow landscapes as family tourists clearly exhibited a higher preference for natural landscapes than individual tourists.

Likewise, with respect to human landscape, this study identified some prominent differences. In terms of gender, females expressed a higher preference for customs and festivals landscapes than males, while males showed a significantly higher preference for heritage landscapes. In terms of different modes of transportation, self-driving visitors showed more preference for customs and festivals, and folk culture; public transport visitors showed more preference for heritage, customs and festivals. This indicates that self-driving and public transport visitors are the main visitors among millennials who are interested in human landscapes. We did not expect a higher preference for food tourism among visitors travelling in a group, which is different from the commonly observed pattern with group tours. In terms of different travel patterns, individual tourists expressed a clear preference for heritage, architecture, customs and festivals, especially, customs and festivals. Compared with the preferences for natural landscapes, individual tourists were one of the main clienteles for human landscapes. More specifically, couples preferred folk culture landscapes. In such landscapes, many photos of the couples were captured using local ethnic elements. It should also be noted that family tourists expressed notably less interest in human landscapes than the other three categories of visitors.

The combination of the landscape preference results for natural and human landscapes likely unravels in the following pattern. Family trips and group self-driving trips prefer landscapes where they can relax, talk, and indulge in outdoor activities. This drives their preferential choice toward meadow landscapes. Family trips take into account the older and the younger members of the family, who enjoy more bonding time and relaxation in a natural environment, thus influencing the preference of this group for architectural and other cultural landscapes. The collation of photographs showed that millennial girls prefer landscapes where people photography can be taken, while boys prefer ornamental landscapes, in line with the gender preference results. Those who travel individually are more attracted to the human landscapes of ethnic minorities in the WSPMA, arousing their desire to explore history, folklore, and ethnic culture.

5. Conclusions

There are three main contributions of this study.

First, our findings broaden the current research results related to mountain tourism [1]. We innovatively studied the natural and human landscape preferences of millennials for mountain tourism based on social media data. The research results showed that the geographic content determined from social media can be used as a reliable substitute for traditional survey methods and used to explore tourists' landscape preferences in mountain tourism. In future research, the drivers behind these preferences and the strengths and weaknesses of each landscape element can also be investigated. This will ultimately improve the competitiveness of mountain tourism from a detailed perspective.

Second, these findings improve the understanding of tourists' preferences for landscapes and reveal useful tourism management implications. In particular, decision/policy-makers and tourism managers need to understand the landscape preferences of tourists [48], the specific subdivision of landscape elements and tourist demographic characteristics can help government and managers conduct more targeted tourism management, planning, and promotion. To our knowledge, the mountain tourism populations and landscape preferences have not been studied to an appropriate extent to date [7].

Third, our findings enrich the research related to social media in landscape preferences. Among the current studies on landscape preferences, landscape studies based on social media data have increasingly become a research hotspot. However, to the best of our knowledge, there are few studies on landscape preferences for mountain tourism through social media [63], so our study complements the research on social media in landscape preferences.

Based on the research of Tieskens et al. [26,27], our results confirm that millennials with different genders, modes of transportation, and travel patterns have their own preferences for mountain landscapes. These differences have important implications in the marketing management and design planning of scenic areas. These can help the government and tourism managers to analyze the advantages of local landscape resources and position products and services according to the aesthetic preferences of different groups for different landscapes. Thus, making tourism promotion and tourism planning more relevant and attractive.

Millennials' preference for natural landscapes can promote a sense of nature conservation, thereby, promoting sustainable behavior towards nature [39,40,49]. At the same time, landscape preferences affect tourism behavior and enhance place attachment, thus, strengthening the willingness to protect these natural areas. Decision/policy-makers and tourism managers should focus on the conservation of natural ecosystems when planning mountain tourism. While carrying out tourism promotion, the unique natural scenery of the WSPMA and nature conservation-related promotion should be performed through the various social media platforms commonly used by millennials. In terms of cultural landscapes, Xu's [64] study confirmed the preference of millennials for cultural landscapes. In response to the low preference for cultural landscapes among tourists on family trips, decision/policy-makers and tourism managers can enrich the form of ethnic activities, so that children and the elderly can be more involved in ethnic cultural activities and experience ethnic cultural features in an interesting way. Children and the elderly can learn about the diverse cultures of the WSPMA through the medium of information reception commonly used by children and the elderly, thus enhancing their interest in cultural landscapes.

On a more detailed level, scenic area operators can create different types of narration (viewpoints or storytelling lines) along the road according to the functional needs of the various visitors to the landscape. For instance, meadow landscapes can be established with space for relaxing, to create a better experience for visitors. In line with the results of Hargittai [65], we found that women are keener to share on social media, so landscape operators can target female groups and create spots that are suitable for photographing people. Tourism managers should use social media to promote high-quality local specialty restaurants to meet the demand for food from people that do not prefer group tours.

6. Limitations

This study had some limitations. First, the size of the samples and collection time were imperfect. Due to the incomplete demographic information from social media data, it was challenging to study education [66] and income [67]. Furthermore, some sample sizes were too small to be analyzed, namely, the transportation by cycling was excluded because its size was too small. Second, there is an inherent limitation in social media data. More specifically, compared with extensive surveys, this method cannot be used to evaluate the impact of social factors on shared content [68], although, social media data are authentic and objective [69].

These limitations do not impact the value of our findings as we strictly focused on the millennials and conducted relevant discussions and research based on their reliance on social media. Overall, our findings certainly contribute to the currently scanty knowledge about the millennial generation in the sustainable development of mountain tourism.

Author Contributions: Conceptualization, M.Y. and S.L.; methodology, M.Y., S.L. and K.D.; software, K.D.; validation, M.Y. and K.D.; formal analysis, M.Y.; investigation, K.D.; resources, M.Y.; data curation, M.Y. and K.D.; writing—original draft preparation, M.Y. and K.D.; writing—review and editing, M.Y. and S.L.; visualization, M.Y. and K.D.; supervision, M.Y.; project administration, M.Y.; funding acquisition, M.Y. All authors have read and agreed to the published version of the manuscript.

Funding: This research was funded by the Sichuan Provincial Social Science Key Research Base (Kangba Cultural Research Center) "Study on the Motivation of Humanistic Tourism along the Sichuan-Tibet Highway in Kang District", grant number: KBYJ2021B011; The Innovation Training Program for University Students of Sichuan Province "Study on the Millennials' Aesthetic Preferences of Mountain Tourism based on Social Media Data", grant number: S202111552126.

Institutional Review Board Statement: The study was conducted according to the guidelines of the Declaration of Helsinki, all the necessary information regarding the study was given. The photos we chose were publicly posted on social media platforms and the study did not expose the users to any harm. Therefore, this experiment was conducted with the approval of the Academic Committee of Sichuan Tourism University, but no submission for review was required.

Informed Consent Statement: Not applicable.

Data Availability Statement: The data presented in this study are available on request from the corresponding author. The data are not publicly available due to privacy.

Acknowledgments: We thank the Sichuan Provincial Social Science Key Research Base (Kangba Cultural Research Center), the Organizer of Innovation Training Program for University Students of Sichuan Province for funding and anonymous reviewers who provided invaluable advice on how to improve the manuscript.

Conflicts of Interest: The authors declare no conflict of interest.

Appendix A

Table A1. Kruskal-Wallis one-way ANOVA independent sample multiple comparison.

Variable	Gender			Transportation			Travel Pattern		
	Item	Mean Rank	SE	Item	Mean Rank	SE	Item	Mean Rank	SE
NL	Male	616.04	0.414	Sd	616.44	0.308	Ind	587.96	0.357
	Female	615.19	0.255	Cy	688.96	0.842	Wf	605.82	0.326
				Ta	631.12	0.398	Wfm	687.26	0.705
				Pt	575.18	0.478	Cp	638.18	0.613
HL	Male	604.22	0.236	Sd	610.32	0.161	Ind	643.00	0.251
	Female	622.01	0.135	Cy	618.22	0.233	Wf	634.05	0.182
				Ta	594.05	0.194	Wfm	478.02	0.143
				Pt	673.23	0.386	Cp	615.60	0.339
F	Male	626.57	0.137	Sd	600.22	0.103	Ind	596.10	0.125
	Female	609.11	0.091	Cy	656.02	0.308	Wf	613.20	0.111
				Ta	644.75	0.150	Wfm	689.90	0.242
				Pt	632.21	0.182	Cp	593.98	0.217
W	Male	541.74	0.123	Sd	591.95	0.095	Ind	651.32	0.134
	Female	658.06	0.093	Cy	829.38	0.349	Wf	605.52	0.107
				Ta	693.58	0.161	Wfm	648.14	0.252
				Pt	565.71	0.192	Cp	563.54	0.193
IS	Male	607.47	0.213	Sd	588.19	0.153	Ind	688.22	0.218
	Female	620.13	0.141	Cy	777.30	0.533	Wf	585.63	0.172
				Ta	674.11	0.256	Wfm	630.57	0.342
				Pt	621.91	0.292	Cp	590.25	0.335
BL	Male	632.47	0.149	Sd	606.46	0.106	Ind	604.64	0.113
	Female	605.71	0.087	Cy	741.50	0.340	Wf	613.32	0.119
				Ta	631.03	0.149	Wfm	671.68	0.247
				Pt	613.08	0.182	Cp	594.37	0.201
Md	Male	622.35	0.134	Sd	681.49	0.109	Ind	488.41	0.076
	Female	611.55	0.090	Cy	523.16	0.154	Wf	627.59	0.107
				Ta	480.54	0.092	Wfm	676.99	0.277
				Pt	544.89	0.118	Cp	736.25	0.232
Wt	Male	665.99	0.107	Sd	615.56	0.099	Ind	659.52	0.184
	Female	586.37	0.089	Cy	580.08	0.265	Wf	605.57	0.081
				Ta	593.27	0.100	Wfm	575.13	0.184
				Pt	656.24	0.163	Cp	611.74	0.201
H	Male	639.89	0.068	Sd	611.29	0.041	Ind	615.95	0.065
	Female	601.43	0.030	Cy	519.00	0.000	Wf	638.90	0.046
				Ta	605.18	0.040	Wfm	531.89	0.024
				Pt	665.66	0.119	Cp	597.76	0.100
CF	Male	596.41	0.046	Sd	625.28	0.037	Ind	676.94	0.051
	Female	626.51	0.032	Cy	756.34	0.187	Wf	601.15	0.040
				Ta	559.19	0.038	Wfm	563.90	0.053
				Pt	640.38	0.067	Cp	608.02	0.069
A	Male	623.02	0.126	Sd	610.51	0.090	Ind	606.90	0.152
	Female	611.16	0.081	Cy	501.96	0.174	Wf	645.27	0.103
				Ta	617.21	0.122	Wfm	497.40	0.073
				Pt	652.45	0.218	Cp	618.82	0.174

Table A1. *Cont.*

Variable	Gender			Transportation			Travel Pattern		
	Item	Mean Rank	SE	Item	Mean Rank	SE	Item	Mean Rank	SE
FC	Male	621.18	0.029	Sd	637.96	0.023	Ind	596.35	0.029
	Female	612.22	0.022	Cy	556.96	0.080	Wf	625.26	0.027
				Ta	563.95	0.035	Wfm	575.29	0.052
				Pt	604.54	0.045	Cp	646.03	0.040
G	Male	606.94	0.028	Sd	606.98	0.018	Ind	607.47	0.022
	Female	620.44	0.016	Cy	559.00	0.000	Wf	625.19	0.024
				Ta	636.27	0.030	Wfm	576.86	0.025
				Pt	629.43	0.041	Cp	625.75	0.036

Note: NL, natural landscape; HL, human landscape; F, forest; W, waterscape; IS, ice and snow; BL, bare land; M, meadow; WT, weather; H, heritage; CF, customs and festivals; A, architecture; FC, folk culture; G, gastronomy; Sd, Self-driving; Cy, Cycling; Ta, Travel agency; Pt, Public transportation; Ind, Individual; Wf, With friends; Wfm, With family members; Cp, Couples.

Table A2. Multiple comparison results.

Variable	(I) Group	(J) Group	Mean Difference (I–J)	Std. Error	Adjusted Sig.	(I) Group	(J) Group	Mean Difference (I–J)	Std. Error	Adjusted Sig.
NL						Ind	Wfm	−99.302	36.301	0.0037
HL						Wfm	Cp	−137.577	38.187	0.002
						Wfm	Wf	156.024	31.037	0.000
						Wfm	Ind	164.974	34.287	0.000
W	Pt	Ta	124.823	33.411	0.001					
	Sd	Ta	−98.870	24.060	0.000					
IS	Sd	Ta	−83.555	23.857	0.001	Wf	Ind	102.589	24.505	0.000
						Cp	Ind	97.964	33.525	0.021
M	Sd	Ta	−196.417	21.865	0.000	Ind	Wf	−139.184	22.395	0.000
	Pt	Sd	133.764	26.320	0.000	Ind	Wfm	−188.580	32.225	0.000
						Ind	Cp	−247.845	30.639	0.000
						Wf	Cp	−108.661	27.408	0.000
H	Ta	Pt	−59.430	21.824	0.019	Wfm	Ind	84.062	22.994	0.002
	Sd	Pt	53.325	18.919	0.014	Wfm	Wf	107.007	20.815	0.000
CF	Ta	Sd	64.642	17.072	0.000	Wfm	Ind	113.042	25.438	0.000
	Ta	Pt	−79.594	23.707	0.002	Wf	Ind	75.793	17.679	0.000
						Cp	Ind	68.926	24.186	0.026
A						Wfm	Ind	109.495	32.049	0.004
						Wfm	Cp	−121.415	35.694	0.004
						Wfm	Wf	147.863	29.011	0.000
FC	Ta	Sd	72.510	14.850	0.000	Wfm	Cp	−70.737	24.254	0.021
G						Wfm	Wf	48.331	16.473	0.020

Note: NL, natural landscape; HL, human landscape; F, forest; W, waterscape; IS, ice and snow; BL, bare land; M, meadow; WT, weather; H, heritage; CF, customs and festivals; A, architecture; FC, folk culture; G, gastronomy; Sd, Self-driving; Ta, Travel agency; Pt, Public transportation; Ind, Individual; Wf, With friends; Wfm, With family members; Cp, Couples. 2. We keep only significant multiple comparison results in the table.

References

1. Tian, J.; Ming, Q. Hotspots, progress and enlightenments of foreign mountain tourism research. *World Reg. Stud.* **2020**, *29*, 1071–1081. [CrossRef]
2. Chen, X.; Qin, J.; Li, X.; Shi, X. Study on Characteristic Development Strategies of Mountain Tourism in High Mountain Canyon Area of Hengduan Mountains in Sichuan—Exploring the Development Path of Mountain Tourism in Western China. *Econ. Ggography* **2012**, *32*, 143–148. [CrossRef]
3. Delitheou, V.; Alexiou, S. Challenges of Tourism Sustainability in Greek Mountain Regions in Decline. *AJHSSR* **2021**, *5*, 448–458. Available online: https://www.researchgate.net/publication/350889823 (accessed on 3 November 2021).

4. Paunović, I.; Jovanović, V. Implementation of Sustainable Tourism in the German Alps: A Case Study. *Sustainability* **2017**, *9*, 226. [CrossRef]
5. Rita, P.; Brochado, A.; Dimova, L. Millennials' Travel Motivations and Desired Activities within Destinations: A Comparative Study of the US and the UK. *Curr. Issues Tour.* **2019**, *22*, 2034–2050. [CrossRef]
6. Denning, A. From sublime landscapes to white gold: How skiing transformed the Alps after 1930. *Environ. Hist.* **2014**, *19*, 78–108. [CrossRef]
7. Giachino, C.; Truant, E.; Bonadonna, A. Mountain Tourism and Motivation: Millennial Students' Seasonal Preferences. *Curr. Issues Tour.* **2019**, *23*, 2461–2475. [CrossRef]
8. Silkes, C.A. Farmers' Markets: A Case for Culinary Tourism. *J. Culin. Sci. Technol.* **2012**, *10*, 326–336. [CrossRef]
9. Kotler, P.; Bowen, J.; Makens, J.; Baloglu, S. *Marketing for Hospitality and Tourism*, 7th ed.; Global Edition; Pearson: London, UK, 2017; p. 527.
10. Bond, N.; Falk, J. Tourism and identity-related motivations: Why am I here (and not there)? *Int. J. Tour. Res.* **2012**, *15*, 430–442. [CrossRef]
11. Falk, J. Identity and the art museum visitor. *J. Art Educ.* **2018**, *34*, 25–34. [CrossRef]
12. Cohen, S. Chasing a myth? Searching for 'self' through lifestyle travel. *Tour. Stud.* **2010**, *10*, 117–133. [CrossRef]
13. Michael, N.; Nyadzayo, M.W.; Michael, I.; Balasubramanian, S. Differential roles of push and pull factors on escape for travel: Personal and social identity perspectives. *Int. J. Tour. Res.* **2020**, *22*, 464–478. [CrossRef]
14. Foot, D.K.; Stoffman, D. *Boom, Bust and Echo 2000: Profiting from the Demographic Shift in the New Millennium*; Macfarlane, Walter & Ross: Toronto, ON, Canada, 1998.
15. Nowak, L.; Thach, L.; Olsen, J.E. Wowing the Millennials: Creating Brand Equity in the Wine Industry. *J. Prod. Brand Manag.* **2006**, *15*, 316–323. [CrossRef]
16. Benckendorff, P.; Moscardo, G.; Pendergast, D. Generation Y and travel. In *Tourism and Generation Y*; Benckendorff, P., Moscardo, G., Pendergast, D., Eds.; CABI: Oxfordshire, UK, 2010; pp. 16–26.
17. Bernardi, M. Millennials, Sharing Economy and Tourism: The Case of Seoul. *J. Tour. Futures* **2018**, *4*, 43–56. [CrossRef]
18. Bonadonna, A.; Giachino, C.; Truant, E. Sustainability and Mountain Tourism: The Millennial's Perspective. *Sustainability* **2017**, *9*, 1219. [CrossRef]
19. Buckley, R. Sustainability issues in mountain tourism. In *Sustainable Mountain Communities*; Taylor, L., Ryall, A., Eds.; The Banff Centre: Banff, CAN, 2003; pp. 169–174.
20. Higham, J.; Thompson-Carr, A.; Musa, G. Activity, People and Place. In *Mountaineering Tourism*; Musa, G., Higham, J., Thompson-Carr, A., Eds.; Routledge: New York, NY, USA, 2015; pp. 1–15. [CrossRef]
21. Cavagnaro, E.; Staffieri, S. A Study of Students' Travellers Values and Needs in Order to Establish Futures Patterns and Insights. *J. Tour. Futures* **2015**, *1*, 94–107. [CrossRef]
22. Hopkins, D. Destabilising Automobility? The Emergent Mobilities of Generation Y. *Ambio* **2017**, *46*, 371–383. [CrossRef] [PubMed]
23. Miller, D.; Merrilees, B.; Coghlan, A. Sustainable Urban Tourism: Understanding and Developing Visitor pro-Environmental Behaviors. *J. Sustain. Tour.* **2015**, *23*, 26–46. [CrossRef]
24. Olsen, J.E.; Thach, L.; Nowak, L. Wine for My Generation: Exploring How Us Wine Consumers Are Socialized to Wine. *J. Wine Res.* **2007**, *18*, 1–18. [CrossRef]
25. Schoolman, E.D.; Shriberg, M.; Schwimmer, S.; Tysman, M. Green Cities and Ivory Towers: How Do Higher Education Sustainability Initiatives Shape Millennials' Consumption Practices? *J. Environ. Stud. Sci.* **2016**, *6*, 490–502. [CrossRef]
26. Tieskens, K.F.; Van Zanten, B.T.; Schulp, C.J.E.; Verburg, P.H. Aesthetic appreciation of the cultural landscape through social media: An analysis of revealed preference in the dutch river landscape. *Landsc. Urban Plan.* **2018**, *117*, 128–137. [CrossRef]
27. Oteros-Rozas, E.; Martín-López, B.; Fagerholm, N.; Bieling, C.; Plieninger, T. Using Social Media Photos to Explore the Relation between Cultural Ecosystem Services and Landscape Features across Five European Sites. *Ecol. Indic.* **2018**, *94*, 74–86. [CrossRef]
28. Kiatkawsin, K.; Han, H. Young Travelers' Intention to Behave pro-Environmentally: Merging the Value-Belief-Norm Theory and the Expectancy Theory. *Tour. Manag.* **2017**, *59*, 76–88. [CrossRef]
29. Rahman, O.; Fung, B.C.M.; Chen, Z. Young Chinese Consumers' Choice between Product-Related and Sustainable Cues-the Effects of Gender Differences and Consumer Innovativeness. *Sustainability* **2020**, *12*, 3818. [CrossRef]
30. Folmer, A.; Tengxiage, A.; Kadijk, H.; Wright, A.J. Exploring Chinese Millennials' Experiential and Transformative Travel: A Case Study of Mountain Bikers in Tibet. *J. Tour. Futures* **2019**, *5*, 142–156. [CrossRef]
31. Xiang, Z.; Gretzel, U. Role of Social Media in Online Travel Information Search. *Tour. Manag.* **2010**, *31*, 179–188. [CrossRef]
32. Sharmin, F.; Sultan, M.T.; Badulescu, A.; Bac, D.P.; Li, B. Millennial Tourists' Environmentally Sustainable Behavior towards a Natural Protected Area: An Integrative Framework. *Sustainability* **2020**, *12*, 8545. [CrossRef]
33. Oliveira, T.; Araujo, B.; Tam, C. Why Do People Share Their Travel Experiences on Social Media? *Tour. Manag.* **2020**, *78*, 104041. [CrossRef]
34. Zhang, Y.; Qiu, J.; Shao, L.; Cui, J.; Meng, X. The structure of pull motivations of rural tourism and the segmentation of rural tourists. *J. Arid. Land Resour. Environ.* **2014**, *28*, 191–196. [CrossRef]
35. Pettebone, D.; Newman, P.; Lawson, S.R.; Hunt, L.; Monz, C.; Zwiefka, J. Estimating Visitors' Travel Mode Choices along the Bear Lake Road in Rocky Mountain National Park. *J. Transp. Geogr.* **2011**, *19*, 1210–1221. [CrossRef]

36. Chen, Y.Y.; Cheng, A.J.; Hsu, W.H. Travel recommendation by mining people attributes and travel group types from community-contributed photos. *IEEE Trans. Multimed.* **2013**, *15*, 1283–1295. [CrossRef]

37. Ana, M.-I.; Istudor, L.-G. The Role of Social Media and User-Generated-Content in Millennials' Travel Behavior. *Manag. Dyn. Knowl. Econ.* **2019**, *7*, 87–104. [CrossRef]

38. Erdeji, I.; Dragin, A. Is Cruising along European Rivers Primarily Intended for Seniors and Workers from Eastern Europe? *Geogr. Pannonica* **2017**, *21*, 115–123. [CrossRef]

39. Fall, L.T.; Lubbers, C.A. Promoting America: How Do College-Age Millennial Travelers Perceive Terms for Branding the USA? In *Global Place Branding Campaigns across Cities, Regions, and Nations*; IGI Global: Hershey, PA, USA, 2016; pp. 270–287. [CrossRef]

40. Loda, M.D.; Coleman, B.C.; Backman, K.F. Walking in Memphis: Testing One DMO's Marketing Strategy to Millennials. *J. Travel Res.* **2010**, *49*, 46–55. [CrossRef]

41. Casalegno, S.; Inger, R.; Desilvey, C.; Gaston, K.J. Spatial covariance between aesthetic value & other ecosystem services. *PLoS ONE* **2013**, *8*, e68437. [CrossRef]

42. Wood, S.A.; Guerry, A.D.; Silver, J.M.; Lacayo, M. Using Social Media to Quantify Nature-Based Tourism and Recreation. *Sci. Rep.* **2013**, *3*, 2976. [CrossRef]

43. Schlieder, C.; Matyas, C. Photographing a city: An analysis of place concepts based on spatial choices. *Spat. Cogn. Comput.* **2009**, *9*, 221–228. [CrossRef]

44. Van Zanten, B.T.; Van Berkel, D.B.; Meentemeyer, R.K.; Smith, J.W.; Tieskens, K.F.; Verburg, P.H. Continental-Scale Quantification of Landscape Values Using Social Media Data. *Proc. Natl. Acad. Sci. USA* **2016**, *113*, 12974–12979. [CrossRef] [PubMed]

45. Richards, D.R.; Friess, D.A. A Rapid Indicator of Cultural Ecosystem Service Usage at a Fine Spatial Scale: Content Analysis of Social Media Photographs. *Ecol. Indic.* **2015**, *53*, 187–195. [CrossRef]

46. Richards, D.R.; Tunçer, B. Using image recognition to automate assessment of cultural ecosystem services from social media photographs. *Ecosyst. Serv.* **2017**, *31*, 318–325. [CrossRef]

47. Tenerelli, P.; Demšar, U.; Luque, S. Crowdsourcing Indicators for Cultural Ecosystem Services: A Geographically Weighted Approach for Mountain Landscapes. *Ecol. Indic.* **2016**, *64*, 237–248. [CrossRef]

48. Hausmann, A.; Toivonen, T.; Slotow, R.; Tenkanen, H.; Moilanen, A.; Heikinheimo, V.; Di Minin, E. Social media data can be used to understand tourists' preferences for nature-based experiences in protected areas. *Conserv. Lett.* **2018**, *11*, e12343. [CrossRef]

49. Kirilenko, A.P.; Stepchenkova, S.O. Sochi 2014 Olympics on Twitter: Perspectives of Hosts and Guests. *Tour. Manag.* **2017**, *63*, 54–65. [CrossRef]

50. Mayer-Schönberger, V.; Cukier, K. Big data: A revolution that will transform how we live, work, and think. *J. Epid.* **2014**, *179*, 1143–1144.

51. Eagle, N.; Pentland, A.; Lazer, D. Inferring Friendship Network Structure by Using Mobile Phone Data. *Proc. Natl. Acad. Sci. USA* **2009**, *106*, 15274–15278. [CrossRef]

52. Krippendorff, K. Content Analysis: An Introduction to its Methodology. by Klaus Krippendorff. *J. Am. Stat. Assoc.* **1984**, *79*, 240. [CrossRef]

53. Li, B. Description of communication content characteristics test of communication research hypothesis content analysis method. *Contemp. Commun.* **1999**, *6*, 39–40.

54. Li, B.; Dong, S.; Xue, M. Ecotourism model and benefit analysis in the marginal areas of ethnic minorities in Western Sichuan. *Rural. Econ.* **2008**, *3*, 51–54.

55. Zhang, J.; Zhang, Y. Trade-Offs between Sustainable Tourism Development Goals: An Analysis of Tibet (China). *Sustain. Dev* **2019**, *27*, 109–117. [CrossRef]

56. Dong, S.; Li, B.; Jin, X. Research on Eco-tourism Development Patterns in the Ethnic Regions of Western Sichuan Province under the Comprehensive Tourism Strategy. *China Popul. Resour. Environ.* **2009**, *19*, 117–122.

57. Konu, H.; Laukkanen, T. Predictors of Tourists' Wellbeing Holiday Intentions in Finland. *J. Hosp. Tour. Manag.* **2010**, *17*, 144–149. [CrossRef]

58. Mohammed Alsawafi, A. Sport Tourism: An Exploration of the Travel Motivations and Constraints of Omani Tourists. *Anatolia* **2017**, *28*, 239–249. [CrossRef]

59. Hanmei, C. Summarization of Human Tourism Resources in Garze Tibetan Autonomous Prefecture. *J. Sichuan Univ. Natl.* **2013**, *22*, 48–54. [CrossRef]

60. Shi, Q.; Wang, Z.; He, L.; Shi, Q.; Anayeti, A.; Liu, M.; Change, S. Landscape Classification System Based on Climate, Landform, Ecosystem: A Case Study of Xinjiang Area. *Shengtai Xuebao/Acta Ecol. Sin.* **2014**, *34*, 3359–3367. [CrossRef]

61. Kimbrough, A.M.; Guadagno, R.E.; Muscanell, N.L.; Dill, J. Gender Differences in Mediated Communication: Women Connect More than Do Men. *Comput. Hum. Behav.* **2013**, *29*, 896–900. [CrossRef]

62. Krueger, R.; Rashidi, T.H.; Vij, A. X vs. Y: An Analysis of Intergenerational Differences in Transport Mode Use among Young Adults. *Transportation* **2020**, *47*, 2203–2231. [CrossRef]

63. Teles da Mota, V.; Pickering, C. Using Social Media to Assess Nature-Based Tourism: Current Research and Future Trends. *J. Outdoor Recreat. Tour.* **2020**, *30*, 100295. [CrossRef]

64. Xu, F.; Morgan, M.; Song, P. Students' Travel Behavior: A Cross-Cultural Comparison of UK and China. *Int. J. Tour. Res.* **2009**, *11*, 255–268. [CrossRef]

65. Hargittai, E.; Litt, E. The tweet smell of celebrity success: Explaining variation in Twitter adoption among a diverse group of young adults. *New Media Soc.* **2011**, *13*, 824–842. [CrossRef]
66. Yang, L. Cultural Tourism in an Ethnic Theme Park: Tourists' Views. *J. Tour. Cult. Chang.* **2011**, *9*, 320–340. [CrossRef]
67. Yang, Y.; Liu, Z.H.; Qi, Q. Domestic Tourism Demand of Urban and Rural Residents in China: Does Relative Income Matter? *Tour. Manag.* **2014**, *40*, 193–202. [CrossRef]
68. Chen, Y.; Parkins, J.R.; Sherren, K. Using Geo-Tagged Instagram Posts to Reveal Landscape Values around Current and Proposed Hydroelectric Dams and Their Reservoirs. *Landsc. Urban Plan.* **2018**, *170*, 283–292. [CrossRef]
69. Luo, Y.; Deng, J. The New Environmental Paradigm and Nature-Based Tourism Motivation. *J. Travel Res.* **2008**, *46*, 392–402. [CrossRef]

 *land*

Article

Comparative Evaluation of Mountain Landscapes in Beijing Based on Social Media Data

Tingting Ding, Wenzhuo Sun, Yuan Wang, Rui Yu and Xiaoyu Ge *

School of Landscape Architecture, Beijing Forestry University, Beijing 100083, China
* Correspondence: gexiaoyu@bjfu.edu.cn

Abstract: An important part of Beijing's ecological pattern, mountain landscapes are also the most important natural tourist destinations in Beijing. The unique mountain environment in Taihang and Yan Mountains attracts Beijing and foreign tourists alike. Tourists publish travel photos and comments on social media, which provides a new opportunity for a systematic evaluation of these mountain parks based on social media data. To fully understand the developmental status of mountain landscapes in Beijing, this paper comparatively evaluates 45 mountain landscapes in Beijing based on social media data. Using big data capture, semantic network analysis, importance-performance analysis (IPA), etc., it explores the composition of tourist groups in mountain parks, the preferences of the tourist groups, and the relationships between park tourists and different influencing factors, and evaluates the recreational experiences of tourist groups. The development of recreational activities was found to be more important to local tourists than scenic sites for foreign tourists. According to gender differences, women were more interested in recreational experiences than men, while men were more interested in the park's landscapes. According to the IPA, tourists were satisfied with the overall recreation offered by mountain landscapes. The perceptual experience was dominated by visual perception, followed by smell; touch, hearing, and taste were of minor importance. Using social media data to analyze mountain landscape resources in Beijing can provide useful insights into the advantages of these landscapes under a variety of site conditions, strengthen local mountain resource development and tourism publicity, integrate tourism management and planning resources in a targeted and attractive manner, and enhance ecological leisure services.

Keywords: mountain landscape; perceived destination image; social media data; Beijing; China; social media data; text analysis; important-performance analysis (IPA); tourism sustainability

Citation: Ding, T.; Sun, W.; Wang, Y.; Yu, R.; Ge, X. Comparative Evaluation of Mountain Landscapes in Beijing Based on Social Media Data. *Land* **2022**, *11*, 1841. https://doi.org/10.3390/land11101841

Academic Editors: Cecilia Arnaiz Schmitz, Nicolas Marine and María Fe Schmitz

Received: 13 September 2022
Accepted: 18 October 2022
Published: 19 October 2022

Publisher's Note: MDPI stays neutral with regard to jurisdictional claims in published maps and institutional affiliations.

1. Introduction

The construction of urban forests is an important measure for adapting to China's national conditions and developmental stage, promoting urban and rural ecological construction, and enhancing residents' ecological welfare. Beijing, as the capital city, is responsible for the construction of the ecological civilization. In 2020, Beijing formulated the Beijing Forest City Development Strategy (2018–2035). As an important mountain resource in Beijing, Taihang-Yan Mountain also plays an important role in the forest city development strategy. The forest city refers to an urban area in which buildings such as offices, houses, hotels, hospitals, and schools are almost entirely covered by plants and trees of different kinds and sizes for protecting the urban ecological environment. In the overall construction of the forest city, the vision of building a healthy mountain forest has been put forward. The Beijing Mountain area is an important water conservation area and ecological barrier in the capital, and it is the main ecological recreation area. By developing and using mountain resources, it is the goal of forest city construction to build an ecological development demonstration area that is suitable for living, industry, and tourism, as well as a model area that displays Beijing's history, culture, and beautiful natural landscape.

Mountain tourism is an important part of contemporary tourism. Because of their biodiversity, rich environmental resources, and rich history and culture, mountain landscapes can satisfy people's desire to be close to nature and pursue health, so mountain tourism is playing an increasingly important role in global tourism patterns [1]. The development and use of a mountain landscape are of great significance in the construction of a forest city, so it is necessary to pay more attention to and study them. Research and analysis of mountain landscape resources play an important role in urban development. Mountain landscape tourism resources are important for attracting tourists, and the richness of the landscapes is an important factor in developing them [2]. From the perspective of tourism, many studies focus on factors related to tourist identity, such as gender, age, mode of travel, and so on. However, further empirical research is required to explore how the factors related to tourists themselves influence broader tourism decisions and results. Therefore, studying what landscapes tourists prefer in mountain parks is of great significance in promoting ecological service function and developing mountain tourism.

As per Cavagnaro et al., modern tourists show a strong interest in topics related to natural resources and like to visit destinations with important natural resources for recreational activities [3–8]. Giachino [9] and others found that modern tourists show seasonal differences in their choices of natural tourist destinations. Selecting a holiday destination is heavily influenced by the image of tourist destinations. The tourist destination image (TDI) consists primarily of impressions (45%), perceptions (27%), beliefs (18%), ideas (18%), and representations (15%). In scientific doctrine, one of the most commonly cited definitions of image is something that can be described as the sum of a person's beliefs, ideas, and impressions of a particular location [10].

Tieskens et al. [11] proved that the analysis of elements of mountain cultural landscapes has a high research value in the exploration of tourists' preference for mountain landscapes. Studies have also shown that analyzing modern tourists' visiting behaviors can promote sustainable mountain tourism development, and their participation is considered necessary for sustaining and improving natural tourism. At the same time, tourists themselves show an important impact on environmental sustainability [12]. Therefore, to promote the sustainable development of tourist destinations, it is of great significance to study the natural and cultural landscape preferences of modern tourists in such destinations. Tourists' preferred activities also differ with age [13], gender [14], mode of transportation [15], and travel mode [16]. However, it remains to be studied in detail whether modern tourists' preferences for natural and cultural landscapes of mountain parks differ due to these factors. Therefore, it is necessary to study the differences in modern tourists' preferences for mountain landscapes when using different travel modes.

The arrival of the era of big data has also provided a new opportunity for the evaluation of mountain park landscapes. In recent years, taking users as the research object, research using big data basically focused on four aspects: mobile phone signaling data, satellite positioning, social media data, and photo analysis with geographical location information [17,18]. The development of modern information technology, especially the popular Internet technologies, such as social media, as a platform for the public to obtain information and publish opinions, has a large number of valuable comments on people, events, products, etc. [19], making it an important source of data that can help evaluate personal emotions, perceptions, opinions, and interests [20]. A large amount of content for evaluation is posted on social media; this data enables textual analysis and sentiment analysis and allows one to study people's preferences for places [21–23]. The research usually uses high-frequency words and semantic network analysis methods to measure and predict users' preferences, such as the travel preferences of outbound Chinese tourists [24], the differences in mental models between tourism marketers and travelers [25], the image perception of specific places [26], and even the services of hotels and other service facilities [27]. One can combine photos with geographical location information with the characteristics of places to effectively predict the number of visits to places and infer people's habits and preferences to help in urban planning and ecological construction [28–33]. The widely used importance-performance analysis (IPA) method was first put forward in 1977 when it was used to analyze product attributes [34]. This method, as a diagnostic model biased toward qualitative research, can help managers identify noteworthy resources and services and provide guidance for landscape planning and construction. Because of its simplicity, intuition, easy operation, and easy interpretation, it has been widely used in landscape architecture in recent years to study the demand characteristics of park visitors [35], recreation situations [36,37], the supply-demand relationship of the cultural ecosystem [38], and so on.

In short, mountain landscapes play an important role in the construction of Beijing's forest city, the promotion of the functions of ecological services, and the development of related tourism resources; social media data are used to understand the landscape preferences of modern tourists in mountain tourism, identify the differences in landscape evaluations by crowds with different characteristics, and evaluate recreational experiences. The aim of this research is to investigate the demographic characteristics and landscape preferences of tourists visiting the 45 mountain landscapes in Beijing by classifying photos based on the associated text uploaded by users to social media. Through an in-depth semantic network analysis of social media comments published by users, the relationships between the evaluations of the mountain landscapes in Beijing and the demographic characteristics and geographical locations of the users will be identified. In addition, the differences in tourists' landscape evaluations based on the different genders and regions will be explored. The landscape characteristics of different mountain systems and 10 administrative divisions are also compared. Finally, through an IPA, tourists' satisfaction in terms of (1) recreation provided by the mountain landscapes in Beijing and (2) experiencing these landscapes via the five senses will be analyzed. This study will provide a theoretical basis for the key points and developmental directions of the construction and improvement of mountain landscapes, leading not only to a new research area but also helping policymakers and tourism managers improve the attractiveness of regional tourism. The Taihang-Yan Mountain area of Beijing is of great significance for the sustainable development of mountain landscape resources.

2. Research Method

Using big data to study tourists' behavior is beneficial for the planning and management of tourist attractions, especially in the field of landscape architecture. In this study, we selected Dianping (https://www.dianping.com, accessed on 10 June 2022), Trip.com Group (https://www.ctrip.com, accessed on 10 June 2022), and Mafengwo (https://www.mafengwo.cn, accessed on 10 June 2022), all of which are the mainstream social media websites for tourism in China. Social media data were used instead of traditional research methods for the following reasons: (1) Social media data can reduce the restrictions related to insufficient sample size, time, place, and self-reporting errors. (2) The photos provided by users record the local environment and experience, which is more authentic. (3) Social media are an important medium for modern tourists to publish and receive tourism information, and the content has important research significance and value. This research is based on the quantitative analysis of photos and evaluation of texts published by tourists on social media. Content analysis is a digital research method for objectively, systematically, and quantitatively analyzing the contents of texts. People upload photos that they like or are interested in on social platforms. Therefore, the contents of the photos were coded and analyzed according to landscape features, and the specific landscape elements that attracted tourists were studied by comparing the frequency of each element. The relationships between various landscape elements and preferences of crowds with different characteristics were analyzed. At the same time, keyword extraction, emotional and semantic network analysis, and IPA were performed on the comments published by tourists on social platforms; then, the differences in landscape evaluations under different factors were explored.

2.1. Research Object

In all, 45 representative mountain landscapes in Beijing were selected as the research objects (Table 1, Figure 1). Generally, in Beijing, the mountains are clear and dangerous, and the terrain rises in steps, forming several levels of viewing platforms. Geological structure and lithology differ greatly, and the landforms are diverse and colorful. The higher the altitude of a scenic mountain, the more natural the scenery is; the lower the altitude, the more anthropic the scenery is. Religious temples are found on most of the tops or foothills of the cultural landscape mountains, indicating that religious culture has a profound influence on the mountains in the suburbs of Beijing. As well as being favored by residents, the temples are also revered by the royal family, illustrating the nexus between religion and politics. Beijing's two mountain spaces, Xishan and Beishan, have significantly different associations with the scenery. A large number of scenic mountains are mostly associated with the content of the Great Wall in Beishan, Beijing, forming a natural cultural landscape similar to the Great Wall. The West Mountains in Xishan are devoid of Great Wall cultural landscapes, with the exception of the enemy towers along the River City.

Table 1. Basic information on mountain landscapes.

Serial Number	Name	Mountain Range	District	Score	Distance to the City Center (km)	Driving Time to the City Center (h)
1	Fragrant Hill Park	Taihang Mountain	Haidian District	4.75	32 km	0.75 h
2	Badachu Park	Taihang Mountain	Shijingshan District	4.80	29 km	0.75 h
3	Ming Tombs National Forest Park	Yan Mountain	Changping District	4.55	54 km	1 h
4	Shangfang Mountain National Forest Park	Taihang Mountain	Fangshan District	4.60	80 km	1.5 h
5	Xishan National Forest Park	Taihang Mountain	Haidian District	4.80	34 km	1 h
6	Beigong National Forest Park	Taihang Mountain	Fengtai District	4.75	20 km	0.5 h
7	Jiufeng National Forest Park	Taihang Mountain	Haidian District	4.60	40 km	1 h
8	Miaofeng Mountain National Forest Park	Taihang Mountain	Mentougou District	4.55	55 km	1.5 h
9	Baishui Temple Forest Park	Yan Mountain	Fangshan District	4.65	67 km	1 h
10	Fahai Temple Forest Park	Taihang Mountain	Shijingshan District	4.00	39 km	0.75 h
11	Yaji Mountain Forest Park	Yan Mountain	Pinggu District	4.80	75 km	1.25 h
12	Laobagoumen National Forest Park	Yan Mountain	Huairou District	4.60	160 km	4 h
13	Dayang Mountain National Forest Park	Yan Mountain	Changping District	3.85	40 km	1 h
14	Jingzhi Lake Forest Park	Yan Mountain	Changping District	4.30	28 km	1 h
15	Yunmeng Mountain National Forest Park	Taihang Mountain	Miyun District	4.55	85 km	1.5 h
16	Xiayunling National Forest Park	Taihang Mountain	Fangshan District	4.20	74 km	1.5 h
17	Tianmeng Mountain National Forest Park	Taihang Mountain	Mentougou District	4.65	40 km	1 h
18	Shuanglongxia Dongshan Forest Park	Taihang Mountain	Mentougou District	4.60	92.2 km	2 h
19	Nanshiyang Grand Canyon Forest Park	Taihang Mountain	Mentougou District	4.50	91.8 km	2 h

Table 1. *Cont.*

Serial Number	Name	Mountain Range	District	Score	Distance to the City Center (km)	Driving Time to the City Center (h)
20	Badaling National Forest Park	Taihang Mountain	Yanqing District	4.65	67.6 km	0.9 h
21	Baihujian Scenic Area	Taihang Mountain	Changping District	4.15	45 km	1 h
22	Yunmeng Mountain Scenic Area	Taihang Mountain	Miyun District	4.55	85 km	1.5 h
23	Bairuigu Scenic Area	Taihang Mountain	Fangshan District	4.55	85 km	2.2 h
24	Guyaju Scenic Area	Yan Mountain	Yanqing District	4.65	92 km	2.5 h
25	Baihua Mountain Scenic Area	Taihang Mountain	Mentougou District	4.65	120 km	3 h
26	Yaji Mountain Scenic Area	Yan Mountain	Haidian District	4.80	42 km	1 h
27	Yangtai Mountain Scenic Area	Taihang Mountain	Pinggu District	4.45	90 km	2 h
28	Fenghuangling Scenic Area	Taihang Mountain	Haidian District	4.70	53 km	0.9 h
29	Shenquanxia Scenic Area	Taihang Mountain	Mentougou District	4.50	55 km	1 h
30	Zhuijiuyu Scenic Area	Yan Mountain	Changping District	4.15	62 km	1 h
31	Shentangyu Scenic Area	Yan Mountain	Huairou District	4.60	65 km	1.2 h
32	Qinglongxia Scenic Area	Yan Mountain	Huairou District	4.65	75 km	2 h
33	Baicaopan Scenic Area	Taihang Mountain	Fangshan District	4.65	120 km	2.5 h
34	Baiyanggou Scenic Area	Taihang Mountain	Changping District	4.30	63.8 km	1 h
35	Jiangjuntuo Scenic Area	Taihang Mountain	Fangshan District	3.80	45 km	1 h
36	Yunfeng Mountain Scenic Area	Yan Mountain	Miyun District	4.55	120 km	3 h
37	Linlong Mountain Scenic Area	Yan Mountain	Huairou District	4.25	75 km	2 h
38	Jiugukou Scenic Area	Yan Mountain	Huairou District	4.10	78 km	1.3 h
39	Xianjugu Scenic Area	Yan Mountain	Miyun District	4.45	125 km	1.7 h
40	Taoyuan Xiangu Scenic Area	Yan Mountain	Miyun District	4.20	101 km	1.5 h
41	Penghewan Scenic Area	Yan Mountain	Miyun District	4.20	105 km	1.24 h
42	Bailongtan Scenic Area	Yan Mountain	Miyun District	3.85	105 km	1.24 h
43	Yunxiugu Scenic Area	Yan Mountain	Miyun District	4.20	175 km	2.32 h
44	Hudongshui Scenic Area	Yan Mountain	Pinggu District	4.45	103 km	1.8 h
45	Qianling Mountain Scenic Area	Taihang Mountain	Fengtai District	4.65	30 km	1 h

Note: Social media has given an overall possible score of 5.

2.2. Data Collection and Statistics

The data collected in this paper are from 45 representative mountain landscapes in Beijing (visited on 10 June 2022) that were manually searched on the website of the Beijing Municipal Bureau of Landscaping and Greening. They included national-level scenic spots and representative tourist destinations of Grade 3A or above. From high to low, China's tourist attractions are classified into five levels, from AAAAA (the highest level) to A (the lowest). After AAAAA and AAAA, AAA (3A) is the third highest level of quality for tourist attractions. In addition, web crawler tools were utilized on Dianping, Trip.com Group, and Mafengwo to crawl for information on the evaluation of the 45 mountain landscapes and photos of the 45 selected mountain landscapes released by tourists. By analyzing users' evaluative social media texts and uploaded and shared travel photos, tourists' demand for recreational activities and their landscape preferences can be understood. From the perspective of data content screening, first, the comments adopted were those published by tourists without any commercial activities, and the language that expressed their feelings and emotions was used as the criterion. As mentioned above, tourists' travel preferences were influenced by gender, mode of transportation, travel mode, and landscape preferences. Therefore, we also collected information about users' gender, transportation mode, and travel mode. Finally, we eliminated comments with prominent advertising information and copyright marks. The sample period was from January 2021 to June 2022. There were 37,572 photos of the 45 mountain landscapes, totaling more than 2.82 million words. Most of the crawled comments were made between January 2021 and June 2022.

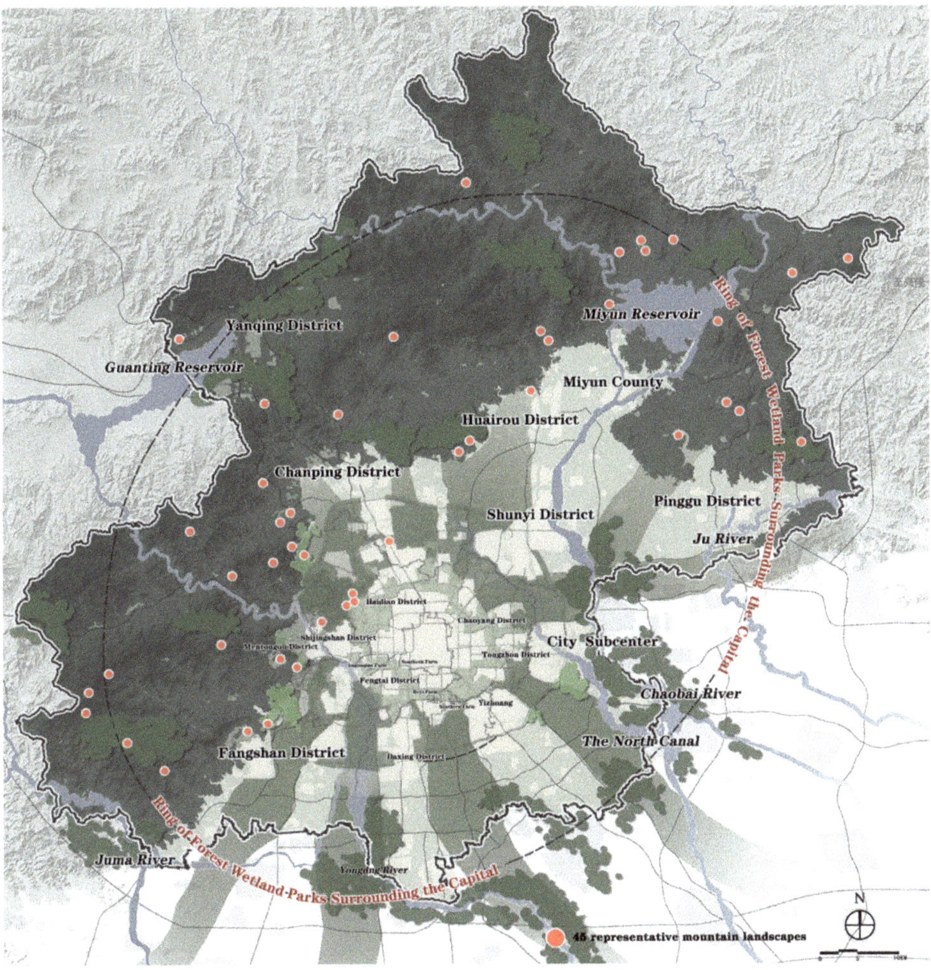

Figure 1. Distribution of the 45 representative mountain landscapes in Beijing.

2.2.1. Data Collection and Preprocessing

The automatically crawled social media data contains a significant amount of noise, so it is necessary to remove the noise data from the text data. To begin with, news information, advertising information, and explanatory texts published by public accounts were removed, and only original content published by individuals was retained. Among the photos posted by users, some photos were primarily based on pictures, and the text description was often too short (for example, only the names and locations of mountain attractions were included); at the same time, there was also content that mentioned mountain attractions; however, the actual description content or evaluation did not follow. A comment that had little relevance to mountain attractions was discarded, as these data were deemed invalid and must be removed. Thus, 20 images were randomly selected from the reserved text for manual verification and removal of invalid information. When the data is invalid, the above operations will be repeated until a valid image with comment data is selected. A total of 31,367 valid images were obtained from 13,990 users, along with 1,631,972 characters.

In terms of data processing, the user information published by social media websites and the keywords in social media comments published by users were extracted, the demographic characteristics of the image users were determined, and the following types of user information were summarized: gender (male and female), mode of transportation (walking; using the subway, a taxi, or a bus; and using a self-driven mode of transport), and travel mode (lone travelers, friends and classmates, families, and couples). Accordingly, the tourism behavior of users was quantitatively analyzed.

2.2.2. Image Recognition and Statistics

On the basis of China's national standard of "Classification, Investigation, and Evaluation of Tourism Resources" (GB/T18972-2017), we divided landscape resources into eight types: physiographic landscape, water landscape, biological landscape, astronomical and climatic landscape, buildings and facilities, historical sites, tourist purchases, and cultural activities (Table 2). On this basis, we quantitatively analyzed 31,367 tourist photos and calculated the frequency of each landscape resource type (Figure 2).

Table 2. Classification of landscape resources.

Main Category	Subcategory	Basic Types
Physiographic landscape	Natural landscape complex	Hills, mesas, valleys, and beaches
	Geological and tectonic traces	Fractured landscape, folded landscape, stratigraphic section, and biological fossil point
	Surface morphology	Hill-shaped landscape, peak-columnar landscape, ravines, and caves
	Natural marks and natural phenomena	Strange natural phenomena and natural landmark
Water landscape	River system	Recreational river sections, waterfalls, and ancient river sections
	Lake and marsh	Recreational lakes, pools, and wetlands
	Groundwater	Springs and buried bodies of water
	Ice and snow area	Snow fields and modern glaciers
Biological landscape	Vegetation landscape	Woodland, single and bushy trees, meadows, and flower fields
	Wildlife habitat	Aquatic animal habitat, land animal habitat, bird habitat, and butterfly habitat

Table 2. *Cont.*

Main Category	Subcategory	Basic Types
Astronomical and climatic landscape	Astronomical landscape	Sun, moon, stars, aurora, and natural or artificial light phenomena
	Weather and climatic phenomena	Clouds, fog rime, rain rime, extreme and special climate displays, and phenological phenomena
Buildings and facilities	Cultural landscape complex	Places for social and commercial activities, military sites and ancient battlefields, places for cultural activities, places for recreation and leisure, places for religious and sacrificial activities, and places for memorials and commemorative activities
	Practical buildings and core facilities	Characteristic blocks, landscape buildings and spaces with viewing functions, bridges, dams, caves, mausoleums, landscape farmland, landscape forest farms, and specialty shops
	Landscape and sketch architecture	Image markers, viewing points, pavilions, platforms, buildings, pavilions, sculptures, archways, forest of steles, porches, tower buildings, landscape trails, flower lawns, fountains, and rock piles
Historical sites	Material cultural relics	Architectural relics and movable cultural relics
	Immaterial cultural relics	Folk literature and art, local customs, traditional costume decoration, and traditional performing arts
Tourist purchases	Agricultural products	Planting, forestry, animal husbandry, aquaculture products, and aquatic products
	Industrial products	Daily industrial product and tourism equipment products
	Handmade arts and crafts	Stationery, fabrics, furniture, ceramics, and paintings
Cultural activities	Personal activity records	Local people and local events
	Festivals and seasons	Religious activities and temple fairs, agricultural festivals, and modern festivals

2.2.3. Text Analysis

A total of 45 representative mountain scenery spots in Beijing were analyzed based on their average star ratings. In order to highlight the differences between different mountain landscapes, the study utilized the tools of "word frequency analysis" and "social network and semantic network analysis" in ROSTCM6 developed by Wuhan University in order to quantify image data in order to generate a collinear network diagram of keywords in tourists' comments on parks and scenic spots, further exploring the core factors that affect mountain landscape evaluations. Using a network diagram, the core elements and deep reasons that affected the evaluation of the mountain landscapes were explored further and the correlation between each element and the evaluation was explored using SPSS tools.

2.2.4. IPA Model Building

To further explore the present situation, problems, and development directions of mountain landscapes in Beijing, the importance-performance analysis (IPA) method was used to analyze tourists' satisfaction and experience in terms of the five senses. IPA, which was proposed by Martilla and James, is used to compare customers' expectations before consumption with their perceived achievements after consumption, and to comprehensively evaluate the performance of each attribute [39]. Since the early 1990s, IPA has been widely used in service industries [40], including service satisfaction evaluations [41], regional attraction analysis [42], tourism policy formulation [43], and scenic spot satisfaction evaluations [44]. In the satisfaction survey, the IPA method requires respondents to evaluate the indicators of the designated survey object in terms of importance and satisfaction in order to form the IPA matrix (Figure 3). The IPA matrix takes tourists' expectations (importance) as the horizontal axis, tourists' satisfaction (performance) as the vertical axis, and the total average as the separation point of the X-Y axis. The space is divided into four quadrants, and the meanings of each quadrant are as follows: the first quadrant is the area of advantage retention, the second quadrant is the area that can be maintained without too much improvement, the third quadrant is the slow improvement area, and the fourth quadrant is the area that needs to be improved. The recreational elements in the related literature on the evaluation of mountain landscape recreation were summed up, the words featured with a high frequency in the evaluation texts were extracted and evaluated, the specific elements worthy of attention in mountain landscapes were integrated, tourists' recreational satisfaction and evaluation factors in terms of the five senses were determined, a vocabulary of recreational satisfaction and elements related to the experience of the five senses was generated (Tables 3 and 4), and each index for each element in the obtained textual data was identified and counted as the result of the importance of each element index in the IPA. The text was classified by emotion, and the frequency of each factor index in positive comments was used as the result for satisfaction. Using the IPA method, based on the results of the textual analysis of social media data, this paper evaluated the satisfaction provided by the mountain landscape in terms of recreation and the experience of the landscape via the five senses and further explored the future direction of development of the mountain landscape.

Figure 2. Images of the classification of mountain landscape resources.

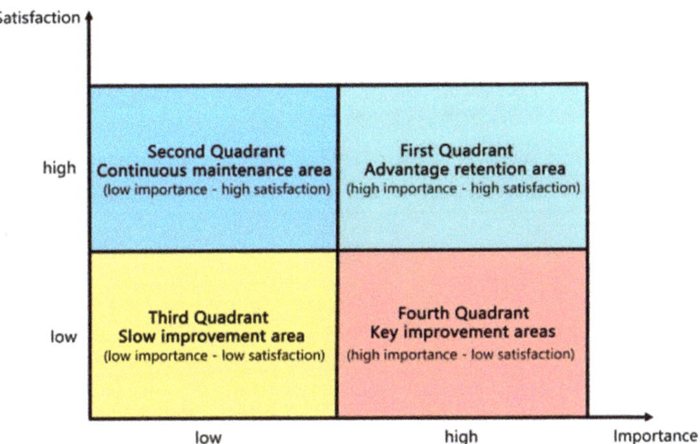

Figure 3. IPA quadrant diagram.

Table 3. Evaluative indexes of tourist satisfaction with mountain landscapes in terms of recreation.

Evaluation Term	Indicators	Indicator Definition
Landscape quality (A)	Natural landscape (A1)	Rivers, streams, and other natural landscapes
	Plant landscape (A2)	Trees, flowers, and other plant landscapes
	Animal landscape (A3)	Squirrels, ducks, hedgehogs, and other animal landscapes
	Astronomical landscape (A4)	Seas of clouds, rimes, rainbows, and other astronomical landscapes
	Historical and cultural landscape (A5)	Temples, ancient temples, ancient buildings, gardens, and other characteristic landscapes
Recreational activities (B)	Outdoor recreational activities (B1)	Hiking, ferrying, picnicking, and other outdoor recreational activities
	Leisure activities (B2)	Taking photos, hiking, walking, and other leisure activities
	Fitness activities (B3)	Sports, hiking, fitness, and other activities
	Humanistic activities (B4)	Burning incense, praying for blessings, and other humanistic activities
Tourism experience (C)	Ticket cost (C1)	Park fares and charges
	Parent-child experience (C2)	Suitability for parent-child activities
	Emotional experience (C3)	Comfort, pleasure, happiness, and other recreational emotions
	Air and environmental quality (C4)	Environmental quality, air freshness, and weather conditions
	Sense of crowded space (C5)	The number of visitors and the degree of space crowding
Infrastructure (D)	Traffic accessibility (D1)	Connectivity of internal and external roads
	Public service facilities (D2)	Parking lots, toilets, trash cans, and other service facilities
	Recreational and entertainment facilities (D3)	Slides, cable cars, cableways, and other recreational facilities
	Navigation signage system (D4)	Guide systems, signage, etc.
	Catering and convenience facilities (D5)	Restaurants, catering, food sales, etc.
	Safety facilities (D6)	Railings, fences, and other safety equipment

Table 3. *Cont.*

Evaluation Term	Indicators	Indicator Definition
Management services (E)	Facility maintenance (E1)	Maintenance and management of public facilities and infrastructure
	Park management services (E2)	Park management, public security maintenance, etc.
	Planning layout (E3)	Park areas, planning and design, route planning, etc.

Table 4. Evaluative indexes of the experiences of mountain landscape tourists in terms of the five senses.

Senses Term	Indicator	Indicator Definition
Vision (F)	Visibility of plants (F1)	Visibility of trees, grass, flowers, etc.
	Visibility of animals (F2)	Visibility of squirrels, ducks, hedgehogs, etc.
	Visibility of natural landscapes (F3)	Visibility of the landscape, rivers, streams, etc.
	Visibility of celestial phenomena (F4)	Visibility of celestial landscapes (sea of clouds, smog, rainbow, etc.)
	Crowd disturbances (F5)	The number of people and the presence or absence of distractions
	Landscape recognizability (F6)	Special sites
	Visibility of roads (F7)	The line, shape, color, etc. of the roads
	Others (F8)	Environmental visibility, etc.
Hearing (G)	Sounds of humans (G1)	Moderate vocals
	Sounds of plants (G2)	Sound of the wind blowing through the plants
	Sounds of animals (G3)	Sounds of birds, insects, and other animals
	Sounds of broadcasts (G4)	Sounds of broadcasts
	Sounds of water (G5)	Sound of flowing water
	Others (G6)	Sounds of wind, rain, etc.
Smell (H)	Smell of air/water (H1)	Smell of the air and water emanating from the landscape
	Smell of plants (H2)	Smell of the scent emanating from the plants
Touch (L)	Feel of sunlight (L1)	Feel of the balance of light and shadow
	Feel of wind (L2)	Feel of the wind environment
	Feel of water (L3)	Feel of water flowing through the landscape
	Feel of temperature (L4)	Feel of the landscape temperature
	Touch of the road (L5)	Feel of the comfort of road contact
	Touch of animals (L6)	Lack of mosquito bites
	Others (L7)	Touch of plants, etc.
Taste (K)	Food sales (K1)	Purchase of food
	Taste of food (K2)	Taste of food, spring water, etc.

3. Research Results and Analysis

3.1. Demographic Analysis of Tourist Groups

Among the 13,990 users who submitted reviews, there were 2882 men (32.4%) and 6014 women (67.6%). It was not possible to distinguish the gender of the remaining users by using public data. It can be seen that women were keener to share their travel experiences on social media.

From the point of view of the mode of transportation, among the review users, 3485 (94.2%) were self-driving tourists, accounting for far more than walking tourists (16, or 0.4%), subway tourists (62, or 1.7%), bus tourists (115, or 3.1%), and taxi tourists (23, 0.6%). Most tourists chose to travel by car, which is probably related to the unique geographical location and landscape characteristics of the mountains. Self-driving is more convenient for reaching the destination and enjoying the beautiful natural scenery along the route.

From the perspective of travel patterns, among the review users, 56.3% chose to travel with their families and only 1.8% chose to travel with their partners. See Table 5 for a statistical analysis of the tourist groups.

Table 5. Statistical analysis of the tourist groups.

	Type	Quantity	Percentage
Gender	Male	2882	32.4
	Female	6014	67.6
Transportation	Walking	16	0.4
	Using the subway	62	1.7
	Using a bus	115	3.1
	Using a taxi	23	0.6
	Using a self-driven vehicle	3485	94.2
Travel mode	Alone	417	12.1
	With friends and classmates	1025	29.8
	With family	1937	56.3
	As a couple	65	1.8
Family	Had children	946	48.9
	Had an elderly person	68	3.5
	Had both children and elderly people	120	6.2
	Had other relatives	803	41.4

3.2. Analysis of the Landscape Preferences of Tourist Groups

3.2.1. Overall Analysis of Tourists' Landscape Preference

Based on the classification of landscape resources summarized in Table 2, 31,367 tourist photos were analyzed quantitatively for landscape elements. Tourists' landscape preferences were determined according to the following order: physiographic landscape (9807, or 31.4%) > buildings and facilities (9000, or 28.8%) > biological landscape (5377, or 17.2%) > water landscape (4090, or 13.1%) > historical sites (1030, or 3.4%) > astronomical and climate landscapes (885, or 2.9%) > cultural activities (781, or 2.6%) > tourism purchases (162, or 0.6%).

3.2.2. Analysis of Landscape Preferences Based on Crowds with Different Characteristics

According to the differences in the gender, transportation mode, and travel mode of the tourists, the landscape preferences of the different groups were statistically analyzed. As far as gender is concerned (Figure 4), men and women tended to prefer the same types of landscapes, and they all showed an obvious preference for landscapes, architecture, and facilities, which is probably because mountain landscapes are dominated by natural landscapes, such as valleys and gullies, and tourists mainly go to scenic mountain spots for sightseeing, so the photos taken are mainly of buildings and places for sightseeing in the landscape. Table 6 presents an analysis of the images shared by people of different genders; men preferred physiographic landscapes, followed by biological landscapes and buildings and facilities, while women preferred water landscapes, followed by astronomical and climatic landscapes and cultural activities. From the perspective of humanistic activities, compared to men, women showed a greater preference for recording their personal activities during travel.

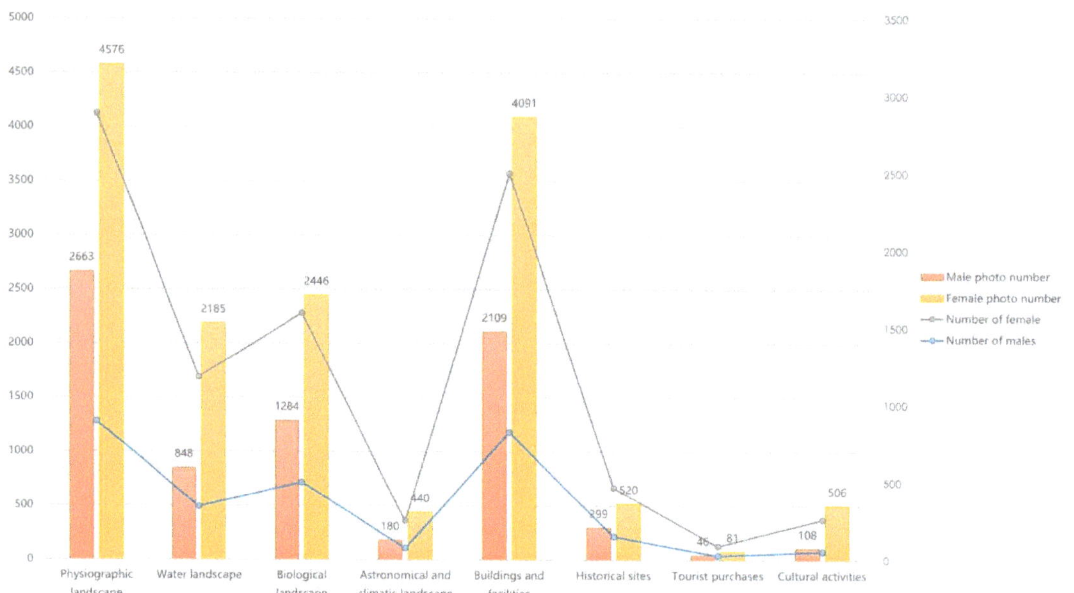

Figure 4. Analysis of tourists' landscape preferences and its difference among different genders.

Table 6. The frequency with which landscape types were visited by each gender.

Types of Landscape Resources	Number of People		Number of Photos		Average Value	
	Male	Female	Male	Female	Male	Female
Physiographic landscape	894	1993	2663	4576	2.98	2.30
Water landscape	345	837	848	2185	2.46	2.61
Biological landscape	501	1094	1284	2446	2.56	2.24
Astronomy and climatic landscape	77	175	180	440	2.34	2.51
Buildings and facilities	826	1674	2109	4091	2.55	2.44
Historical sites	153	313	299	520	1.95	1.66
Tourist purchases	30	61	46	81	1.53	1.33
Cultural activities	56	205	108	506	1.93	2.47

Figure 5 shows the differences in tourists' landscape preferences when using different modes of transportation. Irrespective of the mode of transportation, tourists preferred physiographic landscapes, buildings and facilities, biological landscapes, and water landscapes. However, this choice was more obvious when using a self-driven vehicle. Figure 6 shows the differences in tourists' landscape preferences according to their different modes of travel. Tourists who traveled with their families and partners preferred buildings and facilities, followed by physiographic landscapes, and the degrees of interest were similar between biological landscapes and water landscapes. Tourists who traveled alone or with friends and classmates were more interested in landscapes than in buildings and facilities, and their interest in biological landscapes was also greater than their interest in water landscapes. With regard to the different types of landscape resources, there were certain differences in tourists' travel patterns. As can be seen in Figure 7, tourists who traveled with friends and classmates preferred astronomical and climatic landscapes and tourism purchases.

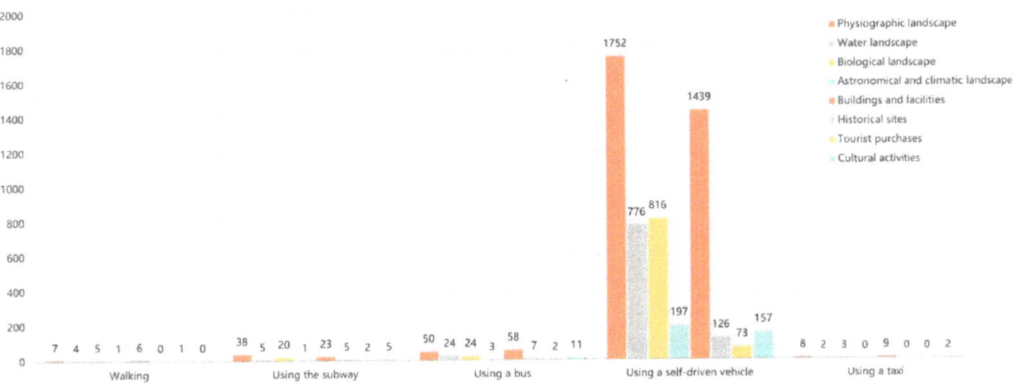

Figure 5. Analysis of tourists' landscape preferences when using different modes of transportation.

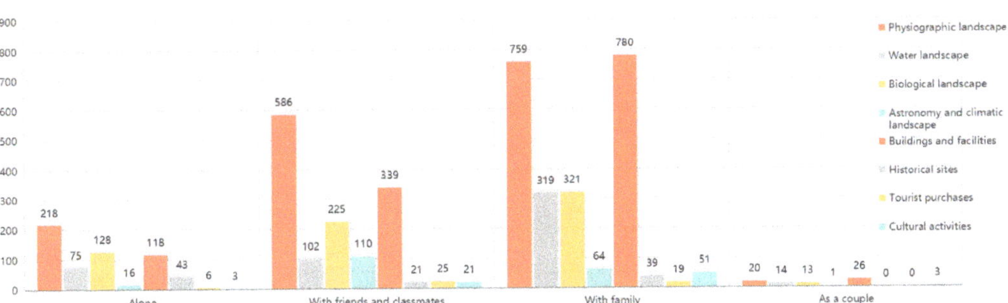

Figure 6. Analysis of tourists' landscape preferences when using different travel modes.

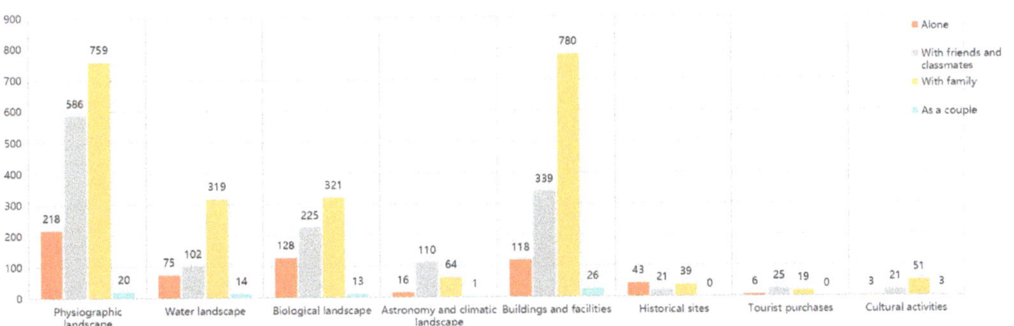

Figure 7. Analysis of tourists' travel mode preferences with respect to different landscape resource types.

3.3. Landscape Evaluation and Analysis of Tourist Groups

To further understand tourists' preferences for recreation types and feelings regarding Beijing's mountain landscapes, word segmentation and word frequency analysis were performed by using ROSTCM6 to evaluate the texts of the review users, and the 30 words related to recreation and emotional experience with the highest frequency were extracted (Figure 8). By and large, tourists tended to show seven recreational behaviors related to mountain landscapes: climbing, rafting, hiking, fitness, camping, picnicking, and sightseeing. Moreover, their emotional experiences were dominated by positive words, such as "suitable", "convenient", "characteristic", and "beautiful", all of which show a positive attitude. It can be seen that tourists loved the overall landscapes and environments of the mountains. The authors of this study extracted the texts of positive reviews by users for semantic network analysis and generated a positive semantic network diagram (Figure 9). The texts of negative user evaluations were also extracted for semantic network analysis, and a negative semantic network diagram was generated (Figure 10). The nodes represent high-frequency vocabulary elements, and the density of connections between elements represents the co-occurrence frequency. The factors of positive evaluations of the mountain landscape were mainly reflected in the beautiful scenery, suitability for outings, traffic accessibility, recreational facilities, and parent-child experiences. The negative factors of the evaluations were mainly reflected in tickets, transportation, infrastructure, and park management services. Although the overall evaluation of the mountain landscape is high, it does not necessarily imply that the parks are well-managed. The management of a forest park involves many complex aspects that are not readily apparent to the public. Negative feedback recorded on social media indicates that tickets, transportation, infrastructure, and park management services may not need to be improved, rather, the meaning implies that the management mode may need to be reviewed.

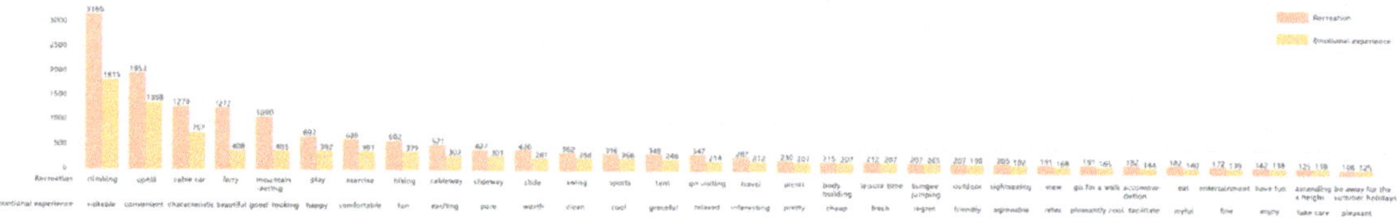

Figure 8. Frequency of the top 30 words related to recreation and emotional experience.

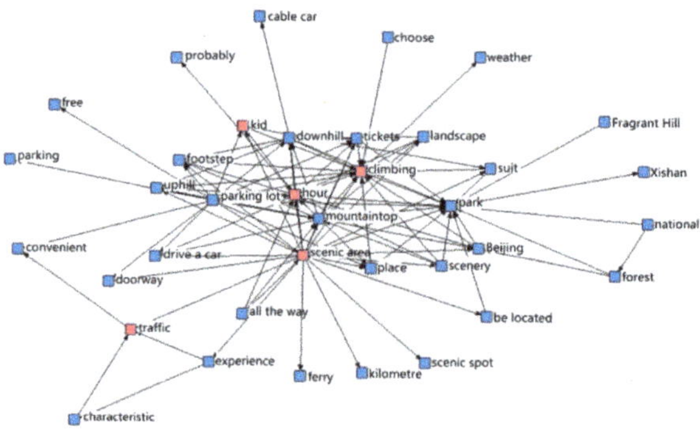

Figure 9. The semantic network of positive evaluations.

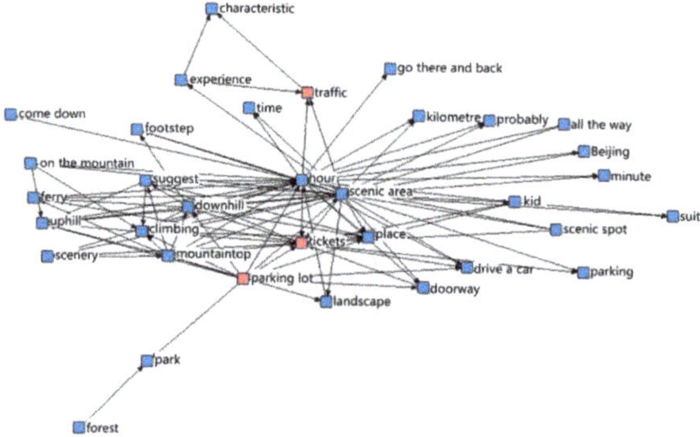

Figure 10. The semantic network of negative evaluations.

3.3.1. Differences in Mountain Landscape Evaluations by Groups with Different Characteristics

The semantic network analysis of the evaluative texts of local and foreign tourists showed (Figures 11 and 12) that the main mountain landscape elements that local and foreign tourists focused on were different. In the user sample, users with undisclosed regional information were excluded. In all, 10,869 local tourists and 1042 foreign tourists were included. Local tourists paid more attention to mountain climbing and recreational experiences, transportation time, infrastructure, fare collection, and parent-child activities. The focus of foreign tourists was mainly on the degree of fame of scenic spots, scenery, scenic environment, consumption cost, and traffic time. It can be seen that local tourists paid more attention to the development of recreational activities, while foreign tourists preferred to register their arrival at the scenic spots.

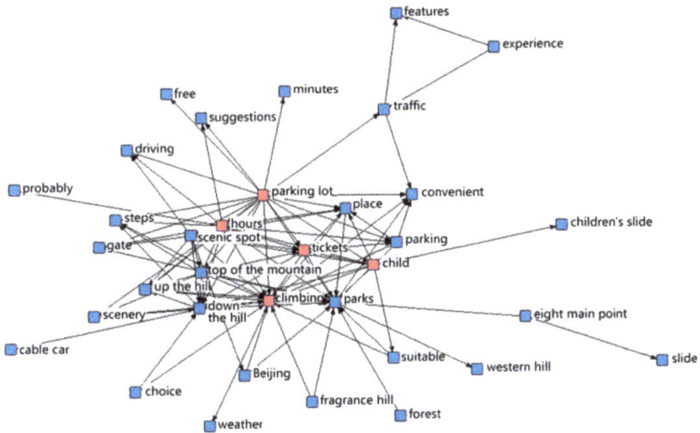

Figure 11. The semantic network of local tourists' evaluations.

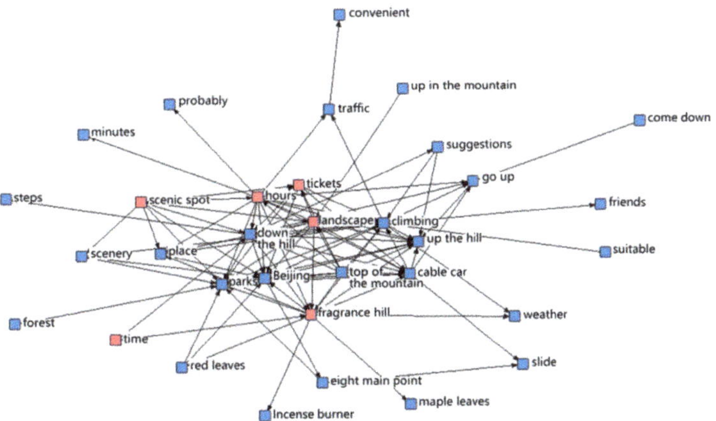

Figure 12. The semantic network of foreign tourists' evaluations.

According to the semantic network analysis of textual data from people of different genders (Figures 13 and 14), male users mainly focused on mountain climbing, infrastructure, and traffic accessibility, and female users paid more attention to the scenic quality, scenic environment, convenient transportation, recreational activities, parent-child experiences, and infrastructure. It can be seen that women paid more attention to recreational experiences than men did.

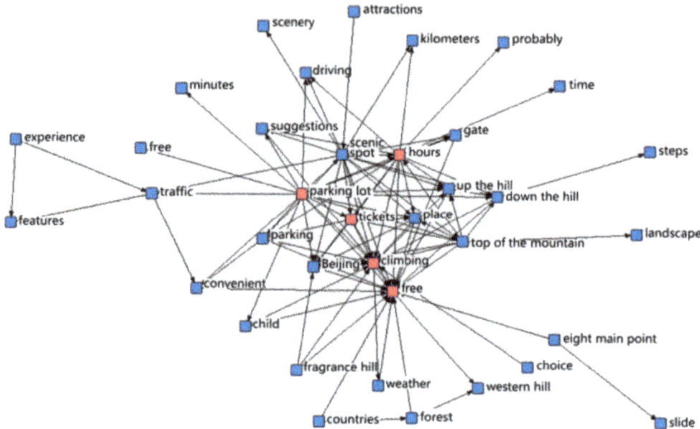

Figure 13. The semantic network of males' evaluations.

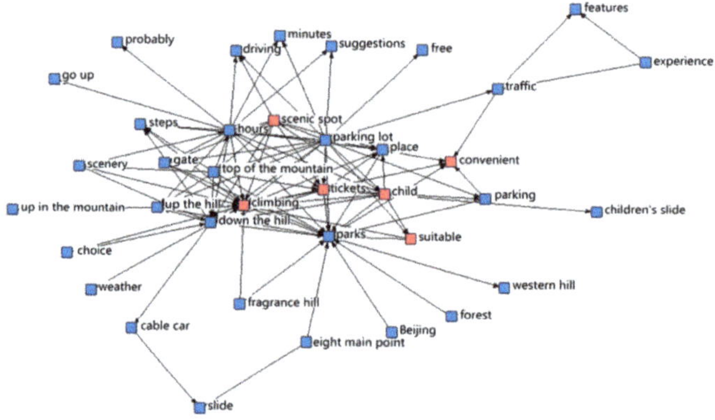

Figure 14. The semantic network of females' evaluations.

3.3.2. Differences in Mountain Landscape Evaluations in Different Geographical Locations

Among the 10 administrative districts in which mountainous landscapes are distributed in Beijing, Haidian District had the highest comprehensive star rating, reaching 4.73 stars, followed by Fengtai District (4.7 stars), Yanqing District (4.65 stars), and Changping District (4.22 stars) (Table 7). The overall evaluation of the mountain landscape was high, which shows that the parks were well-managed and popular with tourists.

For further investigation, the factors that influenced the evaluations of mountainous landscapes—such as the population density, the number of permanent residents, the GDP of the administrative district, the GDP per capita of the administrative district, the average distance from the city center, and the average driving distance from the city center of the administrative district—in each scenic area were analyzed. The correlation analysis using the SPSS tool (Tables 7 and 8) showed that there were no significant correlations between the evaluations of mountainous landscapes and any of these factors. In other words, visitors' evaluations of mountainous landscapes were not related to these external factors, and the internal factors of a scenic area may be more important in influencing the evaluation of a landscape.

The semantic network analysis of the textual evaluations of the mountain landscapes in the Taihang Mountains and Yan Mountains showed that (Figures 15 and 16) the main landscape elements that tourists in the different mountain ranges paid attention to were also different. The analysis showed that the Taihang Mountains mainly have national forest parks, while the Yan Mountains mainly include scenic spots. Tourists paid more attention to the experiences of mountain climbing and recreation in scenic spots when they enjoyed recreation in Taihang Mountains, while tourists visiting the Yan Mountains for recreational activities paid more attention to the traffic accessibility of scenic spots and the time spent in traffic to visit them. The scenic value of the water system landscape in the Yan Mountains was higher, and tourists preferred to go to mountain parks in the Yan Mountains for waterscape viewing and swimming experiences.

Table 7. Factors of different administrative regions were used in the correlation analysis.

	Average Rating Star	Population Density (ppl/km^2)	Number of Permanent Residents (10,000 People)	Per Capita GDP of the District (10,000 CNY)	GDP of the District (1,000,000,000 CNY) [a]	Average Distance from the City Center (km)	Average Driving Time from the City Center (H)
Haidian District	4.73	7515	323.7	26.27310	8504.6	40.2	0.93
Fengtai District	4.70	6628	202.5	9.03506	1829.6	20.0	0.75
Changping District	4.22	1612	216.6	4.94829	1071.8	56.2	1.0
Yanqing District	4.65	173	34.6	5.62138	194.5	79.8	1.7
Fangshan District	4.41	650	131.3	5.78751	759.9	78.5	1.62
Pinggu District	4.57	481	45.7	6.22000	284.1	89.3	1.68
Huairou District	4.44	210	44.1	9.80000	432.6	73.25	2.1
Miyun District	4.32	240	52.7	6.83000	360.3	112.63	1.75
Shijingshan District	4.40	6684	57.0	14.14737	806.4	34	0.75
Mentougou District	4.58	271	39.3	6.84000	268.8	75.67	1.75

[a] One billion yuan.

Table 8. Correlation analysis.

		Average Rating Star	Population Density as per the Latest Yearbook (Ppl/Km^2)	Number of Permanent Residents as per the Latest Yearbook (10,000 People)	per Capita GDP of the District as per the Latest Yearbook (10,000 CNY)	GDP of the District as per the Latest Yearbook (100,000,000 CNY) [a]	Average Distance from the City Center (Km)	Average Driving Time from the City Center (H)
Average rating star	Pearson correlation coefficient	1	0.386	0.221	0.439	0.472	−0.342	−0.124
	Sig. (two-tail)		0.270	0.540	0.204	0.168	0.334	0.732
	Number	10	10	10	10	10	10	10

[a] One hundred million yuan.

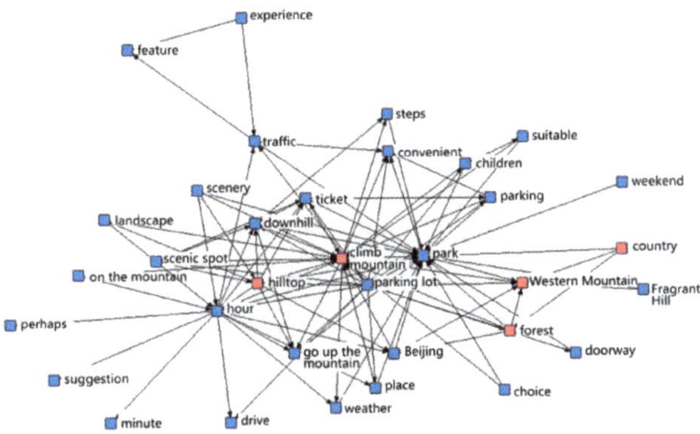

Figure 15. The semantic network for the Taihang Mountain landscape.

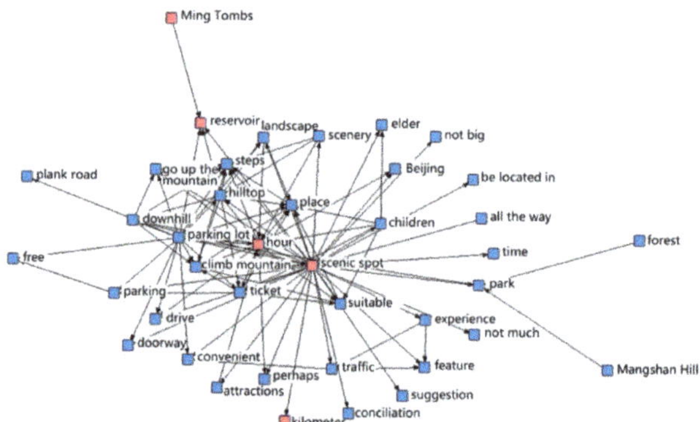

Figure 16. The semantic network for the Yan Mountain landscape.

The semantic network analysis of the textual evaluations of the mountain landscapes in different administrative divisions showed (Figures 17–19) that the mountain landscapes in the different divisions had their own characteristics but they also had similarities. Haidian District has Xishan National Forest Park, which has beautiful scenery, as its main tourist attraction. Shijingshan District's mountain landscape is characterized by Badachu Park, offering rich recreational activities. Fangshan District and Miyun District have similar mountain landscape features, and both are dominated by a natural water system landscape; recreational activities there are characterized by ferrying. The mountain landscape of Fengtai District is mainly characterized by wildlife viewing. Huairou District, Pinggu District, and Yanqing District are rich in the historical sites of mountain landscapes, but they have different characteristics. The mountain landscape in Huairou District is steep and suitable for outdoor activities, such as bungee jumping. The mountain landscape in Pinggu District has a certain religious color. Yanqing District is rich in cave sites with a strong historical background. The Mentougou Mountain landscape is rich in natural scenery, but its characteristics are not obvious. Fangshan District, Miyun District, Fengtai District, Huairou District, and Shijingshan District are all suitable for parent-child activities.

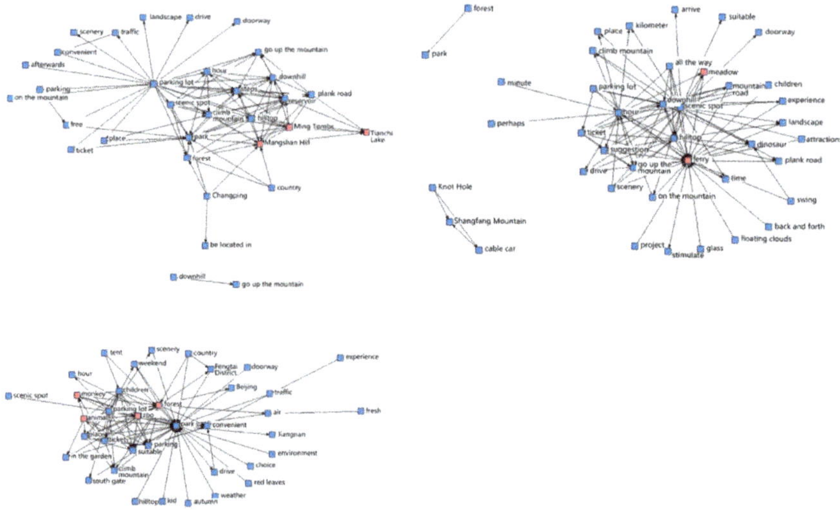

Figure 17. The semantic web of mountain landscapes in various administrative regions (From left to right: Changping District, Fangshan District, and Fengtai District).

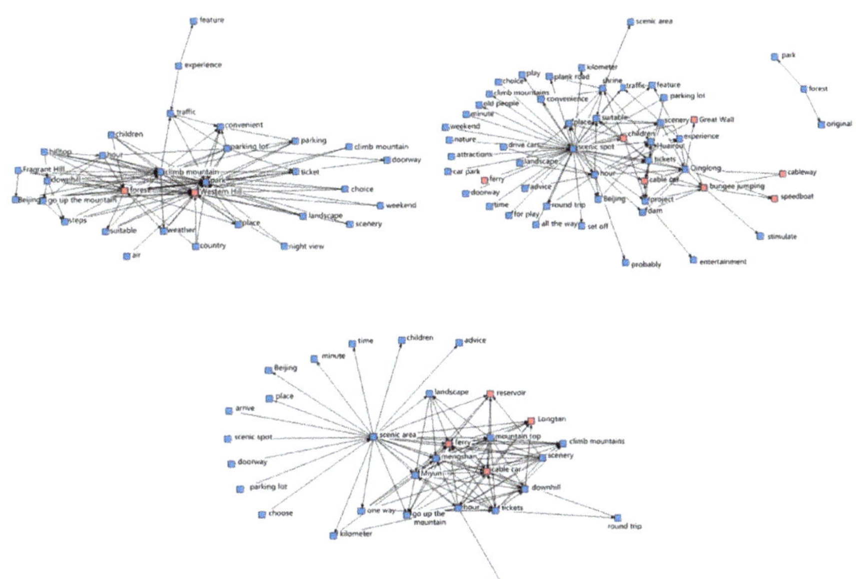

Figure 18. The semantic web of mountain landscapes in various administrative regions (From left to right: Haidian District, Huairou District, and Miyun District).

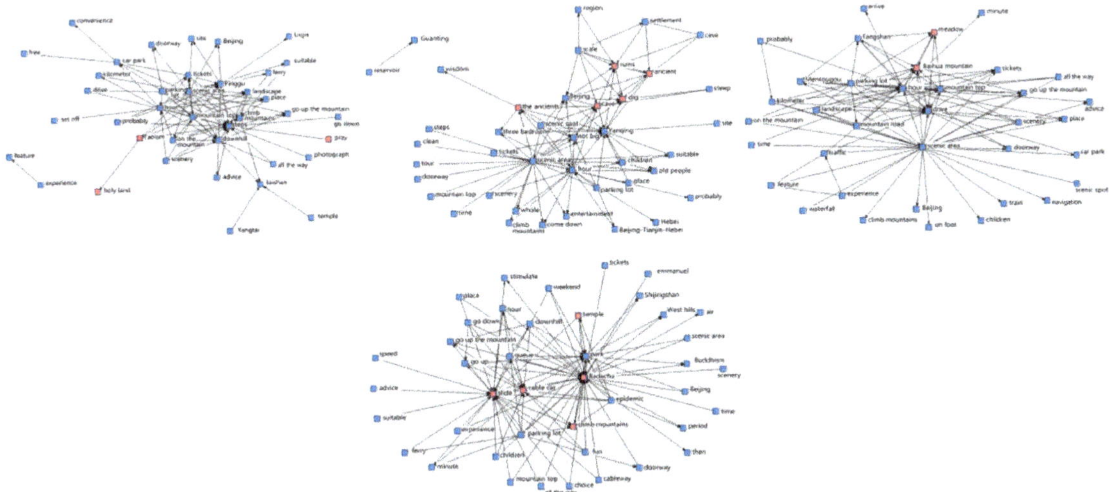

Figure 19. The semantic web of mountain landscapes in various administrative regions (From left to right: Pinggu District, Yanqing District, Mentougou District, and Shijingshan District).

3.4. IPA of Tourist Groups' Recreational Experiences

3.4.1. Analysis of Satisfaction in Terms of Recreation

In a comprehensive IPA of tourists' satisfaction in terms of recreation in mountain landscapes in Beijing (Table 9, Figure 20), most of the indicators were concentrated in the first and second quadrants, indicating that tourists' overall satisfaction with the mountain landscapes was high. The mountain landscapes were shown to have greater advantages in terms of natural landscapes, plant landscapes, outdoor recreational activities, parent-child experiences, and emotional experiences, but public service facilities, traffic accessibility, ticket cost, and park management services needed to be improved.

Table 9. Importance and satisfaction scores of tourists' satisfaction in terms of recreation.

Main Category	Serial No.	Subcategory	Importance	Satisfaction
Landscape quality (A)	A1	Natural landscape	0.1034	0.6669
	A2	Plant landscape	0.0688	0.6982
	A3	Animal landscape	0.0358	0.7259
	A4	Astronomical landscape	0.0068	0.6744
	A5	Historical and cultural landscape	0.0324	0.7017
Recreational activities (B)	B1	Outdoor recreational activities	0.0770	0.6628
	B2	Leisure activities	0.0370	0.6713
	B3	Fitness activities	0.0227	0.6814
	B4	Humanistic activities	0.0020	0.7626
Tourism experience (C)	C1	Ticket cost	0.0527	0.5624
	C2	Parent-child experience	0.0484	0.6826
	C3	Emotional experience	0.1196	0.7087
	C4	Air and environmental quality	0.0318	0.7206
	C5	Sense of crowded space	0.0134	0.5506
Infrastructure (D)	D1	Traffic accessibility	0.1365	0.5885
	D2	Public service facilities	0.0771	0.6132
	D3	Recreation and entertainment facilities	0.0253	0.6641
	D4	Navigation signage system	0.0099	0.5762
	D5	Catering and convenience facilities	0.0082	0.5828
	D6	Safety facilities	0.0268	0.5737
Management services (E)	E1	Facility maintenance	0.0016	0.5093
	E2	Park management services	0.0499	0.6014
	E3	Planning layout	0.0130	0.6738

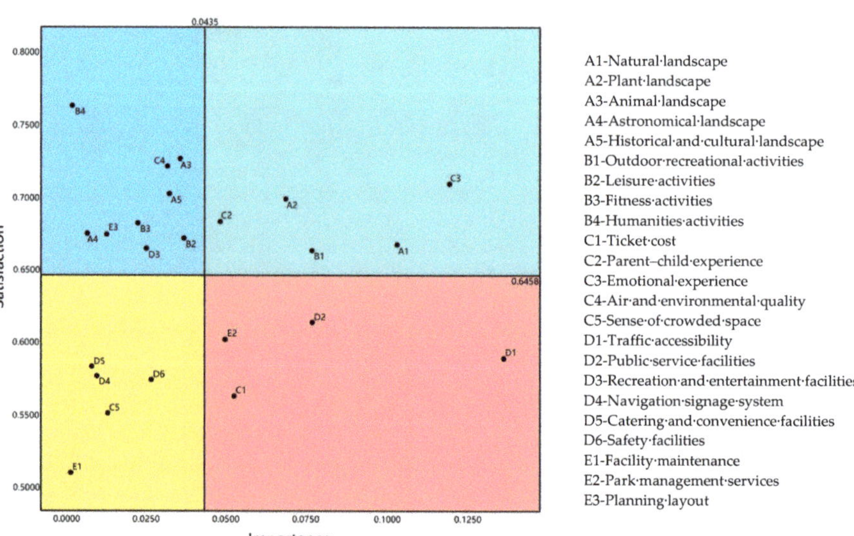

Figure 20. Comprehensive IPA of satisfaction in terms of recreation.

The IPA (Table 10, Figure 21) of mountain landscapes in different mountain ranges showed that tourists were more satisfied with the overall recreation offered by the mountain landscapes in the Taihang Mountains than those offered by the Yan Mountains. Tourists loved the mountain landscapes in the Taihang Mountains in terms of plant landscapes and outdoor recreational activities, while the mountain landscapes in the Yan Mountains were more distinctive in terms of historical and cultural landscapes and leisure activities. In terms of traffic accessibility, public service facilities, ticket cost, and park management services, there was a need for further strengthening and improvement of the facilities. The plant landscapes and the outdoor recreational activities offered in the Yan Mountains had higher importance but a lower satisfaction rating than the average. Therefore, the mountain landscape in the Yan Mountains needs to be improved.

Table 10. Importance and satisfaction scores of tourists' satisfaction in terms of recreation in different mountain systems.

Main Category	Serial No.	Subcategory	Taihang Mountains		Yan Mountains	
			Importance	Satisfaction	Importance	Satisfaction
Landscape quality (A)	A1	Natural landscape	0.0902	0.6655	0.1457	0.6696
	A2	Plant landscape	0.0754	0.7166	0.0479	0.6064
	A3	Animal landscape	0.0409	0.7399	0.0193	0.6309
	A4	Astronomical landscape	0.0078	0.6672	0.0036	0.7241
	A5	Historical and cultural landscape	0.0284	0.6989	0.0452	0.7073
Recreational activities (B)	B1	Outdoor recreational activities	0.0793	0.6750	0.0696	0.6186
	B2	Leisure activities	0.0344	0.6633	0.0456	0.6906
	B3	Fitness activities	0.0246	0.6758	0.0165	0.7078
	B4	Humanistic activities	0.0012	0.7935	0.0044	0.7358
Tourism experience (C)	C1	Ticket cost	0.0521	0.5721	0.0544	0.5328
	C2	Parent-child experience	0.0498	0.6962	0.0437	0.6332
	C3	Emotional experience	0.1176	0.7138	0.1260	0.6933
	C4	Air and environmental quality	0.0331	0.7281	0.0277	0.6921
	C5	Sense of crowded space	0.0149	0.5540	0.0086	0.5314
Infrastructure (D)	D1	Traffic accessibility	0.1363	0.5888	0.1371	0.5874
	D2	Public service facilities	0.0790	0.6139	0.0710	0.6106
	D3	Recreation and entertainment facilities	0.0263	0.6895	0.0219	0.5663
	D4	Navigation signage system	0.0101	0.5667	0.0095	0.6087
	D5	Catering and convenience facilities	0.0075	0.6094	0.0104	0.5219
	D6	Safety facilities	0.0247	0.5774	0.0335	0.5649
Management services (E)	E1	Facility maintenance	0.0507	0.4750	0.0018	0.6047
	E2	Park management services	0.0507	0.6031	0.0474	0.5958
	E3	Planning layout	0.0141	0.6725	0.0093	0.6800

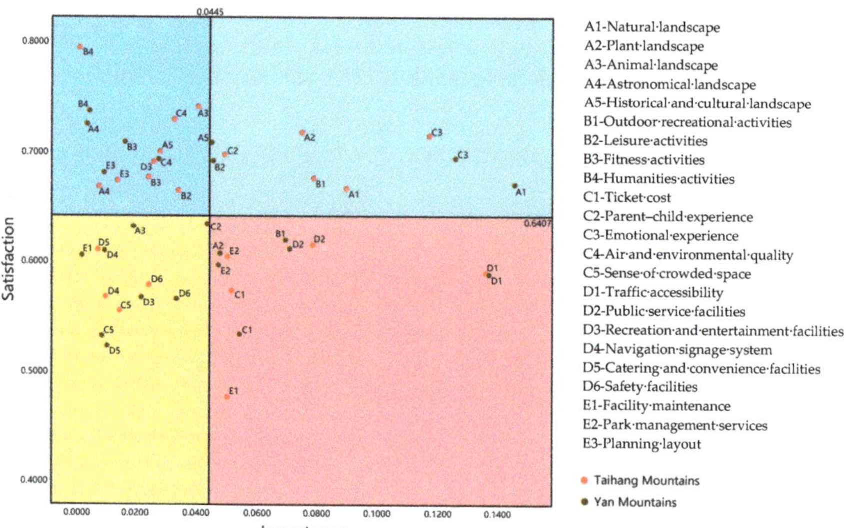

Figure 21. IPA of satisfaction with recreation in different mountain systems.

3.4.2. Analysis of the Experience of the Five Senses

A comprehensive IPA of the experiences of tourists in the mountain landscapes in Beijing in terms of the five senses (Table 11, Figure 22) showed that most of the indicators of the five senses were concentrated in the second and third quadrants, implying that tourists did not pay much attention to the overall sensory experience of the mountain landscapes, and the overall satisfaction of the tourists was high in terms of sight, smell, and touch, but low in terms of taste. Visual and olfactory landscape elements—such as visibility of plants, visibility of animals, landscape recognizability, and smell of air/water—in the first quadrant can continue to be developed; visual and taste elements—such as visibility of natural landscapes, visibility of roads, and food sales—in the fourth quadrant need to be improved and enhanced; and auditory and tactile landscape elements—such as the sounds of plants, sounds of animals, feel of sunlight, feel of water, and touch of the road—in the third quadrant need to be gradually improved.

Table 11. Importance and satisfaction scores of the experience of the five senses.

Main Category	Serial No.	Subcategory	Importance	Satisfaction
Vision (F)	F1	Visibility of plants (F1)	0.2269	0.6973
	F2	Visibility of animals (F2)	0.1179	0.7259
	F3	Visibility of natural landscapes (F3)	0.1538	0.6478
	F4	Visibility of celestial phenomena (F4)	0.0227	0.6743
	F5	Crowd disturbances (F5)	0.0152	0.5548
	F6	Landscape recognizability (F6)	0.1069	0.6962
	F7	Visibility of roads (F7)	0.0547	0.6438
	F8	Others (F8)	0.0355	0.7037
Hearing (G)	G1	Sounds of humans (G1)	0.0111	0.7265
	G2	Sounds of plants (G2)	0.0007	0.6364
	G3	Sounds of animals (G3)	0.0030	0.7500
	G4	Sounds of broadcasts (G4)	0.0052	0.6076
	G5	Sounds of water (G5)	0.0032	0.5859
	G6	Others (G6)	0.0032	0.6701
Smell (H)	H1	Smell of air/water (H1)	0.0582	0.7584
	H2	Smell of plants (H2)	0.0035	0.7570
Touch (L)	L1	Feel of sunlight (L1)	0.0311	0.6474
	L2	Feel of wind (L2)	0.0076	0.7682
	L3	Feel of water (L3)	0.0071	0.5556
	L4	Feel of temperature (L4)	0.0389	0.6747
	L5	Touch of the road (L5)	0.0322	0.6282
	L6	Touch of animals (L6)	0.0175	0.7026
	L7	Others (L7)	0.0013	0.6829
Taste (K)	K1	Food sales (K1)	0.0415	0.6423
	K2	Taste of food (K2)	0.0011	0.6364

For the landscapes of different mountain ranges, the IPA (Table 12, Figure 23) showed that, similarly to the conclusion regarding tourists' satisfaction with recreation, tourists' satisfaction with the mountain landscapes in terms of the five senses was higher for the Taihang Mountains than for the Yan Mountains. The Taihang Mountains were characterized by the visibility of plants, the visibility of animals, landscape recognizability, the smell of air/water, and other visual and olfactory landscape elements that were highly valued, while the tourists in the Yan Mountains attached more importance to the visibility of natural landscapes, landscape recognizability, the visibility of roads, smell of air/water, and the feel of temperature, as well as other visual, olfactory, and tactile landscape elements. That is, tourists in both mountain landscapes attached more importance to visual and olfactory sensory feelings. In terms of sensory elements in urgent need of key improvement, for the Taihang Mountains, more attention should be paid to improving the visibility of natural landscapes, the visibility of roads, and food sales, and for the Yan Mountains, more attention should be paid to the visibility of plants, the visibility of animals, and food sales. Thus, it is clear that both need to improve the sensory experiences of vision and taste.

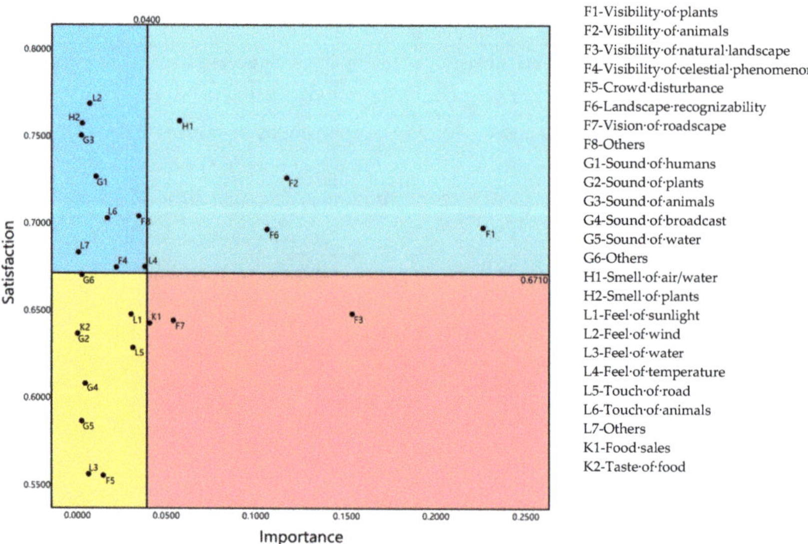

Figure 22. Integrated IPA of the experience of the five senses.

Table 12. Importance and satisfaction scores of the experience of the five senses in different mountain systems.

Main Category	Serial No.	Subcategory	Taihang Mountains		Yan Mountains	
			Importance	Satisfaction	Importance	Satisfaction
Vision (F)	F1	Visibility of plants (F1)	0.2519	0.7166	0.1519	0.6012
	F2	Visibility of animals (F2)	0.1367	0.7399	0.0610	0.6288
	F3	Visibility of natural landscape (F3)	0.1210	0.6418	0.2528	0.6565
	F4	Visibility of celestial phenomena (F4)	0.0265	0.6689	0.0114	0.7126
	F5	Crowd disturbances (F5)	0.0170	0.5703	0.0097	0.4730
	F6	Landscape recognizability (F6)	0.0950	0.6982	0.1428	0.6923
	F7	Visibility of roads (F7)	0.0495	0.6144	0.0704	0.7063
	F8	Others (F8)	0.0360	0.7295	0.0343	0.6221
Hearing (G)	G1	Sounds of humans (G1)	0.0108	0.7631	0.0119	0.6264
	G2	Sounds of plants (G2)	0.0006	0.5385	0.0012	0.7778
	G3	Sounds of animals (G3)	0.0029	0.7727	0.0034	0.6923
	G4	Sounds of broadcast (G4)	0.0023	0.6538	0.0139	0.5849
	G5	Sounds of water (G5)	0.0024	0.6000	0.0058	0.5682
	G6	Others (G6)	0.0033	0.6933	0.0029	0.5909
Smell (H)	H1	Smell of air/water (H1)	0.0607	0.7717	0.0506	0.7106
	H2	Smell of plants (H2)	0.0038	0.7356	0.0026	0.8500
Touch (L)	L1	Feel of sunlight (L1)	0.0319	0.6617	0.0284	0.5991
	L2	Feel of wind (L2)	0.0080	0.7784	0.0063	0.7292
	L3	Feel of water (L3)	0.0059	0.6148	0.0106	0.4568
	L4	Feel of temperature (L4)	0.0367	0.6718	0.0460	0.6818

Table 12. *Cont.*

Main Category	Serial No.	Subcategory	Taihang Mountains		Yan Mountains	
			Importance	Satisfaction	Importance	Satisfaction
Touch (L)	L5	Touch of the road (L5)	0.0328	0.6265	0.0303	0.6336
	L6	Touch of animals (L6)	0.0206	0.7152	0.0084	0.6094
	L7	Others (L7)	0.0013	0.6207	0.0016	0.8333
Taste (K)	K1	Food sales (K1)	0.0416	0.6646	0.0412	0.5746
	K2	Taste of food (K2)	0.0012	0.7037	0.0008	0.3333

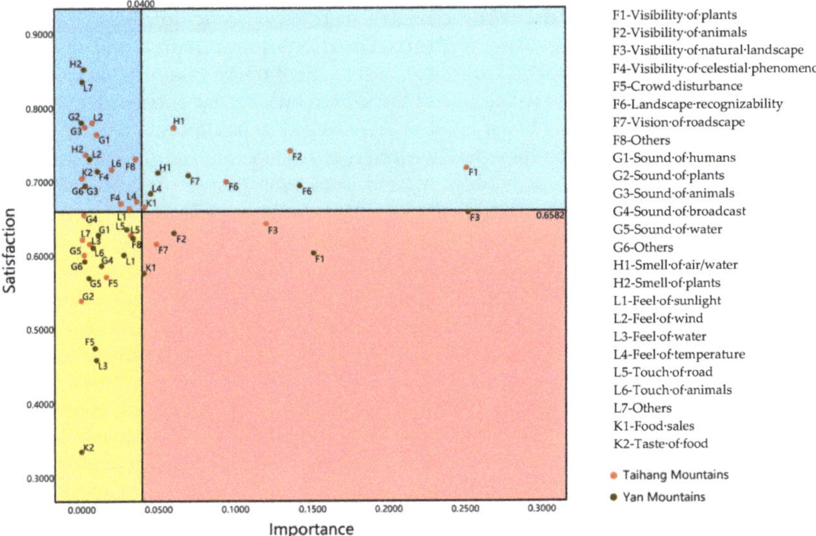

Figure 23. IPA of the experiences of different mountain systems in terms of the five senses.

4. Conclusions

This study investigated the demographic characteristics and landscape preferences of tourists visiting the mountain landscapes in Beijing by classifying photos based on the associated text uploaded by users to social media. In this study, more tourists in mountain landscapes chose a self-driving mode of transportation, which was closely related to the geographical locations of the mountains in the urban countryside. In terms of travel mode, they preferred to travel with friends, classmates, or family members. Regarding landscape preferences, geographical landscapes, buildings, and facilities were the most preferred, followed by biological landscapes and water landscapes; the interest in tourist purchases was the lowest. Tourists' landscape preferences were related to their gender, transportation mode, and travel mode. Men's perception of landscapes was more direct, and they were more inclined toward geographical landscapes and biological landscapes, while women were more emotional, preferring to record the beautiful scenery and personal activities during the trip. The influence of different transportation modes on landscape preferences is not obvious, and the difference is small. Tourists who traveled with friends and classmates preferred astronomical phenomena, climatic landscapes, and tourist shopping. Compared to other sources, the three sources selected in this paper are more representative and provide more images with evaluative significance. The images presented in guides and blogs tend to be more illustrative. For the purpose of studying tourist destination imagery, prescriptive, evaluative, and normative components are valuable conceptualizations. The present study employs both quantitative and qualitative methods of analysis in a quasi-empirical manner. Using this hybrid approach, factual results based on sample data can be developed alongside interpretive results intended to enhance conceptual understanding. As a result of this study, important scientific and practical implications are generated in terms of theoretical frameworks, techniques, and insights, providing a theoretical basis for the key points and developmental directions of the construction and improvement of mountain landscapes.

In this research and analysis of tourists' evaluative texts, tourists' recreational behaviors in mountain landscapes were focused on seven aspects: climbing, rafting, hiking, fitness, camping, picnics, and sightseeing. There were some differences between the positive and negative factors in the overall evaluation of mountain parks in Beijing. The positive evaluation factors of mountain landscapes were mainly focused on the environment, traffic accessibility, recreational facilities, and parent-child experiences, while the negative factors were mainly reflected in tickets, traffic, infrastructure, and park management services. Therefore, in the construction of mountain landscapes, it is necessary to moderately reduce consumption costs, adjust traffic planning, and improve accessibility, infrastructure construction, and the park management and service level.

Local tourists were found to pay more attention to the development of recreational activities, while foreign tourists preferred to visit scenic spots. In terms of gender differences, women paid more attention to recreational experiences than men, while male tourists paid more attention to the park landscapes themselves. Different mountain ranges and administrative divisions of mountain parks affected tourists' landscape evaluations, and the characteristics of mountain landscapes in different geographical locations could also be reflected through the analysis of the evaluative texts.

An IPA of how satisfied tourists were with the recreation offered by the mountain landscapes and their experiences of the landscapes via the five senses was conducted. The results showed that tourists were satisfied with the overall recreation offered by the mountain landscapes. Mountain landscapes offer natural landscapes, plant landscapes, outdoor recreational activities, and parent-child activities, which can be satisfying in terms of tourists' emotional experiences. However, there is an urgent need to improve and upgrade public service facilities, transport accessibility, ticket costs, and park management services. It is necessary to improve park infrastructure construction, optimize the road transport system, improve accessibility, reduce consumption costs, develop management methods, and upgrade park management services. However, in terms of the five senses, tourists did not pay much attention to the overall sensory experience of mountain landscapes. Compared with other aspects, vision played a dominant role in the perceptual experience, followed by smell; the perceptions of touch, hearing, and taste were low. Therefore, more attention should be paid to the creation of the sense of a landscape in the construction of scenic spots.

A comparative evaluation of Beijing's mountain landscapes based on social media data can provide comprehensive information on the advantages of mountain landscapes, thus promoting their development and the construction of a forest city. Accommodation facilities, however, remain an important tourist resource that should be considered. An analysis of tourists' landscape preferences can help the government and managers of scenic spots manage, plan, and promote tourism in a more targeted manner. The analysis of tourists' evaluations can help managers grasp the current situation, advantages, and disadvantages of scenic spots and decide quickly and intuitively on a developmental direction. These data are of great significance for the planning, design, construction, development, and management of different mountain landscapes and can help the government and managers analyze the advantages of local landscape resources, make up for the shortcomings, improve the service level, and environmentally improve the park and landscape quality according to the aesthetic preferences of different tourist groups, making Beijing's mountain landscapes more attractive for sightseeing and enhancing their competitive advantage in the future.

As this research environment is unique, some limitations must be acknowledged. The theoretical and methodological frameworks adopted are useful for conceptualizing specific phenomena within specific cultural contexts (e.g., Chinese information sources). Nevertheless, scholars should exercise caution when extrapolating these findings to other populations. It may be necessary to repeat this experiment with other sources or methods in order to determine more predictable results. It is also recommended to more deeply examine the pattern of target images and the meaning of the images. Researchers can, for example, examine in greater depth the prescribed dimensions of destination imagery for short-haul tourists.

As a result of the nuances of this study and the cases of tourists from various countries visiting Chinese destinations, future research can take several directions. It will be interesting to identify the challenges that local tourism marketers may encounter in gaining access to the information sources preferred by tourists in China and abroad. It is also possible to disseminate the marketing strategies of tourist attractions through various channels, such as official websites, social media, and other promotional channels. Future research on the tourism image of Beijing, China can therefore refer to other news and online platforms at home and abroad to gain a more comprehensive understanding of the produced image. As a final point, it is necessary to clarify how previous experiences at these tourist attractions affect tourists' use of destination images and information sources.

Author Contributions: Conceptualization, T.D.; methodology, T.D.; software, T.D.; validation, W.S., Y.W. and R.Y.; formal analysis, T.D.; investigation, T.D.; writing—original draft preparation, T.D.; writing—review and editing, X.G.; visualization, T.D.; supervision, X.G.; project administration, X.G.; funding acquisition, X.G. All authors have read and agreed to the published version of the manuscript.

Funding: This research was funded by the projects of the National Natural Science Foundation of China, (grant number: 31800606), Beijing Municipal Social Science Foundation of China (grant number: 21JCC094), and Beijing Scientific Research and Postgraduate Education Jointly Construction (grant number: 2015BLU REE01).

Institutional Review Board Statement: Not applicable.

Informed Consent Statement: Not applicable.

Data Availability Statement: All data generated or analyzed during this study are included in this article.

Conflicts of Interest: The authors declare no conflict of interest.

References

1. Tian, M.; Ming, Q. Hotspots, progress and enlightenments of foreign mountain tourism research. *World Reg. Stud.* **2020**, *29*, 1071–1081.
2. Kotler, P. *Marketing for Hospitality and Tourism*, 5th ed.; Pearson Education: Chennai, India, 2018.
3. Higham, J.; Thompson-Carr, A.; Musa, G. Activity, People and Place. In *Mountaineering Tourism*; Musa, G., Higham, J., ThompsonCarr, A., Eds.; Routledge: New York, NY, USA, 2015; pp. 1–15.
4. Cavagnaro, E.; Staffieri, S. A study of students' travellers values and needs in order to establish futures patterns and insights. *J. Tour. Futures* **2015**, *1*, 94–107. [CrossRef]
5. Hopkins, D. Destabilising automobility? The emergent mobilities of generation Y. *Ambio* **2016**, *46*, 371–383. [CrossRef] [PubMed]
6. Miller, D.; Merrilees, B.; Coghlan, A. Sustainable Urban Tourism: Understanding and Developing Visitor pro-Environmental Behaviors. *J. Sustain. Tour.* **2015**, *23*, 26–46. [CrossRef]
7. Olsen, J.E.; Thach, L.; Nowak, L. Wine for My Generation: Exploring How US Wine Consumers are Socialized to Wine. *J. Wine Res.* **2007**, *18*, 1–18. [CrossRef]
8. Schoolman, E.D.; Shriberg, M.; Schwimmer, S.; Tysman, M. Green Cities and Ivory Towers: How Do Higher Education Sustainability Initiatives Shape Millennials'Consumption Practices? *J. Environ. Stud. Sci.* **2016**, *6*, 490–502. [CrossRef]
9. Giachino, C.; Truant, E.; Bonadonna, A. Mountain Tourism and Motivation: Millennial Students' Seasonal Preferences. *Curr. Issues Tour.* **2019**, *23*, 2461–2475. [CrossRef]
10. Marine-Roig, E. Destination image analytics through traveller-generated content. *Sustainability* **2019**, *11*, 3392. [CrossRef]
11. Tieskens, K.F.; Van Zanten, B.T.; Schulp, C.J.; Verburg, P.H. Aesthetic Appreciation of the Cultural Landscape Through Social Media: An Analysis of Revealed Preference in the Dutch River Landscape. *Landsc. Urban Plan.* **2018**, *177*, 128–137. [CrossRef]
12. Kiatkawsin, K.; Han, H. Young Travelers' Intention to Behave Pro-Environmentally: Merging the Value-Belief-Norm Theory and the Expectancy Theory. *Tour. Manag.* **2017**, *59*, 76–88. [CrossRef]

13. Oliveira, T.; Araujo, B.; Tam, C. Why Do People Share Their Travel Experiences on Social Media? *Tour. Manag.* **2020**, *78*, 104041. [CrossRef]
14. Zhang, Y.; Qiu, J.; Shao, L.; Cui, J.; Meng, X. The Structure of Pull Motivations of Rural Tourism and the Segmentation of Rural Tourists. *J. Arid. Land Resour. Environ.* **2014**, *28*, 191–196.
15. Pettebone, D.; Newman, P.; Lawson, S.R.; Hunt, L.; Monz, C.; Zwiefka, J. Estimating Visitors' Travel Mode Choices along the Bear Lake Road in Rocky Mountain National Park. *J. Transp. Geogr.* **2011**, *19*, 1210–1221. [CrossRef]
16. Chen, Y.Y.; Cheng, A.J.; Hsu, W.H. Travel Recommendation by Mining People Attributes and Travel Group Types from Community-Contributed Photos. *IEEE Trans. Multimed.* **2013**, *15*, 1283–1295. [CrossRef]
17. Dang, A.; Zhang, D.; Li, J.; Xu, J. Review of urban and rural landscape planning and design research based on spatio-temporal big data. *Chin. Gard.* **2018**, *34*, 5–11.
18. Dang, A.; Xu, J.; Tong, B.; Li, J.; Qian, F. Research Progress of the Application of Big Data in China's Urban Planning. *China City Plan. Rev.* **2015**, *24*, 24–30.
19. Zhao, Y.; Qin, B.; Liu, T. Text Sentiment Analysis. *J. Softw.* **2010**, *21*, 1834–1848. [CrossRef]
20. Do, Y. Valuating Aesthetic Benefits of Cultural Ecosystem Services Using Conservation Culturomics. *Ecosyst. Serv.* **2019**, *36*, 100894. [CrossRef]
21. Evenson, K.R.; Jones, S.A.; Holliday, K.M.; Cohen, D.A.; McKenzie, T.L. Park Characteristics, Use, and Physical Activity: A Review of Studies Using Soparc (System for Observing Play and Recreation in Communities). *Prev. Med.* **2016**, *86*, 153–166. [CrossRef]
22. Liu, Y.; Bao, J.; Zhu, Y. Research on the Emotional Evaluation Method of Tourist Destination Based on Big Data. *Geogr. Res.* **2017**, *36*, 1091–1105.
23. Shao, J.; Chang, X.; Zhao, Y. A Study on Tourist Behavior Pattern of Huashan Scenic Spot Based on Travel Big Data. *Chin. Gard.* **2018**, *34*, 18–24.
24. Liu, Y.; Bao, J.; Chen, K. A Study on the Emotional Characteristics of Chinese Tourists to Australia: A Text Analysis Based on Big Data. *J. Tour.* **2017**, *32*, 46–58.
25. Pan, B.; Fesenmaier, D.R. Semantics of Online Tourism and Travel Information Search on the Internet: A Preliminary Study. 2002, pp. 320–328. Available online: https://www.researchgate.net/publication/228605286_Semantics_of_Online_Tourism_and_Travel_Information_Search_on_the_Internet_A_Preliminary_Study (accessed on 12 September 2022).
26. Zhang, W.; Du, X. Exploring Mainland Tourists' Perception of Taiwan Province's Tourist Destination Image: Based on the Content Analysis of Online Travel Notes. *J. Beijing Int. Stud. Univ.* **2010**, *32*, 75–83.
27. Xiong, W.; Guo, Y. Text Mining of Online Reviews of Hotel Customers. *J. Beijing Int. Stud. Univ.* **2013**, *35*, 38–47.
28. Sinclair, M.; Ghermandi, A.; Sheela, A.M. A Crowdsourced Valuation of Recreational Ecosystem Services Using Social Media Data: An Application to a Tropical Wetland in India. *Sci. Total Environ.* **2018**, *642*, 356–365. [CrossRef] [PubMed]
29. Fisher, D.M.; Wood, S.A.; White, E.M.; Blahna, D.J.; Lange, S.; Weinberg, A.; Tomco, M.; Lia, E. Recreational Use in Dispersed Public Lands Measured Using Social Media Data and On-Site Counts. *J. Environ. Manag.* **2018**, *222*, 465–474. [CrossRef]
30. Mancini, F.; Coghill, G.M.; Lusseau, D. Using Social Media to Quantify Spatial and Temporal Dynamics of Nature-based Recreational Activities. *PLoS ONE* **2018**, *13*, e0200565. [CrossRef]
31. Kádár, B.; Gede, M. Where do Tourists go Visualizing and Analysing the Spatial Distribution of Geotagged Photography. *Cartogr. Int. J. Geogr. Inf. Geovis.* **2013**, *48*, 78–88. [CrossRef]
32. Carlos, G.P.J.; Gutierrez, J.; Minguez, C. Identification of Tourist Hot Spots Based on Social Networks: A Comparative Analysis of European Metropolises Using Photo-Sharing Services and GIS. *Appl. Geogr.* **2015**, *63*, 408–417.
33. Kisilevich, S.; Krstajic, M.; Keim, D.; Andrienko, N.; Andrienko, G. Event-Based Analysis of People's Activities and Behavior Using Flickr and Panoramio Geotagged Photo Collections. In Proceedings of the 2010 14th International Conference Information Visualisation, London, UK, 26–29 July 2010; pp. 289–296.
34. Martilla, J.A.; James, J.C. Importance-Performance Analysis. *J. Mark.* **1977**, *41*, 77–79. [CrossRef]
35. Gu, X. Research on the Structure, Behavior and Demand Characteristics of Tourists in Shanghai Urban Parks and Their Influencing Factors. Master's Thesis, East China Normal University, Shanghai, China, 2013.
36. Yu, B.; Xie, C.; Yang, S.; Che, S. Corresponding Analysis of Residents' Perceived Satisfaction and Importance of Recreation in Shanghai Urban Community Parks. *Chin. Gard.* **2014**, *30*, 75–78.
37. Fan, Y.; Mao, D.; Zhou, C.; Ye, J.; Chen, L.; Zheng, Y. Evaluation of Recreational Resources in Fuzhou West Lake Park Based on Web Text Analysis. *China Urban For.* **2019**, *17*, 41–46.
38. Wang, M.; Qiu, M.; Wang, J.; Peng, Y. Analysis and Optimization of Supply and Demand Relationship of Cultural Ecosystem Services in Waterfront Space of Suzhou, Shanghai Based on Importance-Performance Analysis. *Landsc. Archit.* **2019**, *26*, 107–112.
39. Liang, H.; Wang, Y.; Liu, M. IPA Analysis of Tourists' Perception of Local Food Experience in Tourist Destinations-Taking Enshi Prefecture, Hubei Province as an Example. *J. Agric. For. Econ. Manag.* **2016**, *15*, 335–342.
40. Chen, X. Revision of IPA Analysis Method and Its Application in Tourist Satisfaction Research. *J. Tour.* **2013**, *28*, 59–66.
41. Chen, P.Z.; Liu, W.Y. Assessing Management Performance of the National Forest Park Using Impact Range-Performance Analysis and Impact-Asymmetry Analysis. *For. Policy Econ.* **2019**, *104*, 121–138. [CrossRef]
42. Go, F.; Zhang, W. Applying Importance-Performance Analysis to Beijing as an International Meeting Destination. *J. Travel Res.* **1997**, *35*, 42–49. [CrossRef]

43. Evans, M.R.; Chon, K.S. Formulating and Evaluating Tourism Policy Using Importance-Performance Analysis. *Hosp. Educ. Res. J.* **1989**, *13*, 203–213.
44. Sever, I. Importance-Performance Analysis: A Valid Management Tool? *Tour. Manag.* **2015**, *48*, 43–53. [CrossRef]

land

Article

Research on the Satisfaction of Beijing Waterfront Green Space Landscape Based on Social Media Data

Siya Cheng, Zheran Zhai, Wenzhuo Sun, Yuan Wang, Rui Yu and Xiaoyu Ge *

School of Landscape Architecture, Beijing Forestry University, Beijing 100083, China
* Correspondence: gexiaoyu@bjfu.edu.cn

Abstract: Urban blue–green space is essential to the normal functioning of the urban landscape ecosystem, and it is also a significant metric for assessing the quality of urban human settlements. In China's territorial space planning, the overall planning strategy's implementation depends on constructing the blue–green space network in the urbanized construction area. This paper used 85 typical riverside parks in Beijing's blue–green space as the research object, collecting and analyzing multiple social media user data. It explored the main factors that influenced people's satisfaction with the landscape design and sensory perception of urban waterfront green space from the perspectives of parks beside different river systems, parks of different types, and parks in different districts. The distinction between urban waterfront green space evaluation was further discussed through variance analysis. The research revealed the following findings: (1) by comparing the total number of park reviews in different seasons, it could be observed that tourists evidently preferred the spring landscape, and the winter landscape construction of waterfront green space needs to be improved. (2) By comparing the review stars of different parks, it could be observed that tourists appreciated parks with multiple functions, excellent recreation facilities, complete management services and parks close to the city center. Functions and services became important influencing factors for park evaluation. (3) There was room for improvement in water ecology in the river landscapes of parks adjacent to various river systems, and people paid more attention to the level of service facilities. (4) According to different categories of parks, people's demand for service facilities, activity organization, cultural displays and other aspects was different. (5) Among parks in different districts, people preferred the distinctive animal and plant landscapes and recreational activities of parks in districts on the outskirts of the city. According to the conclusions, suggestions were made for optimizing and improving Beijing's waterfront green space, providing managers with technical support and a basis for decision-making.

Keywords: riverside park; social media; landscape design satisfaction; sensory perception satisfaction; importance–performance analysis

Citation: Cheng, S.; Zhai, Z.; Sun, W.; Wang, Y.; Yu, R.; Ge, X. Research on the Satisfaction of Beijing Waterfront Green Space Landscape Based on Social Media Data. *Land* **2022**, *11*, 1849. https://doi.org/10.3390/land11101849

Academic Editors: Cecilia Arnaiz Schmitz, Nicolas Marine and María Fe Schmitz

Received: 13 September 2022
Accepted: 17 October 2022
Published: 20 October 2022

Publisher's Note: MDPI stays neutral with regard to jurisdictional claims in published maps and institutional affiliations.

1. Introduction

China's "Notice of the Ministry of Natural Resources on Comprehensively Carrying out Territorial Space Planning Work" issued in 2019 stipulates that "the control scope and balanced distribution requirements of urban structural green space, water bodies and other open spaces within the urban development boundary" [1]. It indicates that, under the new context of implementing the strategy of ecological civilization construction, territorial space planning will constantly reinforce the construction strategy of ecological space development and protection pattern. As a result, the overall planning orientation will shift from urban expansion to ecological constraint [2], and the blue–green space system as the urban ecological space network will receive significant attention.

Traditional ecological space consists primarily of green space made up of parks and green corridors and blue space made up of rivers and wetlands [3]. Changes in the global climate and the acceleration of urbanization have spawned strategies and methods to ensure the safety of urban rainwater, such as sponge cities. The combination of blue space, such as rivers and lakes, and green space has also received more attention [4,5]. Therefore, as an essential component of the urban landscape ecosystem, blue–green space can maintain the regular operation of the urban landscape ecosystem. Moreover, it improves citizens' high quality of life [6–8], which positively affects residents' physical and mental health [9]. Urban livability, social stability, and economic prosperity depend on in-depth research on improving urban blue–green ecological space [6,7]. Currently, ecological functions, such as microclimate regulation [10], rainwater storage [11], and health promotion [12] of waterfront green space in blue–green space, have been extensively studied by scholars domestically and internationally. However, few scholars have placed them in the context of the era of sustainable urban construction and development.

Beijing contains more than 200 rivers that are part of the following 5 major river systems of the Haihe River Basin: the Daqing River System in the southwest, the Yongding River System in the west and south-central region, the Wenyu River-North Canal River System in the central and southeast region, the Chaobai River System in the northeast and eastern region, and the Ju River-Jiyun River System in the eastern region [13]. The urban water system and adjacent green space are intertwined, forming a diverse public open space network that ultimately constitutes Beijing's blue–green space layout. As a vital component of urban blue–green space, riverside parks are important nodes that connect Beijing's natural ecological and cultural recreation corridors and essential objects for people's evaluations of urban blue–green space [14]. To improve the quality of urban blue–green space, it is essential to evaluate the function evaluation and optimization direction of urban riverside parks.

Most current studies that evaluate the green space and park landscape along the waterfront are based on traditional research. Numerous issues, such as high cost, limited content categories, small data sample size, and insufficient collection time, limited the universality and precision of the research results [15–18]. As the popularity of network information technology has increased, so have the opportunities for citizens to participate in the evaluation of urban parks. Many valuable and diverse comments are posted by users on various social media platforms on the Internet [19], which serve as comprehensive data sources for assessing users' emotions and opinions on events [20]. Numerous scholars [21–23] have identified this data information as a tool for public participation in practice due to its diverse content categories, large data sample size, and intense immediacy. Currently, social media data are separated into two categories for various natural landscape studies. On the one hand, it uses photos with geographic location information to predict the number of visits to infer people's travel preferences, in combination with the actual situation of the site. It guides space planning based on public information feedback [24–26]. On the other hand, it collects text data and obtains the image description of the destination [27], the usage profile of the place [28,29], and other elements after conducting a thorough analysis. By using network social media as a platform to conduct various research and understand the city with bottom-up ideas, people can participate in the planning and optimization of urban space. This aspect is more conducive to building a livable environment that meets people's requirements [30–32], thereby increasing people's sense of happiness and fostering a harmonious and stable society.

The realization of urban space's service value heavily depends on people's sensory experiences. Consequently, based on the ecological principle of landscape sensibility, it is possible to fully comprehend the needs of residents by investigating the impact of the status quo of blue–green space on individual residents. The ecological principle of landscape sense aims for sustainable development and studies land use planning, construction, and management from the perspectives of natural factors, physical perception, psychological perception, and the social economy [33,34]. Through the evaluation of the waterfront green space in the urban blue–green space, it is crucial to analyze the optimization strategy of park functions from the two perspectives of the park's landscape design factors and visitors' sensory perception factors.

Using social media data to study the characteristics of people's recreational experiences can quickly and comprehensively understand people's needs and satisfaction levels for various factors of urban riverside parks [35]. It helps city managers create famous waterfront landscapes to maintain and improve urban ecosystem services [36] and closely connect people and urban ecosystem services. However, from the users' perspective, comparative research on the service function evaluation of waterfront green space in Chinese urban blue–green space through quantitative analysis of massive text data is lacking.

Given this research gap, we chose Beijing, with its numerous river systems, as a case study for the investigation. Beijing, the capital of China, is one of the world's megacities. It was founded and flourished on the water. The urban development essentially followed the pattern of "water system–Garden–Imperial City–capital city". The central area of ancient Beijing was a collection of lakes, and the city grew gradually on the ancient Yongding River Ferry [37]. Consequently, the problems faced by Beijing's blue–green space are not only universal but also typical and, in many ways, unique. In addition, the current urban construction in Beijing has changed from incremental development mode to stock renewal mode, and urban renewal will become the hot spot and focus in the future for a long period of time [38]. In recent years, significant exploration and practice of urban renewal has been carried out in Beijing, but it mostly focuses on old residential areas, old buildings, old factories and other types, and less attention is paid to the riverside green space, riverside park and other kinds of blue–green space near old residential areas. As an important part of the sustainable development of urban planning, the renewal and construction of blue–green space is also very necessary.

This paper aimed to explore the core landscape value of waterfront green space in terms of tourists' recreation and perception, and excavate the key points to be improved in the planning and construction of existing parks, in order to play a certain reference role in the function optimization of Beijing riverside parks, so that the waterfront landscape in Beijing's blue–green space can meet the needs of ecological humanities and the requirements of the new era of China. Based on different classification criteria and from the perspectives of recreational elements and perceptional elements in the blue–green space, this paper analyzed the factors that affect the evaluation satisfaction of various riverfront parks by using the text data analysis method of social media, and compared the evaluation results of 85 riverfront parks in the blue–green space of Beijing from multiple perspectives. The following four characteristics were identified:

(1) Compare the number of comments at different times based on the collected social media text data;
(2) Explore the overall evaluation of parks in different water systems, categories and locations based on the weighted average of rating stars;
(3) Study the factors that influence people's satisfaction with landscape design and sensory perception of various parks based on importance–performance analysis;
(4) The differences in the satisfaction of park evaluation factors between parks adjacent to different river systems, parks of different types, and parks in different districts, based on one-way analysis of variance.

2. Literature Review

Cities, water systems and green space are closely related to each other. Water systems are an important organization system of urban open space [39]. Waterfront green space, under the organization of urban water systems, constitutes an urban open space system in which blue and green spaces are interwoven, which results in social and economic benefits, as well as important ecosystem service functions in the city. Since the popularity of urban blue and green space is increasing, a large number of investigations and studies have shown that urban blue and green space has public benefits, such as the positive role of promoting public health. Blue–green spaces have played a role in the health of those living in Japanese megacities by providing a place to reduce stress and mental strain during the coronavirus pandemic [40]. Knight et al. [41] have proven the necessity of improving the ecological quality of blue and green space to enhance residents' life satisfaction.

The evaluation of blue–green space and waterfront green space provided valuable directions and suggestions for blue–green space planning and urban green space renewal. Some scholars have studied and analyzed the environmental quality, aesthetic quality, aging suitability and other aspects of blue and green space and waterfront green space. Mishra et al. [42] proposed the Blue Health Environmental Assessment tool (BEAT) to evaluate the multiple environmental factors that influence people's access to, use of, and health promotion of blue spaces, in order to support evidence-based planning for urban blue space development as a public health resource. Subiza Perez et al. [43] developed a Perceived Environmental Aesthetic Qualifications Scale (PEAQS), which surveyed 331 respondents in 3 locations and summarized the conditions under which people have aesthetic perceptions of blue–green space. Min et al. [44] proposed that environment, function and transportation were the most important factors that affect the elderly's overall satisfaction with waterfront open space through observation of and interviews with the elderly, providing a reference for urban blue and green space to meet the needs of the elderly and improve the service level.

First proposed by Martilla and James in 1977, importance–performance analysis (IPA) was used to compare customers' pre-consumption expectations and post-consumption satisfaction, as well as the customer's evaluation of the performance status of each product attribute [45,46]. The application of IPA to the service industry began in the early 1990s [47], including the evaluation of specific service satisfaction [48], the study of the attractiveness of designated areas [49], and the formulation of tourism policy [50]. In recent years, IPA has been utilized extensively in landscape architecture to examine the characteristics of visitor demand [51] and the recreation situation [16,52]. Liu et al. [53] used IPA to identify the most pressing problems in Shanghai's public spaces on both sides of the Pujiang River and to determine the direction of improvement. In the study of the tourism experience of Haizhu National Wetland Park, Lin et al. [54] used IPA to find out the imperfect aspects of park construction and service, and proposed improvement strategies for the park to improve tourists' satisfaction. Zheng et al. [55] constructed a perception evaluation model based on sentiment analysis and IPA to evaluate the landscape perception of Beijing Yuyuantan Park.

One-way analysis of variance, or ANOVA, is a statistical method widely used to analyze experimental data. In essence, ANOVA is a hypothesis test. It analyzes and com-pares the data fluctuations under a certain influencing factor to infer whether there is a significant difference between the population. If there is a significant difference, it indicates that the influence of this factor is significant [56]. ANOVA is also widely used in landscape architecture to infer whether two things are related. When studying the relationship between landscape configuration characteristics and the urban heat island effect, Connors et al. [57] used ANOVA to test whether different land uses were significantly correlated with different landscape indicators and land surface temperature. Gao et al. [58] used ANOVA to infer whether there were differences in visual behaviors of visitors observing different types of forest landscape space.

The research in this paper conducted IPA on the recreation elements and perception elements of urban waterfront green space based on social media texts, and made scientific and detailed adjustments and improvements to the index system of the IPA of recreation elements, so as to make it more consistent with the spatial characteristics of urban waterfront green space and the evaluation text of tourists. In terms of tourist perception, the evaluation model of tourist perception elements in this study added indicators such as water sound, water and wind contact, which made up for the shortcomings of the previous evaluation index system of landscape perception. In order to further explore whether the scores of riverfront parks are correlated with different classifications of riverfront parks, SPSS tools were used to conduct a one-way analysis of variance. If the results showed correlation, multiple comparison analyses for each index factor were carried out to investigate the differences between index factors in various blue–green spaces in detail.

3. Materials and Methods

3.1. Study Area

According to our survey, at present, there are problems in Beijing riverside park, such as its single function, inconvenient transportation, slow driving system, and lack of waterfront activity space. There is a strong demand for waterfront environment quality improvement and facility function supplement [59]. In addition, the Beijing Urban Master Plan (2016–2035) proposed to build a green space structure of "one screen, three rings, five rivers and nine wedges" in the city [60], in order to strengthen water ecological environment governance, restore the ecological functions of rivers and lakes, and ensure a good urban ecological environment. Therefore, this study selected 85 parks as the research object (Figure 1), which are an important part of the blue–green space in Beijing and are adjacent to the following 5 major water systems: Yongding River, Daqing River, North Canal River, Chaobai River and Jiyun Canal water systems, and compared them with three social networks, Dianping [61], Ctrip [62], and Mafengwo [63].

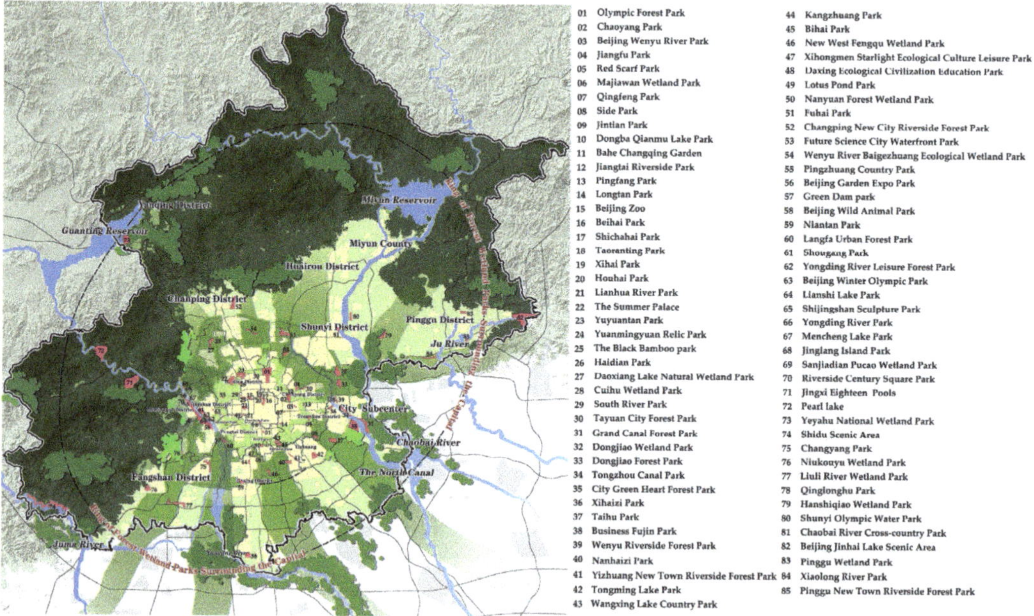

Figure 1. The distribution of riverside parks in 85 blue–green spaces.

In addition, the parks were divided into seven categories based on the park directory information officially released by the Beijing Municipal Bureau of Landscape and Afforestation [64]. Comprehensive parks refer to parks with complete functions, complete facilities and diverse content, which can meet the needs of different people for recreation. Community parks refer to parks where residents in a certain area of residential land carry out daily leisure activities nearby. Historical parks refer to garden scenes that can reflect garden building techniques in a specific historical period and have an impact on the change and cultural development of the city. Specific parks are parks that focus on creating special themes and building specific service content with leisure functions. Gardens are smaller parks that are convenient for nearby residents and workers to use. Ecological parks refer to parks that are located outside the urban construction land and assume multiple functions, such as natural landscape displays and popular science education publicity. Natural parks are areas open to the public in the natural protection area system, which have the functions of leisure and popular science education.

The administrative location of the parks was divided into the following three categories based on the planning documents of the Beijing Municipal government: six central districts (Dongcheng District, Xicheng District, Chaoyang District, Haidian District, Fengtai District and Shijingshan District), inner suburbs (Daxing District, Tongzhou District, Shunyi District, Changping District, Mentougou District, Fangshan District), and outer suburbs (Huairou District, Pinggu District, Miyun District, Yanqing District), to ensure the accuracy and recognition of classification.

3.2. Analysis Process

This paper constructed an evaluation framework of typical blue–green spaces in Beijing based on social media data for the study of landscape satisfaction (Figure 2). Firstly, based on the proportion of the reviews on the three platforms, the weighted average of review stars was carried out to obtain the average rating stars and evaluation result of each park. Then, the basic database text was extracted by crawling and screening the data on the open platform of social media. The text segmentation of the data and the list of high-frequency words were extracted, and the indicator elements were formulated and adjusted as the index system of blue–green space experience. The index system was divided into landscape design index elements and sensory perception index elements, and the frequency analysis was carried out to obtain the results of word frequency importance. The IPA evaluation model was constructed based on the importance analysis results obtained from the analysis of the proportion of positive emotion text, and the difference comparison of evaluation results of various spaces was finally formed.

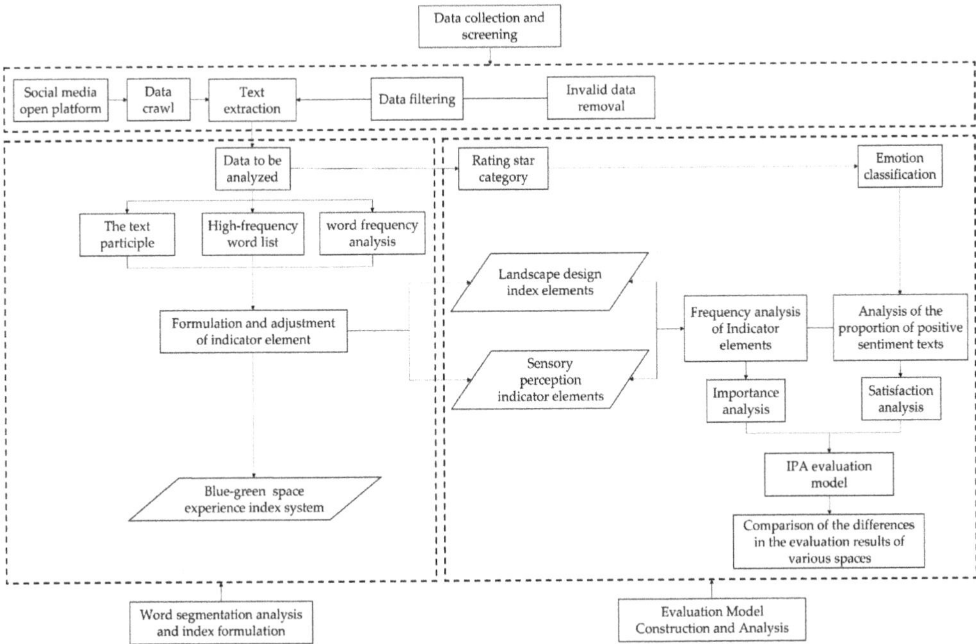

Figure 2. Evaluation framework for the study of landscape satisfaction of typical blue–green spaces in Beijing.

3.3. Data Processing and Index Design

Multiple influential social media platforms in China provided the data for this study, including Dianping, Ctrip, and Mafengwo. The data included the text and image content published by users in a particular space, the gender and region of the rating stars, and other elements. The time period used ranged from 1 January 2006 to 31 December 2021, thereby avoiding the problem of incorrect results due to poor comment information timeliness. In addition, a preliminary screening of the collected information was conducted to obtain valid texts with sufficient information and high approval; finally, 352,837 comments, totaling 51,319,471 words, were obtained. The ratio of foreign visitor reviews to local visitor reviews was 1:6, and the ratio of male visitor reviews to female visitor reviews was 1:3.

Since there is no apparent space between words in written Chinese, we must segment the sequence of Chinese character strings arranged at equal distances to process the information [65]. Therefore, we use tools for word segmentation to analyze the acquired text data and remove any words that lack actual meaning. The research used the ROSTCM6 software to perform word segmentation and word frequency analysis on the text data, and a word frequency table was generated.

This study referred to the park landscape evaluation system [52,66] and the visitor perception evaluation system [67] in the blue–green space recreation evaluation literature. It combined them with the existing high-frequency word data from two aspects of park design and individual perception elements. The high-frequency word data were screened, checked, and classified, and a landscape design factor evaluation system with 4 major indicators and 17 factors (Table 1) and a sensory perception factor evaluation system with 5 major indicators and 18 factors (Table 2) were developed.

Table 1. Evaluation index system of landscape design.

Evaluation Term	Indicators	Indicator Definition
Landscape (A)	Natural ecological environment (A1)	Environmental quality, beautiful scenery, etc.
	Plant landscape (A2)	Trees, leaves, flowers, etc.
	Animal landscape (A3)	Ducks, birds, fish, squirrels, frogs, etc.
	River landscape (A4)	Water, rivers, lakes, ponds, etc.
	Historical and cultural landscape (A5)	Culture, history, royal, red walls, ancient pavilions, the Hall of Abstinence, circular mounds, ancient trees, etc.
Activity (B)	Humanities activities (B1)	Exhibition halls, temple fairs, sacrifices, gardening, etc.
	Country activities (B2)	Boating, camping, mountain climbing, picnics, tents, etc.
	Recreational activities (B3)	Dancing, walking, taking pictures, resting, etc.
	Fitness activities (B4)	Exercise, sports, running, cycling, gym, etc.
Infrastructure (C)	Transportation accessibility (C1)	Highway, subway, bus, driving, walking, distance, location, etc.
	Public service facilities (C2)	Parking lots, restrooms, toilets, trash cans, etc.
	Navigation signage system (C3)	Navigation, maps, explanations, etc.
	Food and beverage facilities (C4)	Catering, restaurants, kiosks, ice cream, commodities, etc.
Management (D)	Consumer spending (D1)	Ticket price, free, charge, consumption, etc.
	Services provided (D2)	Management, attitude, reservations, complaints, quality, maintenance, queuing, etc.
	Planning layout (D3)	Planning, routes, areas, buildings, spaces, etc.
	Science education (D4)	Popular science, exhibitions, learning, knowledge, etc.

Table 2. Evaluation index system of sensory perception.

Senses Term	Indicators	Indicator Definition
Vision (E)	Visual identification (E1)	Vision of special sights
	Vision of plants (E2)	Vision of trees, grass, flowers, etc.
	Vision of water (E3)	Vision of water
	Vision of animals (E4)	Vision of wild ducks, squirrels, birds, etc.
	Vision of humans (E5)	Moderate number of people and no interference
	Vision of roads (E6)	Vision of the line shape, color, etc. of the road
Hearing (F)	Sound of voice (F1)	Moderate voice
	Sound of broadcast (F2)	Sound of broadcast
	Sound of animals (F3)	Sound of birds, insects, etc.
	Sound of water (F4)	Sound of water flow
Smell (G)	Smell of air and water (G1)	Fresh air and good water quality
	Smell of plants (G2)	Smell of plants
Touch (H)	Feel of sunlight (H1)	Feel of the balance of light and shadow
	Feel of wind (H2)	Feel of wind
	Feel of roads (H3)	Comfortable roads
	Feel of water (H4)	Hydrophilic experience
	Contact with animals (H5)	No mosquito bites
Taste (L)	Food available (L1)	Food available

3.4. Evaluation Model and Method

Through the social media text analysis results, this study evaluated the satisfaction of riverside parks in the blue–green space using the IPA method. In the IPA analysis chart, visitor expectations (importance) served as the horizontal axis, visitor satisfaction served as the vertical axis, and the overall average value served as the separation point of the X–Y axis, resulting in a four-quadrant evaluation model (Figure 3). The first quadrant involves the area that continues to strive to develop, including the experience factors that visitors value and are satisfied with. The second quadrant entails the area that does not need much improvement, including the experience factors that visitors think are not significant, but nonetheless make them very satisfied. The third quadrant is the area that needs gradual improvement, including the experience factors that visitors believe are unimportant and unsatisfactory. Finally, the fourth quadrant involves the area that requires major improvement, including the experience factors that visitors value but are unsatisfied with [68]. The visitor expectations (importance) were evaluated by the frequency of each factor; visitor satisfaction was measured by the proportion of positive emotional texts in the text data. First, we determined the corresponding quantity and frequency of each index factor based on the two major index systems, then graded the evaluation text by referring to a Likert-scale satisfaction grading method. The "1–5 star ratings" of the evaluation text corresponded to "very dissatisfied", "dissatisfied", "average", "satisfied", and "very satisfied" [16]. Specifically, evaluation texts with more than 4 stars were considered positive emotional texts. Finally, the number and ratio of positive emotional texts were determined for each index factor to evaluate user satisfaction with the factor.

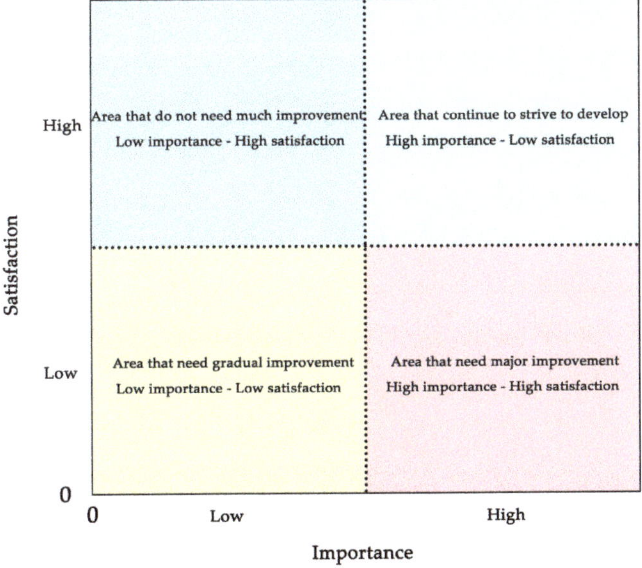

Figure 3. IPA evaluation model diagram with four areas.

3.5. Analysis of Variance and Comparison

In this study, SPSS tools were used to conduct a one-way analysis of variance and multiple comparison analyses for each index factor [56] to investigate the differences between index factors in various blue–green spaces. In addition, the data used in ANOVA enabled further comparative analysis based on the result data obtained from the IPA analysis and all variables were parameterized to facilitate tests used to check their normality and homoscedasticity, with the obtained *p*-values. Therefore, the use of ANOVA was feasible.

4. Results

4.1. Comparative Analysis of the Changes in the Number of Reviews in Different Seasons

Based on the selected reviews with a long time span, the changes in the total number of park reviews from spring (March–May), summer (June–August), autumn (September–November) and winter (December–February) from 2006 to 2021 were compared (Figure 4). As can be observed from the figure, the total number of comments in the four seasons generally increased, and the slower growth or sharp decline in the number of comments in 2020 and 2021 was related to the weakened travel intention of tourists after the outbreak of COVID-19. Due to the increasing needs of tourists for a better life, the use of tourism social media has become more extensive with the increase in park recreation activities, and the sample size of comments has increased. Therefore, the differences in the total amount of comments in the four seasons after 2016 have significantly increased. The total amount of comments in winter was the least, and the total amount of comments in spring, summer and autumn was relatively higher. Tourists' interest in the typical blue–green space in Beijing increased year by year, and they prefer the landscape in spring, summer and autumn, which was related to the climate conditions and the life form of native plants in Beijing. Especially in the past two years, the number of comments was significantly higher in spring, probably because of the rich plant landscape and good water landscape in spring. While maintaining the landscape in spring, summer and autumn, the construction of waterfront green space should enhance the attraction of the park's winter landscape, so as to give full play to its role in social economy and social ecology.

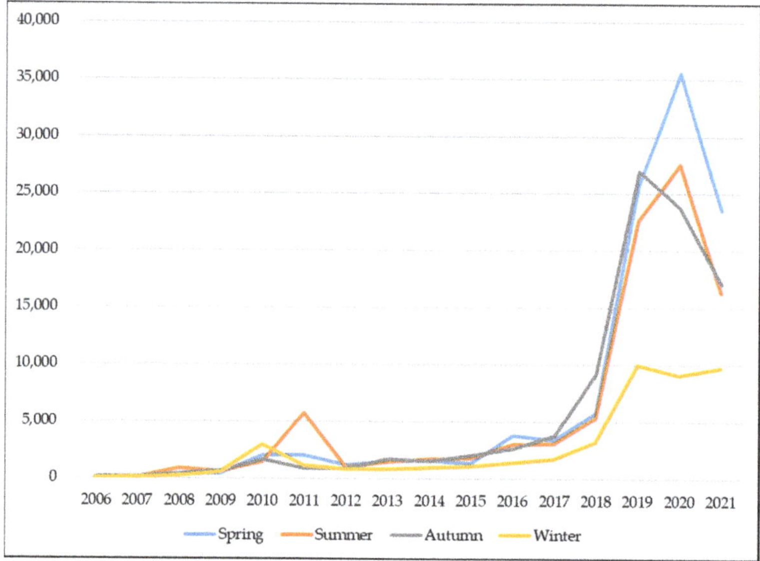

Figure 4. Comparative analysis of the changes in the number of reviews in different seasons.

4.2. Comparative Analysis of Riverside Parks Beside Different River Systems

4.2.1. Rating Star Analysis

According to the statistics (Figure 5), the rating stars of riverside parks from high to low followed the following order: parks beside the North Canal River System, parks beside the Jiyun River System, parks beside the Daqing River System, parks beside the Chaobai River System, parks beside the Yongding River System. The parks beside the North Canal River System receive the highest score, which may be related to their superior geo-graphical location, the large number and variety of parks, the relatively comprehensive coverage of the user demands of all populations, and the large number of samples. The high scores of the parks adjacent to the Daqing River System and the Jiyun River System may be attributable to the late construction date, newer facilities, and the small sample size. Meanwhile, the low scores of the parks adjacent to the Chaobai River System and Yongding River System may be attributable to the fact that the service facilities in the parks are relatively old and do not meet the needs of all users. Therefore, in order to meet the needs of the public, it is recommended that scientific and reasonable space improvements should be implemented.

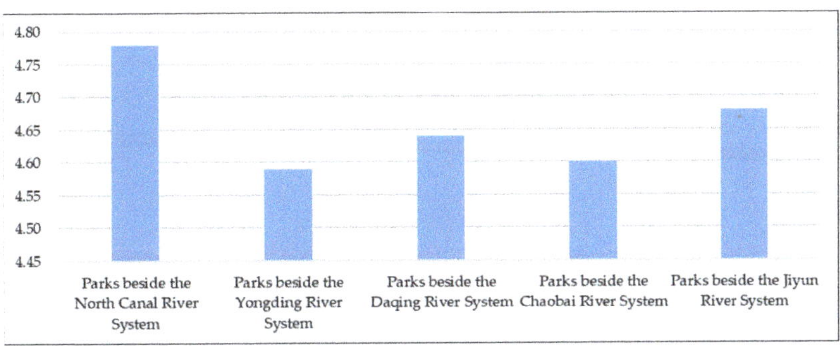

Figure 5. The average stars of riverside parks beside different river systems.

4.2.2. Importance–Performance Analysis Based on Spatial Landscape Design and Visitors' Sensory Perception

From the perspective of spatial landscape design, importance–performance analysis for parks with different river systems (Figure 6) revealed that most of the factors that involved parks beside the North Canal River System were in the first and second quadrants. Meanwhile, only "humanities activities" were in the third quadrant, indicating that people were highly satisfied with all aspects of the parks beside the North Canal River System and were more attentive to the humanities activities. The parks adjacent to the Yongding River System highlighted the following three factors: "river landscape", "country activities", and "public service facilities", which were in the fourth quadrant, indicating that people were satisfied with the Yongding River Park as a whole. However, due to its imperfections, it needed to be improved in the areas of river management, river landscape construction, and park services. Only one factor of the "service provided" in the fourth quadrant for the parks along the Chaobai River System indicated that people were extremely satisfied with all aspects of the parks along the Chaobai River System, but expected the park's service level to be enhanced. Most of the factors of the parks adjacent to the Daqing River System and the Heji Canal river system were in the third and fourth quadrants. It indicates that these parks have significant room for improvement in all aspects, especially regarding the "river landscape", "natural ecological environment", and "consumer spending". These were related to the fact that most of these parks were built relatively late, far from the city center, large in size, it was difficult to refine the landscape construction fully, and they were difficult to manage and maintain.

The importance–performance analysis of parks with different river systems (Figure 7) revealed, from the perspective of visitors' sensory perception, that people were most satisfied with parks adjacent to the North Canal River System. In contrast, their satisfaction with "crowd interference" was slightly lower. It was likely because the parks adjacent to the North Canal River System were closer to the city center, were constructed earlier, and had more comprehensive service facilities, making them more familiar to visitors. As a result, more visitors chose the parks adjacent to the North Canal River System as their travel destination, which affected the park's experience. Most parks adjacent to the Yongding River System were in the first and second quadrants. However, they required improvement in terms of "vision of water", "crowd interference", "vision of roads", and particularly "vision of water". This was likely because the surrounding industrial areas impacted the water quality of the parks. The distribution of factors in the parks adjacent to the Chaobai River System was relatively even, and the "visual recognizability", "vision of water", and "feel of sunlight" needed improvement, consistent with the park's distinctive landscape. Visitors' recreational experiences were hampered by inadequate construction and inadequate humanization of facilities. In addition, the factors of parks beside the Daqinghe River system and the Jihe River system were primarily located in the third and fourth quadrants, and the aspects of "vision of water", "vision of animals", "vision of plants", and "smell of air and water" needed improvement. It indicates that the ecological environment of these parks beside different river systems needed improvement, which may be related to the fact that the Daqing River and Ji Canal watersheds were mostly located in the suburbs, the construction of these parks was late in the year, and their management services were still imperfect.

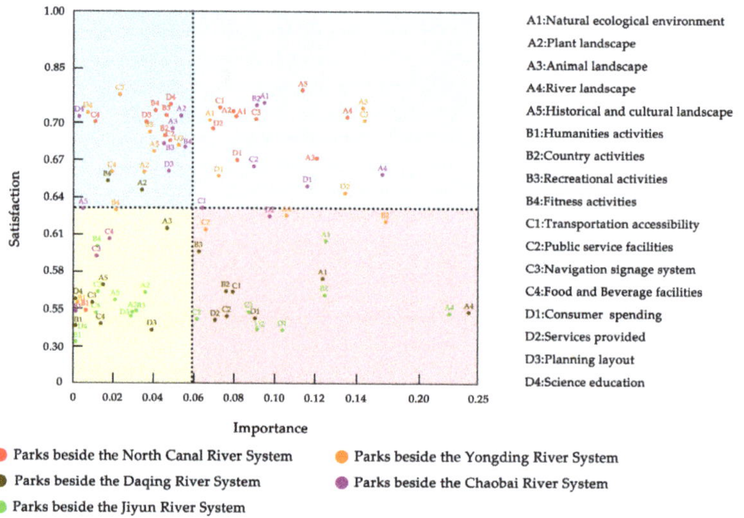

Figure 6. IPA model of riverside parks beside different river systems based on spatial landscape design.

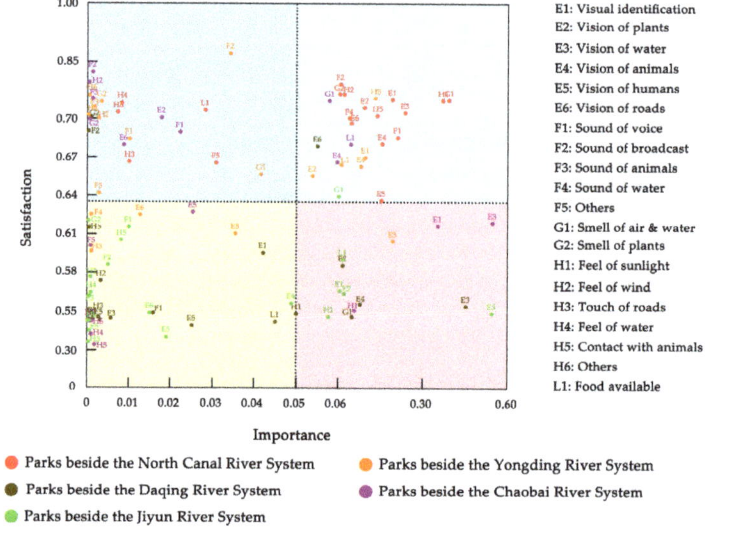

Figure 7. IPA model of riverside parks beside different river systems based on visitors' sensory perception.

4.2.3. One-Way Analysis of Variance and Multiple Comparison Analysis

From the one-way analysis of variance, it was determined that among the 17 landscape design indicators, water parks differed significantly in their satisfaction compared to historical and cultural landscapes, recreational activities, and services provided. Consequently, based on the results of the IPA analysis, several factors were chosen for additional multiple comparison analysis (Table 3).

Table 3. Multiple comparison analysis of landscape design indicators of riverside parks beside different river systems (only some important data are displayed here; all data can be found in the Supplementary File in the Supplementary Materials).

I	J	A5 Mean Dif. (I−J)	Sig.	B3 Mean Dif. (I−J)	Sig.	C2 Mean Dif. (I−J)	Sig.	C4 Mean Dif. (I−J)	Sig.	D2 Mean Dif. (I−J)	Sig.
a	b	0.1405 *	0.0210	0.1116 *	0.0090	0.1172 *	0.0190	0.0695	0.3830	0.1278 *	0.0060
a	e	0.3114 *	0.0090	0.2318 *	0.0050	0.0990	0.2990	0.3734 *	0.0170	0.1862 *	0.0360
b	a	−0.1405 *	0.0210	−0.1116 *	0.0090	−0.1172 *	0.0190	−0.0695	0.3830	−0.1278 *	0.0060
c	e	0.2289	0.1320	0.1237	0.2410	−0.0279	0.8210	0.4162 *	0.0400	0.0449	0.6920

* a (parks beside the North Canal River System); b (parks beside the Yongding River System); c (parks beside the Daqing River System); e (parks beside the Jiyun River System); Sig (short for significant difference, which is a statistical term for the evaluation of data differences in statistics).

The evaluation satisfaction of parks adjacent to the North Canal River System was relatively high. The differences between the North Canal River System and other river systems predominated in each index. The historical and cultural landscape (A5) of the North Canal River System and the Jiyun River System differed significantly. The cultural and historical landscapes of the North Canal River System, the Yongding River System, and the Chaobai River System Park were reported as reasonably good. It may be because most historical parks with a rich cultural heritage are located near the North Canal River System. In contrast, the Daqing River System and the Jiyun River System primarily comprised ecological parks that were recently constructed. The North Canal River System, the Yongding River System, and the Jiyun River System had notably different recreation activities (B3) and service offerings (D2). It may be because the Jiyun River System Park was constructed later in the year, and various garden management service facilities were not yet complete. In addition, it was located far from the city center, transportation was inconvenient, the number of visitors was relatively low, and various activities were not carried out to their full potential.

The results of the one-way analysis of variance revealed that among the 20 sensory perception indicators, there were significant differences in the satisfaction of different water parks in the following 5 aspects: visual identification, vision of animals, vision of humans, odor of plants, and feel of roads. Consequently, based on the results of the IPA analysis, several factors were chosen for additional multiple comparison analysis (Table 4).

Table 4. Multiple comparison analysis of sensory perception indicators of riverside parks beside different river systems (only some important data are displayed here; all data can be found in the Supplementary File in the Supplementary Materials).

I	J	E1 Mean Dif. (I−J)	Sig.	E4 Mean Dif. (I−J)	Sig.	E5 Mean Dif. (I−J)	Sig.	G2 Mean Dif. (I−J)	Sig.	H3 Mean Dif. (I−J)	Sig.
a	b	0.1034 *	0.0443	0.1219 *	0.0058	0.1644 *	0.0281	0.2204 *	0.0106	0.2590 *	0.0072
a	e	0.3138 *	0.0020	0.1164	0.1669	0.0197	0.8904	0.1601	0.3305	0.4131 *	0.0262
b	a	−0.1034 *	0.0443	−0.1219 *	0.0058	−0.1644 *	0.0281	−0.2204 *	0.0106	−0.2590 *	0.0072
b	e	0.2104 *	0.0471	−0.0054	0.9513	−0.1447	0.3427	−0.0603	0.7291	0.1540	0.4285
c	e	0.2563 *	0.0473	−0.0137	0.8997	−0.0901	0.6272	0.2275	0.2858	0.4113	0.0852

* a (parks beside the North Canal River System); b (parks beside the Yongding River System); c (parks beside the Daqing River System); e (parks beside the Jiyun River System); Sig (short for significant difference, which is a statistical term for the evaluation of data differences in statistics).

Regarding visual identification (E1), parks adjacent to the Jiyun River System were significantly less pleasing than those adjacent to the North Canal River System, the Yongding River System, the Daqing River System, and the Chaobai River System. The following reasons were speculated: (1) the parks adjacent to the North Canal River System included historical parks, special parks, and ecological parks, among others. Specifically, the historical parks featured a variety of distinctive garden structures, and the landscape was generally recognizable. (2) Most parks adjacent to the Jiyun River System were natural and ecological parks, which were located far from the city center, had a large area, and had high maintenance and management costs, making it challenging to meet the diverse needs of visitors. In terms of the vision of animals (E4), vision of humans (E5), smell of plants (G2), and feel of roads (H3), the parks adjacent to the Yongding River System provided a lower level of satisfaction than those adjacent to the North Canal River System. It may be because, compared to the parks along the Yongding River System, the parks along the North Canal River System had superior management, more accessible facilities, a more beautiful environment, and greater accessibility regarding traffic. Among all the sensory perception factors, the satisfaction with the vision factor varied the most between parks adjacent to different river systems. In contrast, satisfaction with the other factors was relatively balanced. The parks adjacent to the North Canal River System were the most ideal regarding sensory perception, while the parks adjacent to the Jiyun River System had room for improvement.

4.3. Comparative Analysis of Riverside Parks of Different Types

4.3.1. Rating Star Analysis

According to the statistics (Figure 8), community parks are the parks with the highest rating that were generally welcomed by people. Comprehensive parks and historical parks were given high rating stars and similar scores, indicating that comprehensive parks with diverse content, suitable recreation and supporting management service facilities, and historical parks with profound history and cultural deposits, excellent maintenance and management, and strong scientific and educational significance were favored by most tourists. The scores of specific parks, ecological parks and nature parks are relatively low, indicating that parks with specific content or form and natural landscapes are not able to attract the majority of people, and can be further improved in terms of the theme characteristics and natural landscape. The gardens have the lowest rating, which may be related to the small number of samples, the small scale and the small proportion of residents. The service radius of the park can be further optimized, and public service facilities can be improved to provide residents with a better leisure and recreation experience.

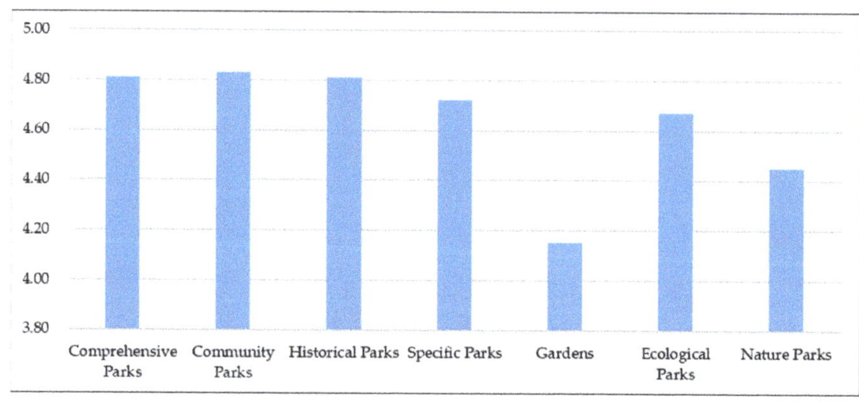

Figure 8. The average stars of riverside parks of different types.

4.3.2. Importance–Performance Analysis Based on Spatial Landscape Design and Visitor Sensory Perception

According to the importance–performance analysis of parks of different types (Figure 9), from the perspective of spatial landscape design, most of the factors of comprehensive parks, community parks, historical parks, and specific parks were in the first and second quadrants. Only "humanities activities" of the comprehensive park was in the third quadrant, indicating that people were highly satisfied with all comprehensive park aspects. Meanwhile, the satisfaction values that involved "public service facilities", "services provided", and "consumer spending" were relatively low; hence, the infrastructure and management services of these parks must be further improved. Only "humanities activities" and "navigation signage system" were in the third quadrant, indicating that people were more interested in community parks. It meant that the guide and identification system of community parks and the types of community activities could be further improved and enriched, as they were closely related to the convenience and comfort of community life. The distribution of the factors of historical parks and the park evaluation score demonstrated that the satisfaction of people's awareness of historical parks was the highest, but "animal landscape" still required improvement. The low level of satisfaction with "services provided" in specific parks may be attributable to the need for more detailed management and maintenance of those parks. Most factors in the ecological park were in the first and second quadrants; a few were in the area where the first and second quadrants intersected with the third and fourth quadrants. Among them, the "river landscape" required improvement, which may have been necessitated by the fact that some natural river channels had been altered by canalization, thereby destroying the original ecological environment. The nature parks and gardens were primarily located in the third and fourth quadrants. In addition to the factors of "animal landscape" and "humanities activities" in natural parks, which were in high-satisfaction areas, other aspects needed improvement. It may be because most natural parks were located far from urban areas, and the management services were not yet ideal. The low level of satisfaction with the park may be attributable to the small sample size.

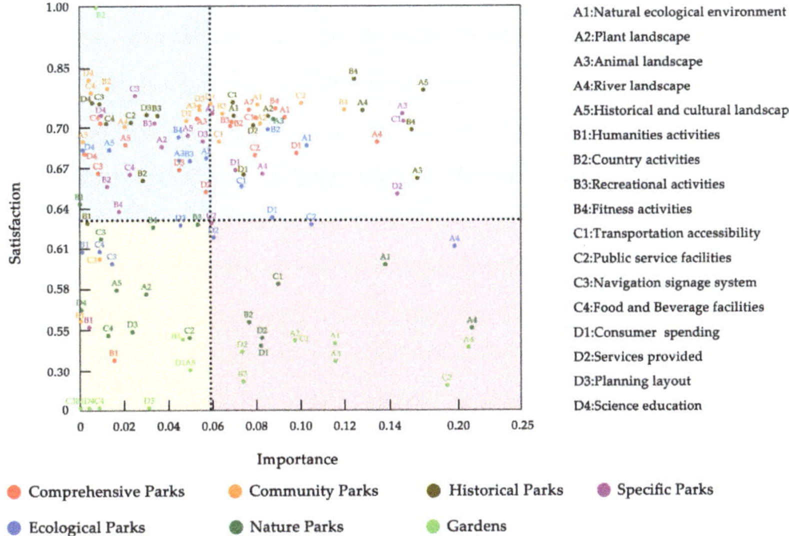

Figure 9. IPA model of riverside parks of different types based on spatial landscape design.

From the perspective of visitors' sensory perception, the importance–performance analysis for parks of different types (Figure 10) revealed that most visual perception factors were in the first and fourth quadrants. It indicates that visual perception was significantly more critical than other sensory perception needs. In terms of different park types, most of the factors of comprehensive parks, community parks, historic parks, and specific parks were in the first and second quadrants, indicating that people were relatively satisfied with the sensory perception of various aspects of these parks. However, there were still factors that needed improvement. Among them, "feel of roads" satisfaction was generally low, particularly in underdeveloped community parks. Most ecological park factors were within the average satisfaction category, but "vision of water" and "vision of animals" needed to be improved significantly. It may be due to the ecological parks' damage to the natural water environment and animal habitats and the length of time needed for restoration. Finally, most garden and natural park factors were in the third and fourth quadrants, indicating that people's satisfaction with all aspects of sensory perception in these parks was relatively low and that all factors required improvement. Again, it was related to the remote and suburban regions with verdant mountains and forests, significant temperature variations, and variable climates.

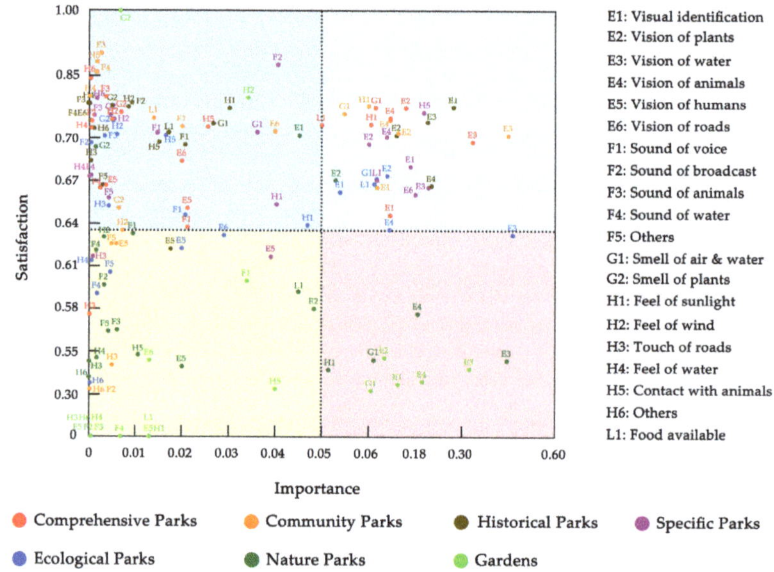

Figure 10. IPA model of riverside parks of different types based on visitors' sensory perception.

4.3.3. One-Way Analysis of Variance and Multiple Comparison Analysis

Except for the natural ecological environment, river landscape, country activities, fitness activities, and navigation signage system, the one-way analysis of variance revealed that among the 17 landscape design indicators, there were significant differences in satisfaction between the parks of different types in all other aspects. Consequently, based on the results of the IPA analysis, several factors were chosen for additional multiple comparison analysis (Table 5).

Comprehensive and historical parks had relatively high overall satisfaction, whereas ecological and nature parks had relatively low overall satisfaction. The differences between each index were most pronounced between comprehensive parks, historical parks, and specific parks, as well as between nature and ecological parks. Especially concerning historical parks and specific parks, each index displayed glaring differences. Comprehensive parks and historical parks with plant landscapes (A2) and animal landscapes (A3) differed significantly from specific parks from the perspective of each index due to their different park type positioning. There were clear distinctions between ecological and historical parks with cultural landscapes (A5). It may be because ecological parks focus more on creating and maintaining natural landscapes. In contrast, historical parks focus more on the continuation and inheritance of historical and cultural landscapes. Significant differences existed between natural and other types of parks regarding consumer spending (D1) and services provided (D2). These differences were attributable to the vast area of natural parks, their distance from the city center, and their high maintenance and management costs.

Table 5. Multiple comparison analysis of landscape design indicators of riverside parks of different types (only some important data are displayed here; all data can be found in the Supplementary File in the Supplementary Materials).

I	J	A2 Mean Dif. (I−J)	A2 Sig.	A3 Mean Dif. (I−J)	A3 Sig.	A5 Mean Dif. (I−J)	A5 Sig.	D1 Mean Dif. (I−J)	D1 Sig.	D2 Mean Dif. (I−J)	D2 Sig.
f	i	0.1272 *	0.0040	0.1598 *	0.0000	0.2169	0.0550	0.0547	0.2600	0.0691	0.1820
	j	0.1004 *	0.0160	0.0993 *	0.0160	0.1501 *	0.0450	0.0204	0.6590	0.0295	0.5500
	k	0.0535	0.3930	0.0690	0.2710	0.3964	0.3560	0.2323 *	0.0020	0.2312 *	0.0030
g	i	0.1007	0.1740	0.1566 *	0.0360	0.2259 *	0.0280	0.0871	0.3000	0.1805 *	0.0460
	j	0.0739	0.3100	0.0961	0.1880	0.1591 *	0.0080	0.0528	0.5230	0.1410	0.1120
	k	0.0271	0.7550	0.0658	0.4490	0.4053	0.3410	0.2647 *	0.0090	0.3426 *	0.0020
h	i	0.1514 *	0.0050	0.1962 *	0.0000	0.2724 *	0.0070	0.0682	0.2520	0.1826 *	0.0050
	j	0.1246 *	0.0160	0.1358 *	0.0090	0.2056 *	0.0020	0.0339	0.5560	0.1430 *	0.0220
	k	0.0777	0.2650	0.1055	0.1310	0.4519	0.2360	0.2458 *	0.0030	0.3447 *	0.0000
i	f	−0.1272 *	0.0040	−0.1598 *	0.0000	−0.2169	0.0550	−0.0547	0.2600	−0.0691	0.1820
	g	−0.1007	0.1740	−0.1566 *	0.0360	−0.2259 *	0.0280	−0.0871	0.3000	−0.1805 *	0.0460
	h	−0.1514 *	0.0050	−0.1962 *	0.0000	−0.2724 *	0.0070	−0.0682	0.2520	−0.1826 *	0.0050
	k	−0.0737	0.2410	−0.0908	0.1490	0.1795	0.9880	0.1776 *	0.0140	0.1621 *	0.0350
j	f	−0.1004 *	0.0160	−0.0993 *	0.0160	−0.1501 *	0.0450	−0.0204	0.6590	−0.0295	0.5500
	g	−0.0739	0.3100	−0.0961	0.1880	−0.1591 *	0.0080	−0.0528	0.5230	−0.1410	0.1120
	h	−0.1246 *	0.0160	−0.1358 *	0.0090	−0.2056 *	0.0020	−0.0339	0.5560	−0.1430 *	0.0220
	k	−0.0469	0.4460	−0.0303	0.6210	0.2463	0.8630	0.2119 *	0.0030	0.2016 *	0.0080

* f (comprehensive parks); g (community parks); h (historical parks); i (specific parks); j (ecological parks); k (nature parks); Sig (short for significant difference, which is a statistical term for the evaluation of data differences in statistics).

From the one-way analysis of variance, it was determined that among the 20 indicators of sensory perception, different types of parks had significantly different levels of satisfaction with each factor. Consequently, based on the results of the IPA analysis, several factors were chosen for additional multiple comparison analysis (Table 6).

Table 6. Multiple comparison analysis of sensory perception indicators of riverside parks of different types (only some important data are displayed here; all data can be found in the Supplementary File in the Supplementary Materials).

I	J	E1 Mean Dif. (I−J)	E1 Sig.	E3 Mean Dif. (I−J)	E3 Sig.	E4 Mean Dif. (I−J)	E4 Sig.	G1 Mean Dif. (I−J)	G1 Sig.	H1 Mean Dif. (I−J)	H1 Sig.
f	i	0.1285	0.3350	0.0605	0.1630	0.1680 *	0.0010	0.1011	0.0640	0.0989	0.9210
	j	0.1081	0.2250	0.0532	0.2000	0.1186 *	0.0090	0.0770	0.1390	0.0655	0.9360
	k	0.2373	0.9690	0.2147 *	0.0010	0.2731 *	0.0000	0.2544 *	0.0020	0.2204	0.8210
g	i	0.1283	0.4950	0.0888	0.2360	0.1514	0.0630	0.1023	0.2760	0.2132 *	0.0450
	j	0.1078	0.5110	0.0816	0.2700	0.1019	0.2020	0.0781	0.3980	0.1799 *	0.0030
	k	0.2371	0.9700	0.2431 *	0.0070	0.2564 *	0.0080	0.2556 *	0.0230	0.3348	0.3660
h	i	0.2030 *	0.0220	0.1418 *	0.0090	0.1829 *	0.0020	0.1342 *	0.0450	0.1548	0.3020
	j	0.1825 *	0.0070	0.1345 *	0.0100	0.1335 *	0.0180	0.1100	0.0900	0.1214	0.0800
	k	0.3118	0.8380	0.2960 *	0.0000	0.2880 *	0.0000	0.2875 *	0.0020	0.2763	0.5820
i	f	−0.1285	0.3350	−0.0605	0.1630	−0.1680 *	0.0010	−0.1011	0.0640	−0.0989	0.9210
	g	−0.1283	0.4950	−0.0888	0.2360	−0.1514	0.0630	−0.1023	0.2760	−0.2132 *	0.0450
	h	−0.2030 *	0.0220	−0.1418 *	0.0090	−0.1829 *	0.0020	−0.1342 *	0.0450	−0.1548	0.3020
	k	0.1088	1.0000	0.1542 *	0.0170	0.1051	0.1280	0.1533	0.0570	0.1215	0.9990
j	f	−0.1081	0.2250	−0.0532	0.2000	−0.1186 *	0.0090	−0.0770	0.1390	−0.0655	0.9360
	g	−0.1078	0.5110	−0.0816	0.2700	−0.1019	0.2020	−0.0781	0.3980	−0.1799 *	0.0030
	h	−0.1825 *	0.0070	−0.1345 *	0.0100	−0.1335 *	0.0180	−0.1100	0.0900	−0.1214	0.0800
	k	0.1293	1.0000	0.1615 *	0.0110	0.1545 *	0.0240	0.1774 *	0.0250	0.1549	0.9820

* f (comprehensive parks); g (community parks); h (historical parks); i (specific parks); j (ecological parks); k (nature parks); Sig (short for significant difference, which is a statistical term for the evaluation of data differences in statistics).

The differences between comprehensive parks and historical parks were negligible, and the differences in various indicators were primarily concentrated between comprehensive parks and other types of parks, with the differences in vision factor being particularly pronounced. Among them, there were significant differences in visual identification between historical and ecological parks (E1). Historical parks focus more on the human environment, whereas ecological parks focus more on the natural environment. There were clear distinctions between nature and other types of parks, except for ecological parks in their vision of water (E3) and vision of animals (E4), which were related to the different construction purposes of natural parks. The protection of natural ecosystems and natural landscapes received the most attention in nature parks, followed by ornamental, cultural, and scientific values. The smell of the air and water in nature parks differed from that of other parks (G1), which may be attributable to the nature parks' distance from the city center, beautiful natural surroundings, and high air and water quality. There were discernible differences between community parks and ecological parks in the quality of sunlight (H1), which may be attributable to differences in the number of visitors, the reason for their visit, and the location of the parks.

4.4. Comparative Analysis of Riverside Parks in Different Districts

4.4.1. Rating Star Analysis

According to the statistics (Figure 11), the average rating stars of riverside parks in six central districts, districts of the inner suburbs and districts of the outer suburbs decrease sequentially. The difference between the lowest and highest average rating star is approximately 0.05, and the overall difference is small. The average rating star of the parks in six central districts is the highest, possibly because they are located in the city center, with long-term operation and sufficient management experience. The satisfaction of the parks in the districts of the inner suburbs is moderate, which may be due to the proximity of the parks to the city center and the variety and content of the parks. The average rating star of parks in districts of the outer suburbs is the lowest, which may be related to the fact that the parks are far away from the city center and are dominated by plant landscapes and ecological recreation activities. Therefore, the positioning of the parks should be improved to enhance the attraction of the parks.

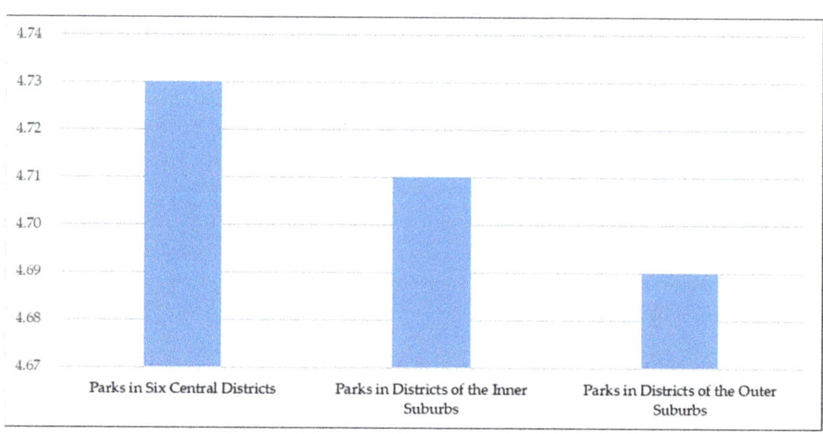

Figure 11. The average stars of riverside parks in different districts.

4.4.2. Importance–Performance Analysis Based on Two Aspects of Spatial Landscape Design and Visitors' Sensory Perception

From the standpoint of spatial landscape design, the importance–performance analysis for parks in different districts (Figure 12) revealed that most park factors within the six central districts were in the first and second quadrants. The highly significant landscape factors were also located in the first and second quadrants. A high level of satisfaction was reported, but "humanities activities" needed to be improved. The residents of the six central districts may have had a higher standard of living and higher expectations and the evaluation criteria for parks may have been more stringent than in other areas.

Consequently, the requirements for developing park cultural activities were more stringent. All the factors of the parks in the districts of the inner suburbs were in the first and second quadrants, indicating that they were relatively acceptable in all respects. The factors of parks in the districts of the outer suburbs were primarily in the third and fourth quadrants, with "animal landscape" and "recreational activities" located in the first and second quadrants, respectively. It may be due to people's desire to be close to nature and to relax in the suburban environment. However, satisfaction with factors such as "consumer spending" and "services provided" was lower, likely because the time and cost of the park activities did not match the satisfaction of the activities available in the parks.

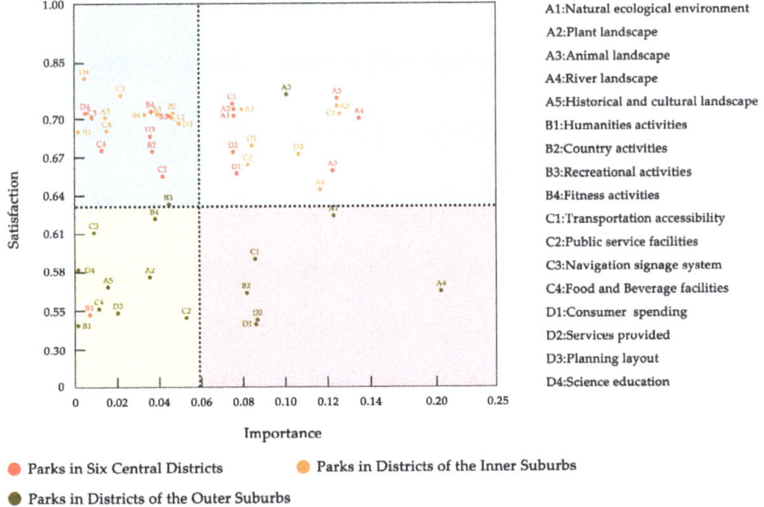

Figure 12. IPA model of riverside parks in different districts based on spatial landscape design.

The importance–performance analysis of parks in different districts (Figure 13) revealed, from the perspective of visitors' sensory perception, that most of the park factors in the six central districts were in the second quadrant. In contrast, the first quadrant contained only visual factors, indicating that the importance and satisfaction of visual factors were high. In addition, the satisfaction of "crowd interference" and "feel of roads" was low due to the park's early construction in the six central districts, its proximity to the city center, and increased visitors. In addition, some facilities have been utilized for an extended period and have not been repaired promptly; therefore, the parks should focus on mitigating the adverse effects of crowd congestion. Most of the factors of parks in the districts of the inner suburbs were in the first and second quadrants, indicating that the parks reported relatively high sensory perception scores. Meanwhile, the satisfaction with "sound of broadcast" was the highest, representing visitors' recognition of the park's broadcast and explanation services. In addition, satisfaction with "the sound of water" and "feel of water" was relatively low. It may be attributable to this region's abundance of water resources and park landscape, which led to people having higher expectations. The sensory perception factors of parks in the districts of the outer suburbs were concentrated in the third and fourth quadrants. As a result, the sensory perception of these parks was unsatisfactory. In contrast, the satisfaction of olfactory factors was relatively high, indicating that the parks in this region possessed more significant advantages regarding the smell of plants, air quality, and water quality. Among the sensory perception factors, the visual factor was the most important, and satisfaction was relatively high. In contrast, the importance and satisfaction of the tactile factor were relatively low. Therefore, park administrators should prioritize enhancing tactile perception.

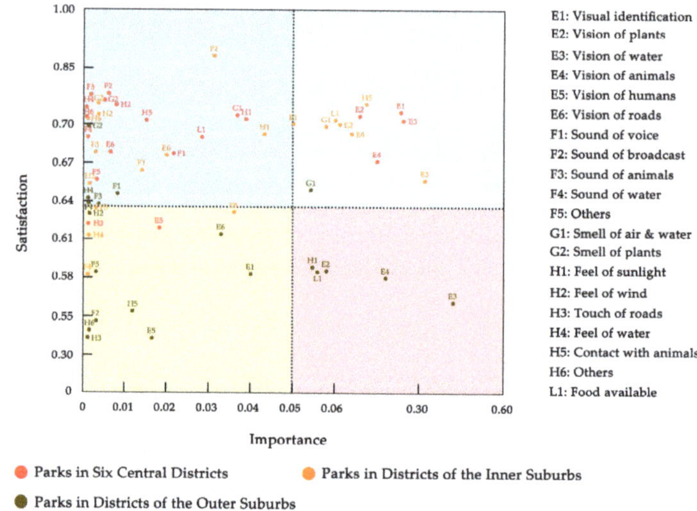

Figure 13. IPA model of riverside parks in different districts based on sensory perception.

4.4.3. One-Way Analysis of Variance and Multiple Comparison Analysis

A one-way variance analysis determined that among the 17 landscape design indicators, there were significant differences in the recreational activities and food and beverage facilities satisfaction of parks in different districts. Consequently, based on the results of the IPA analysis, several factors were chosen for additional multiple comparison analysis (Table 7).

Table 7. Multiple comparison analysis of landscape design indicators of riverside parks in different districts (only some important data are displayed here; all data can be found in the Supplementary File).

I	J	B3		C4	
		Mean Dif. (I−J)	Sig.	Mean Dif. (I−J)	Sig.
l	n	0.1577 *	0.0450	0.2949 *	0.0391
m	n	0.1387	0.0789	0.2818 *	0.0497

* l (parks in six central districts); m (parks in districts of the inner suburbs); n (parks in districts of the outer suburbs); Sig (short for significant difference, which is a statistical term for the evaluation of data differences in statistics).

Regarding recreational activities (B3), the satisfaction of parks in districts of the outer suburbs was significantly lower than that of parks in the six central districts. The difference between the satisfaction of parks in the districts of the inner suburbs and six central districts of the city was insignificant. It may be attributable to the late construction of parks in the districts of the outer suburbs, the low distribution density of parks, and the lack of convenient transportation, all of which impacted the evaluation of park satisfaction. Regarding food and beverage facilities (C4), the satisfaction gap between parks in the districts of the outer suburbs and those in the six central districts and districts of the outer suburbs was more pronounced. It indicates that the management of parks in the outer suburbs was imperfect and that the facilities and services were not fully standardized. In general, the improvement of landscape design factors of parks in different districts was relatively symmetrical. The parks in the districts of the outer suburbs lacked recreational activities (B3) and food and beverage facilities (C4), compared to the parks in the six central districts. Consequently, we should concentrate on enhancing the overall landscape construction and service management of the parks in the districts of the outer suburbs.

The one-way analysis of variance determined that among the 20 sensory perception indicators, the satisfaction of parks in different districts differed significantly, most notably in the perception of plants, the perception of voices, and the perception of roads. Consequently, based on the results of the IPA analysis, several factors were chosen for additional multiple comparison analysis (Table 8).

Table 8. Multiple comparison analysis of sensory perception indicators of riverside parks in different districts (only some important data are displayed here; all data can be found in the Supplementary File in the Supplementary Materials).

I	J	E2 Mean Dif. (I−J)	Sig.	F1 Mean Dif. (I−J)	Sig.	H3 Mean Dif. (I−J)	Sig.
1	m	0.0477	0.4868	0.1028 *	0.0423	0.0759	0.7428
	n	0.0717 *	0.0215	0.1484	0.1622	0.3998 *	0.0057
m	l	−0.0477	0.4868	−0.0103 *	0.0423	−0.0759	0.7428
	n	0.0240	0.8764	0.0457	0.6672	0.3239 *	0.0206

* l (parks in six central districts); m (parks in districts of the inner suburbs); n (parks in districts of the outer suburbs); Sig (short for significant difference, which is a statistical term for the evaluation of data differences in statistics).

Regarding plant vision (E2), relatively low satisfaction with parks in the outer suburbs was reported. It may be because numerous ecological and natural parks in the outer suburbs, which had a large area, were constructed late in the year and had a high maintenance cost regarding the plant landscapes, while some had not yet even formed any plant landscapes. Regarding sound of voice (F1) and feel of roads (H3), parks in the six central districts had the highest satisfaction score, while parks in the districts of the inner suburbs had the lowest satisfaction score. It was due to the long road journey in the outer suburbs, the absence of a convenient road system, and the lack of smooth road surface repair.

Similar to the analysis of landscape design factors, the satisfaction with parks in different districts regarding sensory perception factors was the highest in the six central districts and lowest in the districts of the outer suburbs. Therefore, we should prioritize enhancing the level of park construction in the districts of the outer suburbs. People tend to pay more attention to the ecological quality of nature parks in the districts of the outer suburbs, where the area required for activities is more extensive than parks in the districts of the inner suburbs. Critical factors in the design and construction of nature parks include adequate size and area and adequate and timely management and upkeep.

5. Discussion

The space was based on multiple social media text data collections from 85 typical Beijing riverside parks. First, differences in the park star rating and visitor sentiment of different river systems, types, and districts were analyzed. Second, IPA was used to analyze the primary factors that influenced the satisfaction of landscape design and sensory perception of different river systems, types, and districts. Finally, analysis of variance was used to investigate the differences and causes of the satisfaction of certain factors among various parks.

Liu [53] et al. proposed that infrastructure construction and ecological quality of public space were important influencing factors after studying the factors that affect the recreation satisfaction of riverside public space in Shanghai, and our conclusion is basically consistent with their conclusion. Following the perspective of parks adjacent to different river systems, the study determined that the parks adjacent to the North Canal River System were closer to the city center, constructed earlier, had more comprehensive service facilities, and were more familiar to visitors. Their overall satisfaction was the highest, while the low satisfaction of a few parks was usually due to the imperfect facilities in the parks. For example, there was no parking space, seats, toilets or other basic service facilities in these parks, which made it difficult for visitors to drive there and stay and rest in the park for a long time. Therefore, the level of traffic inside the park could be further optimized, the necessary service facilities could be increased, and the park function system could be improved to reduce the negative impact of heavy traffic on park satisfaction. On the other hand, the parks adjacent to the Yongding River System were close to industrial areas and negatively impacted the natural environment. Therefore, ecological restoration should be bolstered to enhance the quality of aquatic landscapes. At the same time, some large-scale parks, such as Beijing West Eighteen Lakes Scenic Area and Pearl Lake Scenic area, did not keep up with the maintenance and management of the natural landscape and park-supporting service facilities, resulting in low satisfaction with the park. Therefore, more attention should be paid to the areas that must be maintained and restored in the park and one must also protect and improve the key and characteristic landscape of the park, and realize the multi-faceted renewal and construction of the park. Most of the satisfaction factors of the parks around the Chaobai River system were good. There was a need to improve "visual identification", "vision of water", "feel of sunlight", and "services provided". Taking Hanshiqiao Wetland Park as an example, the satisfaction of "water view" was not high mainly because the water system in the park was not strongly connected with the water system outside the park, the water quality was poor, many wetland landscapes were blocked by fences, the management service system in the park was also not comprehensive, and the road system in the park was not complete. To enhance the level of service management, we should focus on the construction of distinctive landscapes and the improvement of the facility systems in the parks, and strengthen the relationship between the internal and external environment of the park. For the parks adjacent to the Daqing River System and the Jiyun River System, the overall satisfaction was relatively average and it is necessary to improve the "river landscape", "natural ecological environment", "consumer spending", "transportation accessibility", "smell of air and water", "plant and animal landscape", and "services provided". Various park management service facilities must be optimized, the types of park activities must be expanded, the ecological landscape of the parks must be enhanced, the modes of travel to each park must be expanded, and systematic facility maintenance and ecological restoration must be performed regularly.

Concerning parks of various types, individuals have varying service requirements. This finding supports previous studies conducted by other scholars in other cities. Wang [29] et al. proposed that different types of parks required different optimization directions. Comprehensive parks should prioritize improving transportation accessibility, overall activity area, park functions, and infrastructure to meet the diverse needs of visitors of varying ages. Through popular science awareness activities, historical parks should focus on displaying and expressing history and culture to their fullest extent. Special parks must be managed and maintained with greater care. The guide sign system and types of community activities, closely related to the convenience and comfort of community life, should be enhanced in community parks. Gardens should expand the distribution area and increase the distribution location's flexibility. Ecological parks should optimize the ecological experience and strengthen the educational function of ecological science. Based on preserving ecology, natural parks should improve various service facilities and increase special attractions to increase their appeal to visitors.

This study also came to a conclusion that previous studies have never mentioned. The parks in the six central districts could be improved in terms of "humanities activities", "vision of humans", and "feel of roads", when compared to parks in other districts. To reduce the adverse effects of crowd congestion, the variety of humanities-related activities should be expanded, and visitors to the park should be better organized. In addition, some damaged structures should be repaired promptly. For parks in the districts of the inner suburbs, the "sound of water" and "feel of water" provided relatively low satisfaction levels. Therefore, the quality of the landscape should be further enhanced. The overall satisfaction of the parks in the districts of the outer suburbs was relatively high; however, there was room for improvement and the following actions should be taken: (1) one must optimize the internal traffic flow of the parks, improve the traffic conditions of the main entrances and exits, and enhance the guidance for self-driving visitors; (2) increase the construction of the parks' distinctive landscape, enhance the level of park management services, reduce park fees, save visitors time and money, and improve the parks' actual accessibility; (3) for natural parks in the outer suburbs, the ecological quality should be enhanced, sufficient area and scale should be ensured, and maintenance should be performed promptly.

In addition, few previous studies have compared the number of reviews in different seasons. Through analysis, this study found that fewer tourists chose to visit riverfront parks in winter. Therefore, as an important part of urban blue–green space, riverside parks can enhance the planting of winter ornamental plants and add winter park activities, such as snow watching and skating, on the basis of ecological priority, so as to enhance the attraction of winter landscapes and further improve the landscape benefits of the park.

In general, the renewal and upgrading of urban parks plays an important role in the renewal and optimization of urban living environments. Managers must implement appropriate strategies to strengthen ecological environment protection, increase the construction of distinctive landscapes, and reduce the cost of tourism to enhance the recreation experiences and happiness of visitors. Simultaneously, the park promotion must coordinate with other urban planning projects to ensure the urban landscape ecosystem's stable operation and realize the city's ecological value and sustainable development. It is the primary direction for improving and optimizing waterfront green space in the urban blue–green space.

6. Conclusions

This study analyzed social media text data for 85 typical riverside parks in blue–green spaces in Beijing, a city with abundant river systems. As an innovative study, this work broke through previous urban blue–green space research limitations. The optimization analysis of waterfront green space in blue–green spaces was discussed from the users' perspective. Concurrently, Internet technology was used to obtain a vast quantity of the most recent information and data. An in-depth comparison was conducted based on the park's landscape design and the sensory perception of park visitors for the online evaluations of parks adjacent to different river systems, parks of different types, and parks in different districts. It was found that the evaluation of waterfront parks was closely related to the nature of land use, the surrounding environment and the visiting time of tourists. Enhancing their ecological function, improving park infrastructure and management services are the main issues facing managers at present.

However, the types of people who utilize social platforms are predominantly elderly, childless, and single. Consequently, there were still some flaws in the research findings. In addition, many other studies were aware of this problem [29,55,69]. In the future, it may be necessary to conduct offline questionnaires that are explicitly designed for the elderly and children to remedy research deficiencies using traditional and modern techniques. We hope that future research can improve the evaluation method of the park system further, propose more scientific and reasonable strategies for the improvement of the functions of waterfront parks, and provide more sustainable suggestions for the optimization of urban blue–green space, which will play a significant role in the maintenance of China's urban ecosystem.

Supplementary Materials: The following supporting information can be downloaded at: 'https://www.mdpi.com/article/10.3390/land11101849/s1, Table S1: List of 85 riverside parks in blue-green space; Table S2: IPA list of riverside parks beside different river systems based on landscape design; Table S3: IPA list of riverside parks beside different river systems based on sensory perception; Table S4: Multiple comparison analysis of landscape design indicators of riverside parks beside different river systems; Table S5: Multiple comparison analysis of sensory perception indicators of riverside parks beside different river systems; Table S6: IPA list of riverside parks in different types based on landscape design; Table S7: IPA list of riverside parks in different types based on sensory perception; Table S8: Multiple comparison analysis of landscape design indicators of riverside parks in different types; Table S9: Multiple comparison analysis of sensory perception indicators of riverside parks in different types; Table S10: IPA list of riverside parks in different districts based on landscape design; Table S11: IPA list of riverside parks in different districts based on sensory perception; Table S12: Multiple comparison analysis of landscape design indicators of riverside parks in different districts; Table S13: Multiple comparison analysis of sensory perception indicators of riverside parks in different districts.

Author Contributions: Data curation and formal analysis, S.C.; methodology and project administration, S.C.; software, Z.Z., W.S., Y.W. and R.Y.; writing—original draft preparation, S.C. and Z.Z.; writing—review and editing, X.G.; funding acquisition, X.G. All authors have read and agreed to the published version of the manuscript.

Funding: This research was funded by the projects of the Beijing Municipal Social Science Foundation of China (grant number: 21JCC094), National Natural Science Foundation of China (grant number: 31800606), and Beijing Scientific Research and Postgraduate Education Jointly Construction (grant number: 2015BLUREE01).

Institutional Review Board Statement: Not applicable.

Informed Consent Statement: Not applicable.

Data Availability Statement: All data generated or analyzed during this study are included in this article.

Conflicts of Interest: The authors declare no conflict of interest.

References

1. Notice of the Ministry of Natural Resources on Comprehensively Carrying out Territorial Space Planning Work. Available online: http://www.gov.cn/xinwen/2019-06/02/content_5396857.htm (accessed on 2 June 2019).
2. Zhang, L. Demarcation and Management of Urban Growth Boundary in Combination with Ecological Security Patterns. Ph.D. Thesis, Zhejiang University, Hangzhou, China, 2018.
3. Bolund, P.; Hunhammar, S. Ecosystem Services in Urban Areas. *Ecol. Econ.* **1999**, *29*, 293–301. [CrossRef]
4. Li, T. New Progress in Study on Resilient Cities. *Urban Plan. Int.* **2017**, *32*, 15–25. [CrossRef]
5. Yu, K.; Li, D.; Yuan, H.; Fu, W.; Qiao, Q.; Wang, S. "Sponge City": Theory and Practice. *City Plan. Rev.* **2015**, *39*, 26–36.
6. Di, W.; Wang, Y.; Chen, F.; Xia, B. Thermal environment effects and interactions of reservoirs and forests as urban blue-green infrastructures. *Ecol. Indic.* **2018**, *91*, 657–663. [CrossRef]
7. Ahmed, S.; Meenar, M.; Alam, A. Designing a Blue-Green Infrastructure (BGI) Network: Toward Water-Sensitive Urban Growth Planning in Dhaka, Bangladesh. *Land* **2019**, *8*, 138. [CrossRef]
8. Yuan, Z.; Shi, T.; Hu, Y.; Gao, C.; Miao, L.; Fu, S.; Wang, S. Urban green space planning based on computational fluid dynamics model and landscape ecology principle: A case study of Liaoyang City, Northeast China. *Chin. Geogr. Sci.* **2011**, *21*, 11. [CrossRef]
9. Pouso, S.; Borja, A.; Fleming, L.E.; Gómez-Baggethun, E.; Uyarra, M.C. Contact with blue-green spaces during the COVID-19 pandemic lockdown beneficial for mental health. *Sci. Total Environ.* **2021**, *756*, 143984. [CrossRef]
10. Yu, Z.; Yang, G.; Zuo, S.; Jrgensen, G.; Vejre, H. Critical review on the cooling effect of urban blue-green space: A threshold-size perspective. *Urban For. Urban Green.* **2020**, *49*, 126630. [CrossRef]
11. Jin, Y.; Zhou, Y.; Shen, J. Research on LID Rainfall Unit Design Method for the Blue and Green Ecological Network System—Based on the Analysis of Mountain Hydrological Characteristics. *Chin. Landsc. Archit.* **2018**, *34*, 83–87.
12. Vaeztavakoli, A.; Lak, A.; Yigitcanlar, T. Blue and Green Spaces as Therapeutic Landscapes: Health Effects of Urban Water Canal Areas of Isfahan. *Sustainability* **2018**, *10*, 4010. [CrossRef]
13. Five Major Water Systems and Major Rivers in Beijing. Available online: http://www.pekingmemory.cn/bdsy/2018/09/28/72.html (accessed on 28 September 2018).
14. Chu, T. Exploration on Park City Design of Riverside Space Under Background of Inventory Land—Taking Urban Design of Riverside Space of Pihe Park in Chengdu as an Example. *Urban Archit. Space* **2022**, *29*, 9–15, 41.
15. Donahue, M.L.; Keeler, B.L.; Wood, S.A.; Fisher, D.M.; Hamstead, Z.A.; Mcphearson, T. Using social media to understand drivers of urban park visitation in the Twin Cities, MN. *Landsc. Urban Plan.* **2018**, *175*, 1–10. [CrossRef]
16. Yu, B.; Xie, C.; Yang, S.; Che, S. Correspondence Analysis on Residents' Perceived Recreation Satisfaction and Importance in Shanghai Urban Community Park. *Chin. Landsc. Archit.* **2014**, *30*, 75–78.
17. Xiao, X.; Du, K. A Study On Recreationists' Satisfaction of Guangzhou City Parks. *Hum. Geogr.* **2011**, *26*, 129–133. [CrossRef]
18. Li, C.; Xu, C.; Zhang, Z.; Gong, L.; Li, B.; Jin, G.; Chen, S. Residents' Satisfaction on Country Parks in Beijing. *J. Beijing For. Univ. (Soc. Sci.)* **2010**, *9*, 68–72. [CrossRef]
19. Zhao, Y.; Qin, B.; Liu, T. Sentiment Analysis. *J. Softw.* **2010**, *21*, 1834–1848. [CrossRef]
20. Do, Y. Valuating aesthetic benefits of cultural ecosystem services using conservation culturomics. *Ecosyst. Serv.* **2019**, *36*, 100894. [CrossRef]

21. Wang, Z.; Fu, H.; Jian, Y.; Salman, Q.; Jie, H.; Wang, L. On the comparative use of social media data and survey data in prioritizing ecosystem services for cost-effective governance. *Ecosyst. Serv.* **2022**, *56*, 101446. [CrossRef]

22. Kim, K.-J. Welfare Activation Strategy for a Urban Park Users. *J. Korea Contents Assoc.* **2012**, *12*, 195–204. [CrossRef]

23. Wang, Z.; Jie, H.; Fu, H.; Wang, L.; Jiang, H.; Ding, L.; Chen, Y. A social-media-based improvement index for urban renewal. *Ecol. Indic.* **2022**, *137*, 108775. [CrossRef]

24. Tieskens, K.F.; Zanten, B.; Schulp, C.; Verburg, P.H. Aesthetic appreciation of the cultural landscape through social media: An analysis of revealed preference in the Dutch river landscape. *Landsc. Urban Plan.* **2018**, *117*, 128–137. [CrossRef]

25. Sinclair, M.; Ghermandi, A.; Sheela, A.M. A crowdsourced valuation of recreational ecosystem services using social media data: An application to a tropical wetland in India. *Sci. Total Environ.* **2018**, *642*, 356–365. [CrossRef] [PubMed]

26. Fisher, D.M.; Wood, S.A.; White, E.M.; Blahna, D.J.; Lange, S.; Weinberg, A.; Tomco, M.; Lia, E. Recreational use in dispersed public lands measured using social media data and on-site counts. *J. Environ. Manag.* **2018**, *222*, 465. [CrossRef] [PubMed]

27. Zhang, G.; Li, J.; Zhang, L. A Research on Tourism Destination Image Perception of Huashan Scenic Spot: Based on Text Analysis of Weblogs. *Tour. Sci.* **2011**, *25*, 87–94. [CrossRef]

28. Jiang, X.; Wu, D.; Wang, X. Investigation about After-use Evaluation of Garden Exhibition Based on Online Comments Data. *Landsc. Archit.* **2018**, *25*, 74–80. [CrossRef]

29. Wang, Z.; Zhao, J.; Peng, Y.; Yue, W. Comparative Evaluation of Guangzhou City Parks: Text Analysis Based on Social Media Data. *Landsc. Archit.* **2019**, *26*, 89–94. [CrossRef]

30. Li, L. Identification Research of Tangible and Intangible Attribute Value of Urban Heritage Based on Deep Learning: A Case Study of Suzhou River. *Urban Dev. Stud.* **2021**, *28*, 104–110.

31. Dwivedi, M. Online destination image of India: A consumer based perspective. *Int. J. Contemp. Hosp. Manag.* **2009**, *21*, 226–232. [CrossRef]

32. Govers, R.; Go, F.M. Projected destination image online: Website content analysis of pictures and text. *Inf. Technol. Tour.* **2004**, *7*, 73–89. [CrossRef]

33. Zhao, J.; Liu, X.; Dong, R.; Shao, G. Landsenses ecology and ecological planning toward sustainable development. *Int. J. Sustain. Dev. World Ecol.* **2016**, *23*, 293–297. [CrossRef]

34. Dong, R.; Liu, X.; Liu, M.; Feng, Q.; Su, X.; Wu, G. Landsenses ecological planning for the Xianghe Segment of China's Grand Canal. *Int. J. Sustain. Dev. World Ecol.* **2016**, *23*, 298–304. [CrossRef]

35. Gao, Y.; Yuan, J.; Yan, M. Study on the influencing factors of the vitality of public space in Wulihe Park in Shenyang. In Proceedings of the 2020/2021 China Urban Planning Annual Conference and 2021 China Urban Planning Academic Season, Chengdu, China, 25–30 September 2021; pp. 1192–1198.

36. Zhao, J.; Yan, Y.; Deng, H.; Liu, G.; Shao, G. Remarks about landsenses ecology and ecosystem services. *Int. J. Sustain. Dev. World Ecol.* **2020**, *27*, 196–201. [CrossRef]

37. Wu, Y.; Fu, H. Trait of Urban Water System Property and its Tourism Utility and Preservation in Beijing. *J. Cap. Norm. Univ. (Nat. Sci. Ed.)* **2004**, *02*, 66–70, 84. [CrossRef]

38. Zhang, F. The new born in the old: Urban renewal in Beijing. *Beijing Plan. Rev.* **2022**, *04*, 164.

39. Zhang, K. The relationship between river and open space in Europe, Cities: Case studies of London and Emshere District Park. *City Plan. Rev.* **2013**, *37*, 76–80.

40. Jo, T.; Sato, M.; Minamoto, T.; Ushimaru, A. Valuing the cultural services from urban blue-space ecosystems in Japanese megacities during the COVID-19 pandemic. *People Nat.* **2022**, *4*, 1176–1189. [CrossRef]

41. Knight, S.J.; McClean, C.J.; White, P.C.L. The importance of ecological quality of public green and blue spaces for subjective well-being. *Landsc. Urban Plan.* **2022**, *226*, 104510. [CrossRef]

42. Mishra, H.S.; Bell, S.; Vassiljev, P.; Kuhlmann, F.; Niin, G.; Grellier, J. The development of a tool for assessing the environmental qualities of urban blue spaces. *Urban For. Urban Green.* **2020**, *49*, 126575. [CrossRef]

43. Subiza-Pérez, M.; Hauru, K.; Korpela, K.; Haapala, A.; Lehvävirta, S. Perceived Environmental Aesthetic Qualities Scale (PEAQS)—A self-report tool for the evaluation of green-blue spaces. *Urban For. Urban Green.* **2019**, *43*, 126383. [CrossRef]

44. Gong, M.; Ren, M.Y.; Dai, Q.; Luo, X.Y. Aging-Suitability of Urban Waterfront Open Spaces in Gongchen Bridge Section of the Grand Canal. *Sustainability* **2019**, *11*, 6095. [CrossRef]

45. Martilla, J.A.; James, J.C. Importance-Performance Analysis. *J. Mark.* **1977**, *41*, 77–79. [CrossRef]

46. Liang, H.; Wang, Y.; Liu, M. Tourists' Perception of Local Food Experience in Tourist Destination by IPA Analysis: A Case Study in Enshi, Hubei. *J. Agro-For. Econ. Manag.* **2016**, *15*, 335–342. [CrossRef]

47. Chen, X. The Modified Importance-performance Analysis Method and its Application in Tourist Satisfaction Research. *Tour. Trib.* **2013**, *28*, 59–66.

48. Chen, P.; Liu, W. Assessing management performance of the national forest park using impact range-performance analysis and impact-asymmetry analysis. *For. Policy Econ.* **2019**, *104*, 121–138. [CrossRef]

49. Go, F.; Zhang, W.J. Applying Importance-Performance Analysis to Beijing as an International Meeting Destination. *J. Travel Res.* **1997**, *35*, 42–49. [CrossRef]

50. Evans, M.R.; Chon, K.S. Formulating and Evaluating Tou Rism Policy Using Importance-Performance Analysis. *J. Hosp. Tour. Res.* **1989**, *13*, 203–213. [CrossRef]

51. Gu, X. Research for Vistors' Structure, Behavior, Demand Characteristics and Influencing Factors in Shanghai Parks. Master's Thesis, East China Normal University, Shanghai, China, 2013.
52. Fan, Y.; Mao, D.; Zhou, C.; Ye, J.; Chen, L.; Zheng, Y. Recreational Resources Evaluation of Fuzhou West Lake Park Based on Internet Text Analysis. *J. Chin. Urban For.* **2019**, *17*, 41–46.
53. Liu, Q.; Pan, Y.; Zhang, Z.; Wang, X. Recreational Satisfaction of Typical Riverfront Public Spaces in Shanghai Based on IPA Analysis. *J. Chin. Urban For.* **2021**, *19*, 29–34.
54. Lin, P.; Chen, L.L.; Luo, Z.S. Analysis of Tourism Experience in Haizhu National Wetland Park Based on Web Text. *Sustainability* **2022**, *14*, 3011. [CrossRef]
55. Zheng, T.; Yan, Y.; Zhang, W.; Zhu, J.; Wang, C.; Rong, Y.; Lu, H. Landsense assessment on urban parks using social media data. *Acta Ecol. Sin.* **2022**, *42*, 561–568.
56. Dai, J.; Yuan, J. Comparison of one-way analysis of variance and multiple linear regression analysis. *Stat. Decis.* **2016**, *9*, 23–26. [CrossRef]
57. Connors, J.P.; Galletti, C.S.; Chow, W.T.L. Landscape configuration and urban heat island effects: Assessing the relationship between landscape characteristics and land surface temperature in Phoenix, Arizona. *Landsc. Ecol.* **2012**, *28*, 271–283. [CrossRef]
58. Gao, Y.; Zhang, T.; Zhang, W.K.; Meng, H.; Zhang, Z. Research on visual behavior characteristics and cognitive evaluation of different types of forest landscape spaces. *Urban For. Urban Green.* **2020**, *54*, 126788. [CrossRef]
59. Wang, Y.; Liu, J.; Shao, L.; Tang, Y. Guiding Factors and Orientations for Improving Urban Waterfront Space: A Case Study of Haidian Section of Qinghe River in Beijing. *World Archit.* **2022**, 24–33. [CrossRef]
60. Li, F.; Li, K.; Li, X. Research on Scenario Planning of Beijing Second Green Belt Country Park Ring. *Landsc. Archit.* **2021**, *28*, 58–64. [CrossRef]
61. Dianping Official Website. Available online: https://www.dianping.com/ (accessed on 2 February 2022).
62. Ctrip Official Website. Available online: https://www.ctrip.com/ (accessed on 2 February 2022).
63. Mafengwo Official Website. Available online: https://www.mafengwo.cn/ (accessed on 2 February 2022).
64. Notice of Beijing Municipal Bureau of Landscape and Afforestation on the Issuance of the Measures for the Classification and Classification of Parks in Beijing. Available online: http://yllhj.beijing.gov.cn/zwgk/fgwj/gfxwj/202206/t20220621_2747422.shtml (accessed on 2 February 2022).
65. Liu, Q.; Jia, H. A View of Chinese Word Automatic Segmentation Research in the Chinese Information Disposal. *Comput. Eng. Appl.* **2006**, *175–177*, 182.
66. Wang, M.; Qiu, M.; Wang, J.; Peng, Y. The Supply-demand Relation Analysis and Improvements Based on Importance-Performance Analysis of Cultural Ecosystem Services in Waterfront Areas Along the Suzhou Creek in Shanghai. *Landsc. Archit.* **2019**, *26*, 107–112. [CrossRef]
67. Zheng, T.; Yan, Y.X.; Lu, H.; Pan, Q.; Zhu, J.; Wang, C.; Zhang, W.; Rong, Y.; Zhan, Y. Visitors' perception based on five physical senses on ecosystem services of urban parks from the perspective of landsenses ecology. *Int. J. Sustain. Dev. World Ecol.* **2020**, *27*, 214–223. [CrossRef]
68. Xu, B.; Shi, Q.; Zhang, Y. Evaluation of the Health Promotion Capabilities of Greenway Trails: A Case Study in Hangzhou, China. *Land* **2022**, *11*, 547. [CrossRef]
69. Hausmann, A.; Toivonen, T.; Fink, C.; Heikinheimo, V.; Kulkarni, R.; Tenkanen, H.; Di Minin, E. Understanding sentiment of national park visitors from social media data. *People Nat.* **2020**, *2*, 750–760. [CrossRef]

land

Article

Urban Forest Tweeting: Social Media as More-Than-Human Communication in Tokyo's Rinshinomori Park

Diego Martín Sánchez [1,*] **and Noemí Gómez Lobo** [2]

1   GIPC—Cultural Landscape Research Group, ETSAM Madrid School of Architecture, Universidad Politécnica de Madrid, 28040 Madrid, Spain
2   CAVIAR—Quality of Life in Architecture Research Group, Department of Architecture, University of the Basque Country (UPV/EHU), 20018 Donostia, Spain
*   Correspondence: d.martins@upm.es

Abstract: Urban parks are places that have significant impact on the physical and mental health of citizens, but they are also for safeguarding biodiversity and thus fostering human–nature interactions in the everyday landscape. The exploration of these spaces through social media represents a novel field of research that is contributing to revealing patterns of visitor behavior. However, there is a lack of comparable research from a non-anthropocentric perspective. What if we could use social media as a more-than-human communication medium? This research aims to reveal the possibility of communicating the urban forest's voice through the examination of the official Twitter account of a metropolitan park in Tokyo. To this end, an analysis of the content of the messages is carried out, focusing on the narrative voice from which the message is told, the protagonists, the action performed, the network of actors deployed, and the place where it occurs. It is found that the majority of these messages are delivered from a non-human perspective, where plants, animals, or meteorological agents behave deploying complex networks of more-than-human interaction. The current study reveals the latent potential of non-humans as possible agents within the realm of social media, which can mediate the relationships between humans and their environment. It introduces a layer that can be incorporated into future lines of research, as well as provides a model case that illustrates a good practice in the management and communication of urban green spaces.

Keywords: twitter; more-than-human; urban forestry; urban park; Tokyo

Citation: Martín Sánchez, D.; Gómez Lobo, N. Urban Forest Tweeting: Social Media as More-Than-Human Communication in Tokyo's Rinshinomori Park. *Land* **2023**, *12*, 727. https://doi.org/10.3390/land12040727

Academic Editors: Cecilia Arnaiz Schmitz, María Fe Schmitz and Sarel Cilliers

Received: 31 December 2022
Revised: 19 March 2023
Accepted: 20 March 2023
Published: 23 March 2023

1. Introduction

The climate emergency is driving a growing academic momentum in the more-than-human dimension across disciplines such as urban planning, environmental studies, communication, and urban geography. The anthropocene is the new geologic epoch that recognizes human activity as having a major impact on geological time. Nevertheless, the management of the planet is intertwined with forces conceived as both human and natural, as different species have evolved co-constitutively with human communities [1]. As urban populations continue to grow, an exponential number of publications are exploring the importance of natural environments within cities. This ethical–political approach urges abandoning the conception of cities as the antithesis of nature and considering them as part of a "metabolic" relationship between society and nature, the city as an ecological space [2]. The current study engages with such context exploring the perspective of the urban forest itself. To aid a conceptualization that transcends the human as the reference, this introduction provides a theoretical overview of the ecofeminist framework and the philosophical background of co-producing urbanism with other species. Then, the literature on the intersection between social media and green spaces, and specifically Twitter, are reviewed as novel tools for understanding the relationship with nature in the city.

1.1. An Ecofeminist Perspective: Framing Communication beyond the Human

Transcending the anthropocentric worldview, ecofeminist thought adds a layer of consciousness that contributes to creating a reality with greater awareness of recording actions beyond the human. Haraway calls for creating a kinship with all forms of life, including those agents of the environment that are co-shaping and co-creating our habitats [3]. Communication scholars such as Plec highlight the "role of communication in the construction and transformation of human relationships with the more-than-human world" [4]. Rethinking conventional modes of social interaction, she criticizes the modernist paradigm that denigrates narratives linked to other beings on earth.

Until this day, rhetoric based on technological progress points to humans as the only ones capable of communicating. However, other forms of life constitute our environment and constantly speak to us. Plec suggests "internatural communication", an alternative and inclusive significance that embraces other possibilities beyond human life forms. Puig de la Bellacasa and Bennett bring to the forefront of the discussion notions of care that, in conjunction with the co-production of the public sphere and its eco-social realities, can engender new models of coexistence beyond human exceptionalism [5,6]. From multispecies geography, Pitt proposes to engage more directly with plants and how they perform to clarify "not just what plants mean to people but how they contribute to gardens" [7].

1.2. Co-Producing Urbanism with Other Species

New planning theories analyze multi-species use and occupation of space, recognizing the capacity for action of other non-human entities. Various bodies co-produce our urban spaces, and these bodies have diverse material dimensions. Their ecologies condition and question conventional city planning that fabricate policies based on the culture-nature separation. Urban planning continues to address ecology from a human angle, focusing on establishing measures solely according to the human population [8]. Conceiving human beings as part of a larger web implies generating new vocabularies and practices challenging the dominant ways of knowing, towards more inclusive trans-species urbanisms [9].

The creation of multispecies planning tools and narratives would support the realization of diverse urban futures that are more socially and environmentally just [10]. Empathetic imaginaries of everyday practices and trans-species encounters emerge from examining the city, vividly expressed by Gruen's "entangled empathy" to create more-than-human urban environments [11]. This notion urges us to reconsider what counts in urban planning, going beyond "homo urbanis" as a benchmark for ethical action. It is necessary to assume responsible interactions of care attuned to the urban ecosystem inhabited by diverse beings [12]. To analyze the interactional phenomena between nature and humans, urban green spaces present a unique field of work to provide new theories and methodologies.

1.3. Social Media for Researching: Urban Green Spaces

Urban planning research has consistently focused on green spaces as essential elements for human well-being [13]. Some of the main avenues of scholarship regard their impact on health, sense of place, and community interactions—or their usage, in terms of when and how people interact, as well as identifying barriers to their accessibility [14,15]. Emerging technologies are advancing techniques for assessing human interaction with urban green spaces. The evaluation of these relationships can be from an observational approach or through technological devices that operate with big data. Social networks are a clear example of the application of crowdsourcing methodologies, as they allow anyone connected to the Internet to produce information. Recent studies on the intersection between the performance of nature and user experience in urban parks demonstrate that social media is potentially an effective evaluation tool for public park management [16]. User-generated content serves as a vehicle for conveying visitor feedback to park staff, accessing opinions about the use of a place, emotions, or the movement of people in a space [17].

Since 2018, among the emerging social networks in the research of urban green spaces, Twitter appears as a powerful source to understand citizens' feelings, satisfaction levels, and behavior [18]. It allows communication of free statuses and messages with other users connected to the network, with a limit of 280 characters per text, having open access to different accounts. The ease of getting public data and a large number of users have presented Twitter as a source valid and desirable [19]. Twitter data have contributed to gaining insight into the sentiment variation between urban and natural land cover types [20]. Geo-referencing tweets at the urban park scale also allowed linking well-being to specific spatial and user characteristics [21]. Adding a cross-linguistic perspective, other scholars have compared human demands and emotions in different cultural contexts [22].

There is a body of research using Twitter data to understand liveliness in public parks, associating density, diversity, accessibility, travel demand, and human behaviors. To assess the quality of urban green spaces, Twitter data are discussed concerning the morphology of the built environment, pedestrian routes, and the variation in demographic characteristics of visitors, whether gender, age, or ethnicity [23]. Some studies link large-scale geotagged data to reveal temporal and spatial patterns and infer activity purposes [24]. Others identify Twitter as a source for comprehending spatiality in urban parks and the ability to bond people from various neighborhoods [25].

Most recently, in the wake of COVID-19, there has been a proliferation of studies using Twitter to examine the interaction with urban green spaces during the pandemic. Researchers found the potential of Twitter as a source of real-time information to enable quicker engagement in emergencies, as well as a snapshot of the challenges faced by park administrators. Some focused on the contested opinions on regulations in different countries, detecting the frequency of topics, and manually sorting the opinions by the research team [26]. Other studies collected park-specific values, emphasizing the importance of users' behavior in informing management strategy [27].

1.4. Twitter, An Emerging Tool for Exploring Biodiversity

Twitter-based biodiversity research includes studies that utilize geolocation methodologies similar to those observed in the analysis of urban parks. For instance, this includes applying them to assess the risk posed by visitor pressure in protected areas [28], or to clarify user sentiment and characteristics in national parks [29]. Further research focuses on the evaluation of certain keywords to establish indicators of interest [30], to illuminate the main actors around specific topics [31], or to determine the popularity of species both endangered [32] and familiar [33]. Several studies have also explored the utility of Twitter as a tool for the identification and monitoring of both invasive [34] and endemic species [35]. When it comes to monitoring, Twitter serves as a powerful resource for getting citizens' views on park policies. It can magnify controversies or spark public discourse about biodiversity management [36].

However, as previously reviewed, digitized information generated through devices, applications, sensors, or platforms is mainly person-centered. Lupton questions this human-focused approach, reclaiming the collection of digital technologies as a registry of knowledge forms about different aspects of life, including non-human ones [37]. The current study aims to add to the social landscape of cities by exploring urban green spaces from the perspective of the non-human. It contributes to filling the gap in the literature that uses Twitter as a data source to explore the more-than-human world that speaks to us. Understanding data as phenomena endowed with vitality, the critical curation of Twitter can create an awareness of the natural realm of our parks, contributing to a better sense of the biodiversity in the city.

2. Materials and Methods

2.1. Case Study

In this scenario, Japan arises as a fruitful context that transcends the anthropocentric worldview, presenting relational and ontological models that consider human and non-human beings as interconnected and trans-agential. Shinto heritage and folkloric traditions bridge boundaries between the human and the non-human, considering them together as participants in collective life [38]. Japanese animistic cosmology overturns classical Western dichotomies between what constitutes otherness, culture, nature, subject, and object. Japan subverts the hyper-stratified categories of knowledge in dualistic assumptions that conceive humans as different and superior to other beings [39].

Despite being a country with an impressive presence of nature, two-thirds of its land is forest, and the anonymous urban green spaces that serve populated neighborhoods of its major city, Tokyo, often go unnoticed [40]. Close-to-home parks, elements that provide contact with nature and allow recovery in case of emergencies, have proven to be essential in planning [41]. Rinshinomori Park, a neighborhood-scale green space presents an ideal case study to explore the intersection with nature in a dense urban fabric. Nestled in central Tokyo, Rinshinomori Park has a great diversity of flora and fauna across its 12 hectares. Additionally, it is also selected for its background as a center of experimentation in forestry, which has produced a particularly biodiverse forest, with century-old tree species and a significantly high canopy density. The park workers, who take care of the relations between natural resources and citizens, are often invisible to the public eye [42]. However, they have found in social media a way to narrate the spatial and temporal rhythms of green space management and the life of other beings that inhabit the urban forest.

The Rinshinomori Park Twitter account, @ParksRinsi, presents a single voice, that of the park, which in turn brings together multiple perspectives. Other Tokyo Metropolitan Park Administrations also manage similar Twitter accounts, some of which started tweeting even earlier, have gained a larger number of followers, or are more prolific in their posts. However, by voicing tweets from the perspective of different urban forest actors, the Rinshinomori Park Twitter account is representative of how social media can serve as a more-than-human praxis. Park staff act as mediators, tweeting to give voice to those agents who do not usually communicate, such as plants or animals that inhabit the urban forest. They make visible all the beings beyond the human, including them, but also other phenomena, such as meteorology or latent emergencies related to catastrophes. Through consistent analysis of the short texts and images, it is possible to know who is exercising the action and what is being represented, combining quantitative and qualitative evaluation. The park's Twitter account, as the object of study in this article, is a novel case study for reporting on environmental issues that transcend the human-use perspective, and illustrate how the park is a multi-species habitat.

2.2. Methodology

The existing body of scholarship provides several methods to deal with Twitter data. A spatial–temporal investigation based on the geolocation of tweets is among the most common ones [43], which can also be cross-referenced with other demographic [44] or ecological data [28]. Other publications complement this analysis by focusing on the content of the message, which can examine the appearance of certain keywords [45] or the feelings and emotions evoked in the tweets [46]. Besides the studies purely based on sentiment analysis [29], other works focus on the network generated by the messages [31], on their popularity, or on the engagement they foster [47]. This article grounds the exploration of urban forestry as more-than-human practice through the examination of the contents posted on the official Twitter account of Tokyo's Rinshinomori Park: @ParksRinsi. The investigation follows content analysis methodologies, focusing on semantic aspects, distinguishing more-than-human terms [48], the analysis of the narrator's perspective, and identification of the actors' network evoked in each message.

In comparison to the outreach levels of other parks within Twitter, such as Central Park in New York with more than 200 thousand followers, @ParksRinsi has a local impact profile, with around three thousand followers to date of this article, a number similar to the rest of the Tokyo metropolitan parks. The account has been active since March 2016, posting its first and representative message on Friday, April 1 of that same year: "Cherry blossoms are in full bloom". The current study will examine all tweets published since then, until 25 June 2022, the date on which the first abstract of this article is prepared.

Among the total 1333 messages published during this period, 311 retweets and 14 replies were excluded, to focus on the 1008 tweets with original content (Figure 1). All tweets were downloaded using Twitter API and sorted by date of publication. Then, each message was translated from Japanese to English using DeepL API, and Google Translate for cross-checking. The accompanying media—images and videos—were also downloaded to support the text analysis. The content of each tweet was analyzed manually according to the following categories: *narrative voice, main actor, action, actor-network*, and *place*. The *narrative voice* is considered as the grammatical tense in which the action takes place—past, present, future—in conjunction with the grammatical person—first, second or third, and singular or plural; it is taken as an indicator to register the plurality of points of view registered in the different tweets. Whether the *main actor* in each tweet is human or non-human is recorded as a critical factor for identifying the type of relationship that unfolds. The *action* described in the tweet is also documented, noting whether it corresponds to an anthropocentric view. Then, the number of actors mentioned in each tweet is noted as the *actor-network count*, regardless of their human or non-human character, to know the extent of the network implied by the tweet. Finally, it is recorded whether the text mentions the specific *place* within the park, since in some tweets it emerges as a determining factor that conditions the message. Once this analysis is done, the patterns of more-than-human relationships found are extracted based on the main actor and the network of actors involved in the tweet.

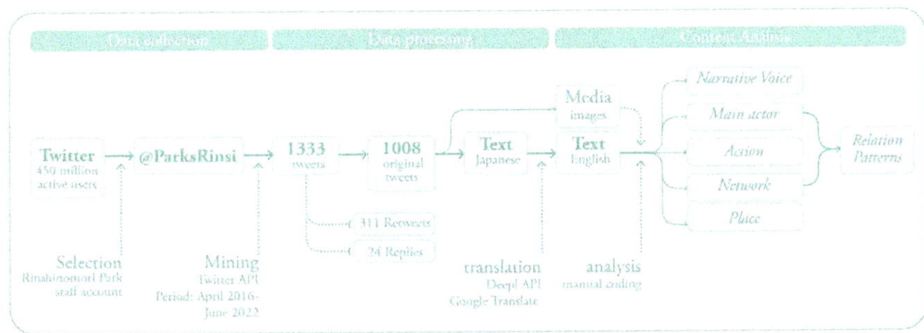

Figure 1. Flow diagram of the methodology.

2.3. Analysis Example

If we take as an analysis example (Figure 2) the tweet of Thursday, 20 January 2022, we can observe that the message is narrated in the present tense and in the first person singular from the point of view of a non-human main actor: the aerial root. The central action is to see, to notice, and a network of five actors is deployed: the swamp cypress itself, the winter, the observer, the other two parks in Tokyo where it is possible to find the same phenomenon, and the place within the park itself—the pond.

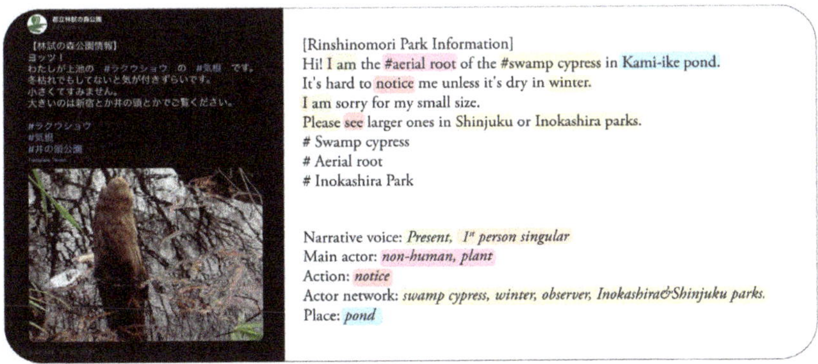

Figure 2. Analysis example. Original tweet in Japanese, English translation and categories.

3. Results

3.1. Content Analysis Categories

Narrative voice. As noted above, the narrative voice refers to the time and the point of view from which the action is narrated. It is crucial when dealing with an urban forest, where multiple temporalities differ depending on the perspective of each constituent subject. The analysis shows that the narration can happen in the present, such as the appearance of a certain flower, or it can herald future events, like upcoming neighborhood events or potential weather disasters, as well as recall the historical past of Rinshinomori when it used to be a public research facility (Figure 3). One of the characteristics found in the Rinshinomori Park account is that every possible narrative voice can be found along its timeline.

Figure 3. Narrative voice. This tweet talks about the history of the park through the trees. Original Japanese text and English translation.

Main actor: Identifying the protagonist was key to this analysis and required a detailed breakdown of all messages. A total of 1097 protagonists were found (Table 1), which implies that in a small but significant portion of tweets we can observe two protagonists. As established earlier, two main categories were distinguished: human and non-human. This division reveals that the majority of the messages, 57.2%, are performed by a non-human protagonist, a significant result that reinforces the more-than-human communication approach of this account.

Table 1. Main actors count.

Human Main Actor				Non-Human Main Actor							
Professional		Non-Professional		Plants		Climate		Animal		Others	
staff/municipality	385	neighbors	49	plant	160	rain	39	bird	31	festivity/event	23
		children/students	20	tree	129	temperature	27	insect	35	facility/furniture	13
		volunteers/association	16	flower	22	wind/typhoon	21	reptile	9	park	9
-	-			fruit/seeds	18	season	20	fish	5	covid	10
		-	-	leaves	17	sun/sky	14	amphibians	3	pond	5
				fungi	9	ice/snow	7	mammal	1	-	-
(81.9%) 385		(18.1%) 85		(56.6%) 355		(20.4%) 128		(13.4%) 84		(9.6%) 60	
(42.8%) 470						(57.2%) 627					
				1097							

Regarding the Human group, several recurring actors emerge: park staff, neighbors, volunteers and associations, or students and children. Within this group, there is a clear division between professionals and citizens, with 81.9% of the human protagonists featuring park workers. This outcome is partially explained by the fact that the Twitter account is managed by the park administration, but also underscores the existing barrier for citizens to access natural resources. Still, it is noteworthy that 18.1% of the human main actors are neighbors, volunteers, or students. If we consider the Non-human group, the following actor subgroups can be distinguished: plants, animals, climate and meteorological agents, spatial elements, and others. Among these, the botanical realm appears most frequently as the non-human protagonist of the tweets (56.6%), which is logical given that there are more than 6100 large trees and most of the park's surface is covered by dense vegetation of great biodiversity. Of the remaining subgroups, 20.4% are related to meteorological and climatic elements, an understandable aspect when trying to inform potential visitors about the park's status. Similarly noteworthy is the 13.4% of animal protagonists, where birds and insects are the most representative. This might respond to the interest shown from the neighbors' side, since there is a community of enthusiastic wild bird photographers and, especially in summer, local children go out in search of beetles and cicadas to capture (Figure 4).

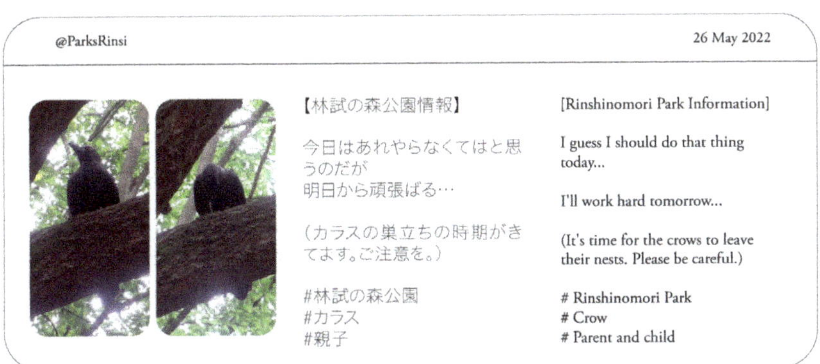

Figure 4. Tweet from the perspective of a crow. Original Japanese text and English translation.

Action. Considering the entirety of actions that appear in the park messages, there is a great diversity of outcomes. As would be expected, verbs associated with human activities appear repeatedly, for instance, those activities performed by the park staff that relates to the practice of urban forestry such as plant, care, prune, harvest, or repair, (Figure 5) or those related to the management of this kind of public space such as clean, organize, research, test, open, prohibit, make, protect, or inform. Others express the daily life of the park's neighbors such as visit, enjoy, like, celebrate, sing, photograph, exhibit, sell, etc., as well as their active involvement with the urban forest's inhabitants such as identify, participate, cook, help, or build.

Figure 5. Tweet expressing actions of care carried by the park workers. Original Japanese text and English translation.

Nevertheless, the frequent appearance of verbs related to non-human practices is also remarkable, such as the actions of the plant kingdom such as bloom, fall, take root, grow, sprout, or produce; or those performed by the animal community such as fly, hatch, eat, sound, nest, attack, breath, step, or feed; as well as the meteorological ones such as rain, blow, shine, change, rise, or destroy. It is worth mentioning the appearance of verbs that reflect a sensorial perception of the park such as hear, smell, touch, bathe, or play, which contrast with the strictly visual experience that has dominated 20th-century urban planning.

Actor-network. By counting the actors that appear in each tweet, besides the protagonist, the aim is to reveal the plurality of the network involved in each message (Figure 6). It is noteworthy that the vast majority of the tweets, almost 80%, deploy a network of at least two actors, thus establishing multidirectional relationships and diversifying their scope. This result is interesting when overlaid with the previous main actor category, indicating that most of the messages dealing with the urban forest deploy a more-than-human actor-network.

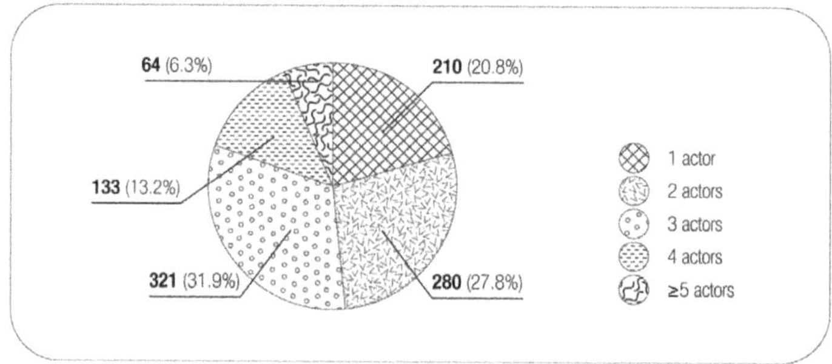

Figure 6. Percentage of the tweets according to the number of actors involved.

Place. Only one-third of the tweets include the location within the park as part of the message. By recording all these places, the following sets emerge: access, connections, open spaces, water bodies, service centers, and community gardens (Figure 7). All these sites have in common that they can be visited by neighbors, which indicates that the appearance of the place in a tweet is a human-oriented message, although most of the park's surface is covered by a mass of trees.

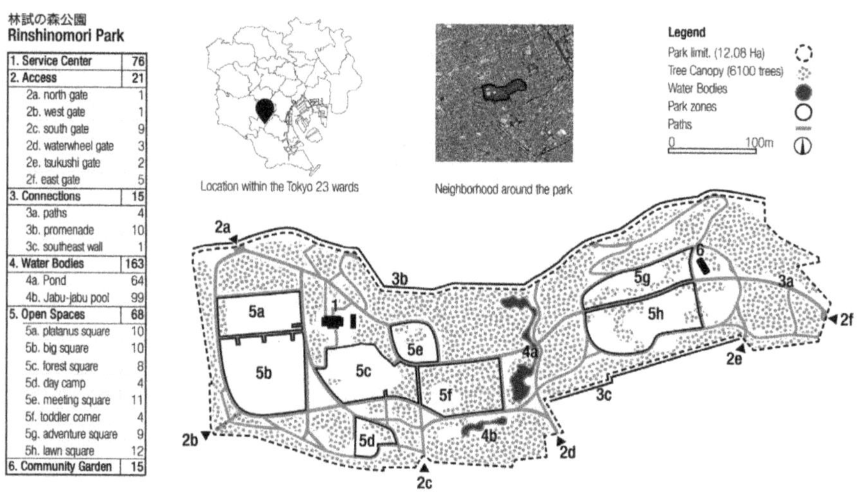

林試の森公園
Rinshinomori Park

Place	Count
1. Service Center	76
2. Access	21
2a. north gate	1
2b. west gate	1
2c. south gate	9
2d. waterwheel gate	3
2e. tsukushi gate	2
2f. east gate	5
3. Connections	15
3a. paths	4
3b. promenade	10
3c. southeast wall	1
4. Water Bodies	163
4a. Pond	64
4b. Jabu-jabu pool	99
5. Open Spaces	68
5a. platanus square	10
5b. big square	10
5c. forest square	8
5d. day camp	4
5e. meeting square	11
5f. toddler corner	4
5g. adventure square	9
5h. lawn square	12
6. Community Garden	15

Figure 7. The number of times a specific place inside the park is mentioned in a tweet.

The most recurrent set is the water bodies. The jabu-jabu pool is a small outdoor pool for children which, besides operating exclusively in summer, is tremendously popular with the residents. The pond is a small ecosystem, and although it is not accessible to humans, it is a highly appreciated place in the neighborhood because it contains great biodiversity that changes throughout the year. The service center is one of the most repeated places, as it is where the park staff work, and also functions as an information and citizen center.

3.2. More-Than-Human Communication Patterns

Based on the categories of the main actor and the composition of the actor-network involved in the tweet, the following communication patterns were found (Figure 8):

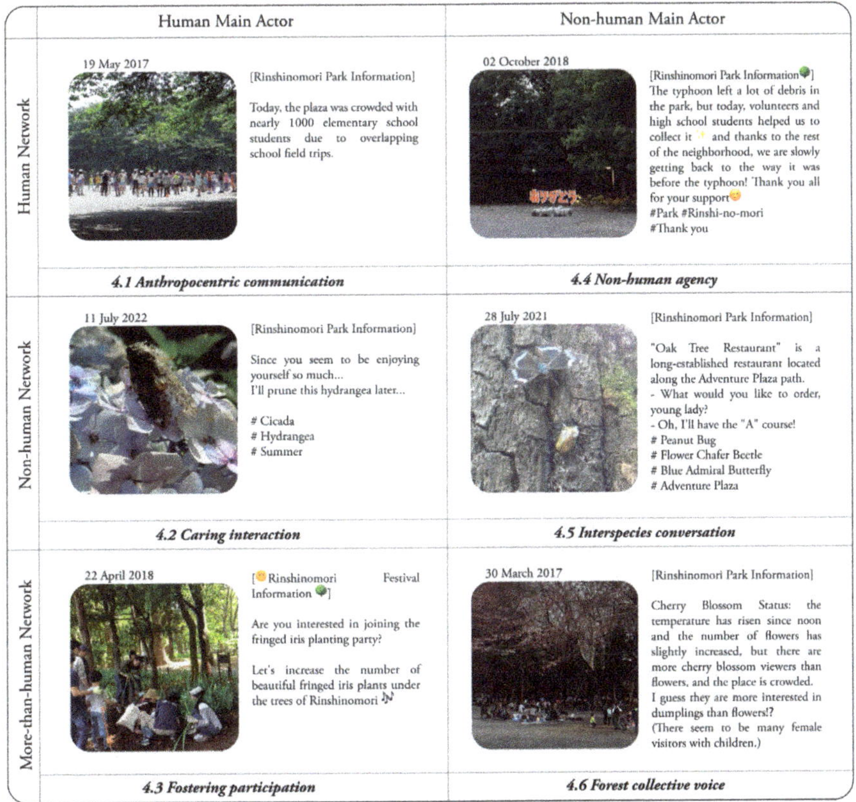

Figure 8. More-than-human communication patterns. *Anthropocentric communication*: relation between only human actors; *Caring interaction*: communication from humans to non-humans; *Fostering participation*: a human protagonist that involves a more-than-human community; *Non-human agency*: the non-human actor that deploys a human network; *Interspecies conversation*: tweets depicting a strictly non-human conversation; *Forest collective voice*: an event triggered by a non-human, engenders a more-than-human actor-network.

Anthropocentric communication: This pattern reflects the communication from humans to humans, and therefore represents the conventional relations on Twitter. In our case study, this pattern is rather uncommon because it is easy for non-human actors from the park to appear in the post. Still, event announcements, upcoming activities, rules, and administrative information are some examples of this pattern.

Caring interaction: Another type of pattern observed is that in which humans address non-humans. Considering that non-humans cannot understand this kind of message, it may still be a patronizing view in which a human speaks to a plant or animal from an intellectually superior position. However, the value resides in the fact that most of this group reports unnoticed everyday care practices performed by park workers. These messages also create a first layer of attention to other beings by placing them as an audience.

Fostering participation: Other tweets are transmitted from a human protagonist involving a more-than-human community of actors. This pattern contains messages that call for citizen participation in natural events that are occurring in the park. In this way, more-than-human interaction within a public space is encouraged or conditioned through a social network.

Non-human agency: As previously established, we discover a majority of tweets where the conventional sequence is inverted and non-humans have agency in the message. Within this condition, a pattern of tweets emerges where non-humans are the protagonists of an action involving humans. Paradigmatic instances are weather changes or possible natural disasters, which affect the community of neighbors and workers.

Interspecies conversation: In this particular pattern, humans are completely excluded from the message and purely non-human actors are recorded in the tweet. In these messages, forest communication is represented as we would perceive it if we were able to translate codes that are foreign to our species. Some of these registered interspecies conversations are flowering changes due to the advance of the seasons, or animals that share their eating habits based on fruits from the park.

Forest collective voice: A final pattern represents the highest expression of actor inclusiveness in the message. The main actor is non-human acting within a more-than-human network. Not only has the agency of the action been changed, shifting it to the non-human, but a community that is both human and non-human is involved in the message.

4. Discussion

Our contribution to the existing body of research is twofold: on the one hand, to reinforce Twitter as a fertile ground for research and, at the same time, to value the Rinshinomori park Twitter account as an example of good practice to foster more-than-human relations that can be extrapolated to the maintenance and administration of similar urban green spaces.

4.1. Twitter as a Research Tool

As reviewed in the introduction, the existing literature regarding the use of Twitter in the context of urban green spaces and biodiversity interaction is an emerging area of research. Thus, a large proportion of the existing work aims to demonstrate its credibility as a valid tool in the investigation of parks [49,50]. While other research has a more practical outlook, we are proposing the application of this instrument as potential support in the management of urban landscapes and ecosystems [51].

Already, an extensive body of scholarship is relying exclusively on the use of Twitter [52] or its Chinese counterpart Weibo [53] as the only material of study. However, it is equally common to find contributions that combine its exploration with other social networks such as Flickr [28] and Instagram [49], with demographic data [44], or with internet searches and digital newspapers [30]. Other authors propose to supplement its utilization with traditional sources such as field work [48], semi-structured interviews [49,50], or surveys [51].

The worldwide coverage provided by this social network allows researchers to analyze diverse case studies, situated on a broad range of spatiotemporal scales. Analogous to the present investigation, which responds to the specificity of a single urban green space, there are previous studies circumscribed to a single urban park [50,51]. Others use Twitter to analyze several parks in the same metropolitan area [52], based on the comparison of different cities in the same country [29], between countries [49], or even on a global scale.

Defining aspects when using Twitter as study material are the actors selected and how information is mined for subsequent analysis. The possibility of accessing a global network has resulted in a tendency to deal with a large number of anonymous users [43]. To process an enormous amount of data, scholars frequently apply filters according to geolocation or specific content in the messages. To a lesser extent, there are fewer studies that use a single Twitter thread [48], or the content generated by a reduced number of specialized accounts [47]. Building along this line, the present study is the first that relies on the detailed analysis of the content generated by a single account, contributing with a novel approach to the existing body of research.

4.2. Rinshinomori Park Twitter Account as Good Practice for More-Than-Human Communication

The data processing procedure is consistent with existing research methodologies, with the automated steps first and then ending with authors screening tweets [27]. The vast amount of information provided by social networks and big data has led to a proliferation of automatic analysis methodologies [28] assisted by machine learning, which yield quantitative analyses. However, in this paper, as in others [45], the information is coded manually to perform a qualitative analysis that needs a fine-grained semantic understanding that would otherwise be very difficult to carry out automatically. On a different note, and in line with other studies [35], this article highlights the use of images associated with tweets as analysis tools, which are complementary to the textual content.

Twitter is an instrument frequently used by citizens to convey their demands to public decision-makers, resulting in a rich repository of policy-related opinions. For this purpose, previous studies systematically analyzed the textual content of the collected tweets following sentiment analysis modeling techniques [54]. The current study shares the intention to ensure that research results inform park managers. However, rather than inferring the specific needs of the public, the aim is to give visibility to non-human actors in a public park.

Other scholars have measured "urban life" by tracking the relationship between spatial structures and human behaviors via Twitter. Adopting a locally sensitive analysis, they have added data on points of interest and walking behaviors to the methodology, revealing hidden correlations between the shape of streets and their liveliness [55]. The current study adds to the responsibility of the ongoing reconstruction of planning theories by adding a hidden layer in the public sphere of members beyond the human realm.

Human-centered perspectives feature plants as mere background constituents of landscapes without treating them as individual entities [7]. As mentioned above, recent studies have used Twitter data as a source of information to examine the interaction between people and green city spaces [43]; however, the present study advances the discussion by using Twitter data to understand the "perspective of the urban forest" itself.

Previous research centers on events or uses where the human being is the focus [22]. In the current study, Twitter is explored as a means of creating a cybercommunity where voices of the more-than-human are recorded, moving from Twitter as "a data source to inform human use of urban green spaces" to a data source that informs about how the park behaves. Thus, times and events related to nature can be understood from its viewpoint, an alternative filter that disrupts the anthropocentric view of the urban environment.

While the use of Twitter intends to enable real-time response to urgent situations, allowing administrations and city planners to adapt their communication skills [54], the current study complements this approach by using Twitter data to understand the perspective of the urban forest as perceived by humans. Twitter analysis on how to communicate the voices of the more-than-human increases environmental awareness, not only among citizens but also among the actors involved in the design and management of public space. In addition to enhancing physical and mental health on a daily routine, such messages can serve to encourage greater citizen participation in the maintenance of urban forests, with better involvement in public maintenance policies.

Participatory observation elicits empathy and produces awareness, even if it does not lead to full understanding. The interpretation of animal communication has implications for the communicative relationships that mediate the public [56].

4.3. Limitations

Although the intent of our study was not to carry out a demographic sampling, but rather to show Twitter as a potential channel for raising public awareness of biodiversity and getting the park's voice heard, it may suffer from the biases pointed out in the previous literature. Twitter is a biased medium regarding gender, age, or nationality. The receptors of the "forest tweeting" would be the habitual users of this type of social network, young people and men, with an underrepresentation of other population groups such as elderly women or low-income earners [26]. In addition, the fact that the park staff are the ones that are tweeting brings the dilemma that arises when humans use their voice to humanize animals, treating them as people or speaking for them [57].

When it comes to language interpretation and translation, previous studies already found limitations in capturing nuances about sentiment and understanding perception. Automated sentiment analysis of big data was not sufficient, applying manual coding analysis, defining categories, and using a mixed qualitative and quantitative review [41]. We share with other studies the limitation that scarce information in short texts may lead to misinterpretation by the researchers, although human ratification brings a nuanced understanding of subtle language expressions as irony. As an improvement, the role of covariates such as seasonal rhythms could be included in the analysis to achieve high accuracy [20].

5. Conclusions

Taking a critical position nourished in the previous theoretical body around the role of social media in urban studies, this article has insisted on changing Twitter's conventional human perspective to a more-than-human one that allows us to generate a different world-view to inhabit the city as a shared space with other beings. For this purpose, the Twitter account of Tokyo's Rinshinomori Park was carefully selected as a case study because it is considered a valuable practice in communicating the voice of the urban forest. It was decided to meticulously analyze all the content published to date in a qualitative manner, in search of those characteristics that would allow us to understand how this practice is being performed. Following the analysis, the appearance of non-humans as protagonists in more than a half of the tweets is evidenced, thus conditioning the activities that are communicated. It is also revealed how, when tweeting, this urban forest deploys complex actor-networks, which connect a plurality of diverse agents and which manifest themselves in different patterns of more-than-human communication. This exploration has illustrated flora and fauna's agencies as producers of the urban milieu, revealing the park's voice in its own right. These tweets tell stories of interactions between various actors so that other social media users can unveil a latent layer that was otherwise invisible.

Additionally, the current study provides an innovative, replicable methodology that is consistent with the existing literature that uses Twitter as research material. It reveals the latent potential of non-humans as possible agents within the realm of social media, which can mediate the relationships between humans and their environment. This introduces a layer that can be incorporated into future lines of research, such as the possibility of combining it with citizen science methodologies based on sentiment analysis or geolocation, which would identify which particular park areas and attributes are more visited and represented to extract behavioral use patterns related to biodiversity. It could also be combined with social media engagement studies to detect species which appeal to the neighbors. This paper also illustrates a model case that could be extrapolated to other contexts as a good practice in the management and communication of urban green spaces. Joining the efforts of other researchers, we expect that the more-than-human perspective can serve as a basis for future planning guidelines that encourage biodiversity as a constituent part of park design. Through broadcasting the park's voice, the dynamics of the natural environment can be better understood, and Twitter emerges as a potential tool for urban design and planning. In this way, more-than-human interaction within a public space is encouraged through a social network.

Author Contributions: Conceptualization, D.M.S. and N.G.L.; methodology, D.M.S. and N.G.L.; investigation, D.M.S. and N.G.L.; data curation, D.M.S. and N.G.L.; writing—original draft preparation, D.M.S. and N.G.L.; writing—review and editing, D.M.S. and N.G.L.; visualization, D.M.S. and N.G.L.; supervision, D.M.S. and N.G.L.; funding acquisition, D.M.S. and N.G.L. All authors have read and agreed to the published version of the manuscript.

Funding: This research was funded by the European Union—Next Generation EU Margarita Salas Grant and by the project LABPA-CM: CONTEMPORARY CRITERIA, METHODS and TECHNIQUES FOR LANDSCAPE KNOWLEDGE AND CONSERVATION (H2019/HUM5692), funded by the European Social Fund and the Madrid regional government.

Institutional Review Board Statement: Not applicable.

Informed Consent Statement: Not applicable.

Data Availability Statement: Not applicable.

Conflicts of Interest: The authors declare no conflict of interest. The funders had no role in the design of the study; in the collection, analyses, or interpretation of data; in the writing of the manuscript; or in the decision to publish the results.

References

1. Human Animal Research Network Collective (Ed.) *Animals in the Anthropocene, Critical Perspectives on Non-Human Futures*; Sydney University Press: Sydney, Australia, 2015.
2. Bruce, B. Environmental issues: Writing a more-than-human urban geography. *Prog. Hum. Geogr.* **2005**, *29*, 635–650.
3. Haraway, D. *When Species Meet*; University of Minesota Press: Minneapolis, MN, USA, 2008.
4. Plec, E. (Ed.) *Perspectives on Human-Animal Communication: Internatural Communication*; Routledge: New York, NY, USA, 2013.
5. Puig de la Bellacasa, M. *Matters of Care: Speculative Ethics in More Than Human Worlds*; University of Minnesota Press: Minneapolis, MN, USA, 2017.
6. Bennett, J. *Vibrant Matter: A Political Ecology of Things*; Duke University Press: Durham, NC, USA, 2010.
7. Pitt, H. On showing and being shown plants-a guide to methods for more-than-human geography. *Area* **2015**, *47*, 48–55.
8. Houston, D.; Hillier, J.; MacCallum, D.; Steele, W.; Byrne, J. Make kin, not cities! Multispecies entanglements and 'becoming-world' in planning theory. *Plan. Theory* **2018**, *17*, 190–212.
9. Maller, C. *Healthy Urban Environments: More-Than-Human Theories*; Routledge: London, UK, 2018.
10. Kirksey, S.; Helmreich, S. The emergence of multispecies ethnography. *Cult. Anthropol.* **2010**, *25*, 545–576.
11. Gruen, L. *Entangled Empathy: An Alternate Ethic for Our Relationships with Animals*; Lantern Books: New York, NY, USA, 2015.
12. Steele, W.; Wiesel, I.; Maller, C. More-than-human cities: Where the wild things are. *Geoforum* **2019**, *106*, 411–415.
13. Lafortezza, R.; Carrus, G.; Sanesi, G.; Davies, C. Benefits and well-being perceived by people visiting green spaces in periods of heat stress. *Urban For. Urban Green.* **2009**, *8*, 97–108.
14. Low, S.; Taplin, D.; Scheld, S. *Rethinking Urban Parks: Public Space and Cultural Diversity*; University of Texas Press: Lake Austin Centre, TX, USA, 2009.
15. Daily, G. (Ed.) *Nature's Services: Societal Dependence on Natural Ecosystems*; Island Press: Washington, DC, USA, 1997.
16. Schwartz, R.; Hochman, N. *The Social Media Life of Public Spaces: Reading Places Through the Lens of Geo-Tagged Data*; Routledge: New York, NY, USA, 2014.
17. Huang, Y.; Li, Z.; Huang, Y. User Perception of Public Parks: A Pilot Study Integrating Spatial Social Media Data with Park Management in the City of Chicago. *Land* **2022**, *11*, 211.
18. Zsolt Farkas, J.; Hoyk, E.; Batista de Morais, M.; Csomós, G. A systematic review of urban green space research over the last 30 years: A bibliometric analysis. *Heliyon* **2023**, *9*, e13406. [CrossRef]
19. Roberts, V.H. Using Twitter data in urban green space research: A case study and critical evaluation. *Appl. Geogr.* **2017**, *81*, 13–20.
20. Lin, Y.; Wood, S.A.; Lawler, J.J. The relationship between natural environments and subjective well-being as measured by sentiment expressed on Twitter. *Landsc. Urban Plan.* **2022**, *227*, 104539. [CrossRef]
21. Chen, S.; Liu, L.; Chen, C.; Haase, D. The interaction between human demand and urban greenspace supply for promoting positive emotions with sentiment analysis from twitter. *Urban For. Urban Green.* **2022**, *78*, 127763. [CrossRef]
22. Plunz, R.A.; Zhou, Y.; Carrasco Vintimilla, M.I.; Mckeown, K.; Yu, T.; Uguccioni, L.; Sutto, M.P. Twitter sentiment in New York City parks as measure of well-being. *Landsc. Urban Plan.* **2019**, *189*, 235–246. [CrossRef]
23. Chen, Y.; Song, Y.; Li, C. Where do people tweet? The relationship of the built environment to tweeting in Chicago. *Sustain. Cities Soc.* **2020**, *52*, 101817. [CrossRef]
24. Niu, H.; Silva, E.A. Understanding temporal and spatial patterns of urban activities across demographic groups through geotagged social media data. *Comput. Environ. Urban Syst.* **2023**, *100*, 101934. [CrossRef]
25. Chuang, I.-T.; Benita, F.; Tunçer, B. Effects of urban park spatial characteristics on visitor density and diversity: A geolocated social media approach. *Landsc. Urban Plan.* **2022**, *226*, 104514. [CrossRef]

26. Sainz-Santamaria, J.; Moctezuma, D.; Martinez-Cruz, A.; Téllez, E.; Graff, M.; Miranda-Jiménez, S. Contesting views on mobility restrictions in urban green spaces amid COVID-19—Insights from Twitter in Latin America and Spain. *Cities* **2023**, *132*, 104094. [CrossRef]

27. Huang, J.-H.; Floyd, M.F.; Tateosian, L.G.; Aaron Hipp, J. Exploring Public Values through Twitter Data Associated with Urban Parks Pre- and Post-COVID-19. *Landsc. Urban Plan.* **2022**, *227*, 104517. [CrossRef]

28. Hausmann, A.; Toivonen, T.; Fink, C.; Heikinheimo, V.; Tenkanen, H.; Butchart, S.H.; Di Minin, E. Assessing global popularity and threats to Important Bird and Biodiversity Areas using social media data. *Sci. Total Environ.* **2019**, *683*, 617–623. [CrossRef]

29. Mangachena, J.R.; Pickering, C.M. Implications of social media discourse for managing national parks in South Africa. *J. Environ. Manag.* **2021**, *285*, 112159. [CrossRef]

30. Cooper, M.W.; Minin, E.D.; Hausmann, A.; Qin, S.; Schwartz, A.J.; Correia, R.A. Developing a global indicator for Aichi Target 1 by merging online data sources to measure biodiversity awareness and engagement. *Biol. Conserv.* **2019**, *230*, 29–36. [CrossRef]

31. Bruzzese, S.; Ahmed, W.; Blanc, S.; Brun, F. Ecosystem Services: A Social and Semantic Network Analysis of Public Opinion on Twitter. *Int. J. Environ. Res. Public Health* **2022**, *19*, 15012. [CrossRef] [PubMed]

32. Kidd, L.R.; Gregg, E.A.; Bekessy, S.A.; Robinson, J.A.; Garrard, G.E. Tweeting for their lives: Visibility of threatened species on twitter. *J. Nat. Conserv.* **2018**, *46*, 106–109. [CrossRef]

33. O'Reilly, K.E. It's Beginning to Look a Lot Like # 25DaysofFishmas: Communicating Freshwater Biodiversity Using Social Media. *Fisheries* **2022**, *47*, 395–405. [CrossRef]

34. Daume, S.; Galaz, V. "Anyone Know What Species This Is?-Twitter Conversations as Embryonic Citizen Science Communities. *PLoS ONE* **2016**, *11*, e0151387. [CrossRef] [PubMed]

35. Schuette, S.; Folk, R.A.; Cantley, J.T.; Martine, C.T. The hidden *Heuchera*: How science Twitter uncovered a globally imperiled species in Pennsylvania, USA. *PhytoKeys* **2018**, *96*, 87–97. [CrossRef] [PubMed]

36. Pickering, C.M.; Norman, P. Assessing discourses about controversial environmental management issues on social media: Tweeting about wild horses in a national park. *J. Environ. Manag.* **2020**, *275*, 111244. [CrossRef]

37. Lupton, D. *Data Selves. More-Than-Human Perspectives*; Polity Press: Cambridge, UK, 2020.

38. Jensen, C.B.; Blok, A. Techno-animism in Japan: Shinto cosmograms, actor-network theory, and the enabling powers of non-human agencies. *Theory Cult. Soc.* **2013**, *30*, 84–115. [CrossRef]

39. Plumwood, V. Nature in the Active Voice. In *The Handbook of Contemporary Animism*; Harvey, G., Ed.; Acumen Publishing: Cambridge, UK, 2013; pp. 441–453.

40. Martín, D.; Tsukamoto, Y.; Gómez, N. Tokyo Metropolitan Parks as Urban Forestry Assemblages: Reframing more-than-human commons in. the city. *J. Asian Archit. Build. Eng.* **2021**, *21*, 2636–2651.

41. Marchi, V.; Speak, A.; Ugolini, F.; Sanesi, G.; Carrus, G.; Salbitano, F. Attitudes towards urban green during the COVID-19 pandemic via Twitter. *Cities* **2022**, *126*, 103707. [CrossRef]

42. Martín, D.; Gómez, N. Rethinking Urban Forestry through Resources Accessibility. *AIJ J. Technol. Des.* **2021**, *27*, 1074–1079. [CrossRef]

43. Liu, Q.; Ullah, H.; Wan, W.; Peng, Z.; Hou, L.; Qu, T.; Ali Haidery, S. Analysis of Green Spaces by Utilizing Big Data to Support Smart Cities and Environment: A Case Study About the City Center of Shanghai. *ISPRS Int. J. Geo-Inf.* **2020**, *9*, 360. [CrossRef]

44. Hamstead, Z.A.; Fisher, D.; Ilieva, R.T.; Wood, S.A.; McPhearson, T.; Kremer, P. Geolocated social media as a rapid indicator of park visitation and equitable park access. *Comput. Environ. Urban Syst.* **2018**, *72*, 38–50. [CrossRef]

45. Daume, S. Mining Twitter to monitor invasive alien species—An analytical framework and sample information topologies. *Ecol. Inform.* **2016**, *31*, 70–82. [CrossRef]

46. Schwartz, A.J.; Dodds, P.S.; O'Neil-Dunne, J.P.M.; Ricketts, T.H.; Danforth, C.M. Gauging the happiness benefit of US urban parks through Twitter. *PLoS ONE* **2022**, *17*, e0261056. [CrossRef]

47. Arts, I.; Duckett, D.; Fischer, A.; Van Der Wal, R. Communicating nature during lockdown–how conservation and outdoor organisations use social media to facilitate local nature experiences. *People Nat.* **2022**, *4*, 1292–1304. [CrossRef]

48. Miyazaki, Y.; Teramura, A.; Senou, H. Biodiversity data mining from Argus-eyed citizens: The first illegal introduction record of Lepomis macrochirus macrochirus Rafinesque, 1819 in Japan based on Twitter information. *Zookeys* **2016**, *569*, 123–133. [CrossRef]

49. Tenkanen, H.; Di Minin, E.; Heikinheimo, V.; Hausmann, A.; Herbst, M.; Kajala, L.; Toivonen, T. Instagram, Flickr, or Twitter: Assessing the usability of social media data for visitor monitoring in protected areas. *Sci. Rep.* **2017**, *7*, 17615. [CrossRef]

50. Johnson, M.L.; Campbell, L.K.; Svendsen, E.S.; McMillen, H.L. Mapping Urban Park Cultural Ecosystem Services: A Comparison of Twitter and Semi-Structured Interview Methods. *Sustainability* **2019**, *11*, 6137. [CrossRef]

51. Sim, J.; Miller, P. Understanding an Urban Park through Big Data. *Int. J. Environ. Res. Public Health* **2019**, *16*, 3816. [CrossRef]

52. Kovacs-Györi, A.; Ristea, A.; Kolcsar, R.; Resch, B.; Crivellari, A.; Blaschke, T. Beyond Spatial Proximity—Classifying Parks and Their Visitors in London Based on Spatiotemporal and Sentiment Analysis of Twitter Data. *ISPRS Int. J. Geo-Inf.* **2018**, *7*, 378. [CrossRef]

53. Ullah, H.; Wan, W.; Ali Haidery, S.; Khan, N.U.; Ebrahimpour, Z.; Luo, T. Analyzing the Spatiotemporal Patterns in Green Spaces for Urban Studies Using Location-Based Social Media Data. *ISPRS Int. J. Geo-Inf.* **2019**, *8*, 506. [CrossRef]

54. Cui, N.; Malleson, N.; Houlden, V.; Comber, A. Using social media data to understand the impact of the COVID-19 pandemic on urban green space use. *Urban For. Urban Green.* **2022**, *74*, 127677. [CrossRef] [PubMed]

55. Huang, J.; Cui, Y.; Chang, H.; Obracht-Prondzyńska, H.; Kamrowska-Zaluska, D.; Li, L. A city is not a tree: A multi-city study on street network and urban life. *Landsc. Urban Plan.* **2022**, *226*, 104469. [CrossRef]
56. Head, L.; Atchison, J. Cultural ecology: Emerging human- plant geographies. *Prog. Hum. Geogr.* **2009**, *33*, 236–245. [CrossRef]
57. Adams, T.E. Animals as Media: Speaking through/with Nonhuman Beings. In *Perspectives on Human-Animal Communication: Internatural Communication*; Plec, E., Ed.; Routledge: New York, NY, USA, 2013; pp. 17–35.

 *land*

 MDPI

Article

Contributions of Social Media to the Recognition, Assessment, Conservation, and Communication of Spanish Post-Industrial Landscapes

Ángeles Layuno Rosas * and Jorge Magaz-Molina *

School of Architecture, University of Alcalá, 28801 Alcalá de Henares, Spain
* Correspondence: angeles.layuno@uah.es (Á.L.R.); jorge.magaz@edu.uah.es (J.M.-M.)

Abstract: The paper aims to draft how phenomena such as abandonment, territorial disarticulation, environmental pollution, socioeconomic imbalances, and heritage consideration issues that surround landscapes where industrial activity has ceased are reflected on social media in Spain. The research focuses on the most popular social media platforms in Spain: Instagram, Facebook, and Twitter. A manual sample strategy was conducted to ensure an individualized approach to user-generated content. Sampling was carried out separately for three aspects: (1) keywords at a general level, (2) terms used to define industrial landscapes, and (3) recognition of significant industrial landscapes related to governmental facilities built in the 20th century, wherein we take into account three potential profile types: (i) individuals; (ii) NGOs/associations and/or public administrations; and (iii) academics. The results show that social media platforms are widely used as tools to disseminate information about industrial landscapes, but the contributions of each platform are uneven and incomplete in relation to the reality of post-industrial landscapes. However, it is worth recognizing the added value that their possible interaction brings as a reference for current civic debates. How social media contributes toward mitigating the difficulties of recognition, comprehension, and protection of post-industrial landscapes is emphasized in our conclusions.

Keywords: social media; industrial landscape; post-industrial landscapes; assessment; conservation; communication

Citation: Layuno Rosas, Á.; Magaz-Molina, J. Contributions of Social Media to the Recognition, Assessment, Conservation, and Communication of Spanish Post-Industrial Landscapes. *Land* **2023**, *12*, 374. https://doi.org/10.3390/land12020374

Academic Editors: Cecilia Arnaiz Schmitz, María Fe Schmitz and Nicolas Marine

Received: 31 December 2022
Revised: 25 January 2023
Accepted: 27 January 2023
Published: 30 January 2023

1. Introduction

The European Landscape Convention (Florence, Council of Europe, 2000) [1] advanced the idea of "landscape" as "any part of the territory as perceived by the population, whose character is the result of the action and interaction of natural and/or human factors...", summarizing some earlier statements and assertions contained in previous international UNESCO charters and documents. Ideas such as the need to renew heritage concepts through a cross-cutting approach comprise a symbiotic set of natural and cultural elements (tangible and intangible) in which a social group recognizes its identity and is committed to transmitting it to future generations in an improved and enriched way (Conference of Stockholm, 1998) [2].

The European Landscape Convention emphasizes the management of landscape as a favorable economic resource for communities as well as an element of identity, both in spaces of exceptional beauty or in the most ordinary and degraded ones. Hence, the document opens the way to improving and preserving the landscape from cultural, environmental, ecological, and social perspectives. As such, a debate has emerged around industrial and post-industrial landscapes as a type of cultural landscape subject to heritage protection.

The processes of de-industrialization and the consequent possibility of heritage assessment of landscapes resulting from the obsolescence of industry are current phenomena occurring on a global scale [3]. The renovation of both the industrial structure of Southeast Asia [4–6] and decarbonization programs for the electricity industry in Western countries [7–9] has impelled actions of integration of post-industrial legacy in cultural, urban, and territorial strategies. Given the growing sensitivity toward post-industrial landscapes, it is worth recognizing the spreading acceptance of "factory esthetics" and the "industrial ruin" by disciplines derived from "industrial archeology" [10–13]. However, do these actions, aimed at preserving the industrial heritage and landscape, respond to the social demand of recognizing productive landscape as part of the community's cultural heritage? Is it possible to consider a more comprehensive approach to citizens' sensitivities beyond the administrative procedures or debates covered in the academic media?

As with most landscapes, the approach to industrial landscapes places us at a complex crossroad due to (1) its subjective nature (emotional, mental, or cultural for each individual or social group, sometimes showing polarities in terms of what an inhabitant/visitor perceives and experiences); (2) its heterogeneous classification (depending on each productive sector, the orographic conditions, the climate–environment context, etc.); and (3) its hybrid tangible–intangible dimension and evolutionary and dynamic nature. In this frame, post-industrial landscapes also include those landscapes that lack monumentality, which cannot be catalogued but are part of economic identity and social networks, despite the difficulties of being recognized and comprehended.

On the other hand, if landscape is a cultural construct based on the perception of individuals and society, the final goal of landscape assessment should be to satisfy the needs and interests of the community. This makes it possible to establish a comparison between the objective and technical visions of a landscape and that based on subjective and identity aspects held by societies, which constitute the builders and managers of landscapes whose essence is always to evolve as dynamic and developing organisms [14]. In this context, it is worth reviewing the discussion of the cultural geographer John Brinckerhoff Jackson in his work A Sense of Place, a Sense of Time [15] regarding the need to learn to read the landscape. JB Jackson identified a dichotomy that is still valid among two perspectives: the esthetic approach based on perceptual–visual aspects to landscape endowed with design and artistic qualities versus the phenomenological approach around the development of places based on the uses and customs of their citizens, appropriate for a more autochthonous reading of the landscape and to the recognition of its historical evolution. David Lowenthal's lessons [16] also have a place in this issue: encouragement of the participatory and creative attitude of citizens in relation to their landscapes, the importance of landscape management policies from local and social identities based not only on the exploration of its growing economic value but also on the reaffirmation of its affective dimension or the collective memory.

The dichotomy inherent to the landscape, between the livable and the visitable, and the contrast between the social memory and the tourist image of landscapes [12], is linked to the selective and subjective character expressed in the reading of the landscape. This phenomenon is especially evident when it comes to highlighting partial images that exclude other realities contained within the same place, or other readings that should be considered. We enter fully into the definition of the social landscape; the landscape as a social and identity construction, not only for the visitor but, above all, for the inhabitant, who becomes an individual and a collective in the midst of a scenario of interaction and otherness among various collectives. The diversity of parallel and non-exclusive symbolic constructions, as well as the spaces in which the collectivity is self-represented, evolves upon all these bases. The new ways for virtually inhabiting, comprehending, and registering a landscape, thanks to the use of new technologies such as Google Maps, Street View, OpenStreetMap, Flickr, or Mapillary, have transformed the traditional essence of the act of viewing a landscape. Social media information plays an important role as an instrument to activate strategies leading to the assessment, protection, and rehabilitation of post-industrial landscapes.

Regarding these arguments, international reference guidelines on industrial heritage also integrate, in their methodological corpus, the attention to citizen participation and the importance of dissemination, as stated in the Nizhny Tagil Charter on Industrial Heritage (2003), derived from the International Committee for the Conservation of the Industrial

Heritage [17]. Likewise, the Dublin Principles for the conservation of industrial heritage sites, structures, areas, and landscapes (ICOMOS/TICCCIH, 2011) [18] confers substantial value to the tasks of documentation and understanding of attributed values to industrial sites, structures, and landscapes by both specialists and communities. Furthermore, the need to support communication aspects through various channels using new technologies is emphasized.

These premises connect with the concept of "digital cultural participation in heritage", in which new formulas for citizen involvement toward safeguarding cultural heritage are proposed [19–22] at the same time that they can be inscribed under article 27 of the Universal Declaration of Human Rights. In the European framework, the Faro Convention on the Value of Cultural Heritage for Society [23] encourages participation in cultural heritage activities and fosters the development of heritage-related technologies and digital content, where industrial landscapes should also be recognized. In the Spanish-speaking context, we can highlight, among others, the results of the debate promoted by the 102nd issue of the PH journal [24] that gathered 30 contributions. The PAYSOC project of the Andalusian Historical Heritage Institute has also explored approaches to recognizing cultural landscapes through virtual ethnography [25].

Social media platforms, as the new transnational scope virtual agoras, play a key role in disseminating information, concerns, and interests in contemporary society, as well as in the construction of collective imaginaries, esthetics, and global narratives. There are several academic studies on the role of social media platforms in the dissemination and conservation of cultural and natural heritage [26,27]. However, there are few that specifically determine their contributions to industrial heritage [28–32]. In this sense, the extent to which user-generated content contributes to the growing recognition of the industrial landscape as a heritage resource, similarly to the identification of its elements and values, does not have a consolidated scientific literature. This article is the first to approach the role of social media in the collective cultural construction of the Spanish industrial landscape.

Taking, as a case study, several examples linked to the industrial landscapes generated by the state policy of Francoism (1939–1975) in the 1950s and 1960s, we analyze (1) the characterization of the industrial landscape based on information disseminated through social media (e.g., institutional, partner, or individual channels) to identify the cultural, esthetic, or identity values attributed by community and (2) critical reactions to their distortion or reuse for exploitative purposes.

Considering the post-industrial landscape as part of the identity of regions and their inhabitants [33] as much as the difficulties for their assessment, protection, and management, we aim to conduct a critical analysis of the role of social media as a tool in the identification, assessment, and protection of obsolete industrial landscapes in Spain. How, when, and where are these obsolete industrial sites considered as landscapes of heritage value? How do social networks portray the challenges surrounding the regions where industrial activity has ceased and where issues of abandonment, ruin, territorial disarticulation, environmental pollution, and socioeconomic problems arise? What role do heritage approaches play on public social media debates around deindustrialization? Based on these premises, this study proposes a critical approach on how user-generated content collected on social media can contribute to the construction and dissemination of information in an active and interactive way. In this context, the prevalence of social (i.e., marginalized landscapes) and academic stereotypes (i.e., sublimated landscapes) based on assessments of visual experience stand out, as opposed to other considerations such as environmental, heritage, or social factors.

The role of social media platforms in the cultural revitalization of industrial landscapes is also explored. Despite adverse circumstances, the cultural, historical, technical, or esthetic values attributed to these regions do not disappear, and these are susceptible to sustaining cultural reactivation initiatives capable of offering economic alternatives to the development of inhabited communities [34].

Through the analysis of the information disseminated through social media as a support for opinions and concerns, we seek to answer the following questions:

- What is meant by industrial post-industrial landscapes with regard to the different agents involved?
- Could social media lead to a proactive role in the re-activation of these landscapes to foster the interaction of all involved agents?
- Can user-generated content contribute to the identification of morphological, esthetic, or sociocommunity parameters that suggest guidelines for articulating processes of obsolete industrial landscapes valorization?
- What presence do memory and historical elements of landscape, both geographical and anthropic, have on social media?

2. Materials and Methods

As is well known, social media refer to any digital tools that allow their users to quickly create and share content. Given that this is the first approach to the role of social media in the recognition of the industrial landscape in Spain, and lacking reference papers, we propose to approach the subject, here, by working on some quantitative and qualitative strategies to lay the basis for subsequent, more detailed studies.

In social media, a wide range of websites and applications are used. In this first approach, we selected three of the most used social media platforms in Spain [35]: Instagram, a photo-sharing app; Facebook, a news, information, and audiovisual content-sharing app; and Twitter, an app for sharing short written messages.

Data collection was carried out from mid-September to the end of December 2022 with the aim of outlining the scope of terms related with the industrial landscape, and the profiles and forums in which these concepts are handled, to identify different content offered by the various social media platforms.

Search engine tools and metric analyses of these platforms were consulted directly for keyword and/or hashtag tracking. We employed a manual study strategy in order to ensure an individualized approach to user-generated content and identifying the type of profile of origin and precise context, because this is the most effective way to determine the level of involvement of different culture services of Spanish Public Administrations, which have recently incorporated social media.

To this end, several keywords were monitored following previous studies by Campillo, Ramos, and Castelló [36] and Mariani, Di Felice, and Mura [37] that focused on audience parameters (followers and publications) and the level of user interaction ("likes" and "shares" on Facebook; and reactions, retweets, and comments on Twitter). Sampling was carried out separately for three aspects: (1) keywords at a general level, (2) terms used to define industrial landscapes, and (3) recognition of significant industrial landscapes. In both sampling and results, we took into account three potential profile types: (1) individuals; (2) groups, that could be NGOs/clubs/associations, and/or public administrations; and (3) academics.

2.1. Sampling 1

The first task involved tracking keywords in Spanish relating to both industrial (industrial) and post-industrial (postindustrial) landscapes (paisajes) and heritage (patrimonio). This preview of the subject explored the presence of landscape and industrial heritage references on social media platforms of individuals, public administrations, and NGOs.

2.2. Sampling 2

A second task involved looking for industrial landscape images uploaded to social networks by the three types of profiles mentioned: personal, NGO/association/club and institutional, and academic.

The images were then analyzed using landscape characterization studies, based on López-Sánchez et al. [38], being applied to the three main categories of parameters (e.g., morphological, esthetic–perceptual, and social) that were further subdivided into associated subparameters to draft 17 attributes (Table 1).

Table 1. Landscape characterization of sampling #2.

Parameters for Characterizing the Landscape	Associated Subparameters
Morphological analysis *(Identification of elements)* Territorial/urban/architectural/ infrastructures/populations/facilities/historical/geomorphology/ environmental problems	Architecture Town planning Heritage (authenticity, integrity) Delimitation Environmental problems
Esthetic–perceptive *(Qualitative assessment, from experts or users)* Monumental/anti-monumental/degraded/presence of stereotypes	Esthetic interest Scale Historicity
Social *(Assessment of the work memory,* *experts or users)* Intangible heritage/perception of local population and users/ landscape as economic resource/tourism	Sense of identity Collective memory Personal experiences Functionality Documentary contributions Tourist resources Distortion or commodification Critical reactions to transformation or rehabilitation of heritage elements

2.3. Sampling 3

In a third step, we developed several case studies. Twelve prominent sites were selected (Table 2), linked to industrialization actions promoted by the Spanish government as part of a program of the National Institute of Industry (INI) created in 1941. These hubs were the object of intense public propaganda campaigns in the mid-20th century and the sites had been the subject of previous studies in terms of historical and heritage aspects [39–42].

Table 2. Studies sites of sampling #3.

Industry or Productive Space	State	Location	Company Town	State	Location
INI—Instituto Nacional de Industria (National Institute of Industry)	Active	Urban (Madrid)	**Parque Marqués de Suances.** Canillejas (Madrid)	Inhabited	Periurban
ENASA—Empresa Nacional de Autocamiones (National Truck Company)	Active	Periurban (Madrid)	**Ciudad Pegaso.** Barajas (Madrid)	Inhabited	Periurban
ENDESA—Empresa Nacional de Elecrtricidad, SA (National Electricity Company)	Undergoing decarbonization	Rural (León–Aragón–Galicia–Almería)	**Compostilla.** Ponferrada (León)	Inhabited	Periurban
ENSIDESA—Empresa Nacional Siderdúrgica, SA (National Steel Company)	Active	Periurban (Asturias)	**Llaranes.** Áviles (Asturias)	Inhabited	Periurban
Minas de Rodalquilar (Rodalquilar Mines)	Abandoned	Rural (Almería)	**El Arteal.** Níjar (Almería)	Abandoned	Rural
ENCASUR—Empresa Nacional Carbonífera del Sur (National Company of Coal from the South)	Decommissioned and environmentally restored	Rural (Ciudad Real–Córdoba)	**Poblado Asdrúbal** (Puertollano–Ciudad Real)	Abandoned	Periurban

In addition to the National Institute of Industry's headquarters in Madrid, 5 of the 500 industrial establishments distributed throughout the Iberian Peninsula were selected. The intention was to offer a varied representation of different industrial sectors, regions, and current status. Their company towns were also selected.

We tracked the ongoing echo, on social media platforms (Facebook, Instagram, and Twitter), of these industrial facilities and their adjoining worker settlements. The goal of this sampling was to approach the recognition of the esthetic landscape values attributed to these spaces and, particularly, to the attention they attracted on Instagram. Through our selection of sites, we also sought to trace relationships that remain between the habitats of 20th-century workers and the original productive facilities.

Given the widespread use of company acronyms for the selected facilities, we chose to track these shortened terms instead of the long official names. Company towns were instead sampled on social networks for their popular names.

For these industrial hubs, which bring together productive spaces and workers' habitats, we propose an analysis of the contents with the aim of assessing their degree of recognition. The sampled images are classified according to the following parameters: people, document, industrial profile, railway stamp, machinery and utensils, urban stamp, architecture, dismantling, ruin, renewable, heritage event, or other.

With all the information collected in the three samplings, we conducted a comparative analysis to determine:

- the differences among the three selected profiles;
- what are the most common variables;
- the differences in criteria and content between the different social media platforms.

3. Results

3.1. Sampling #1

Keyword sampling pointed to an uneven diffusion of concepts, as well as the generic use of terms such as "industrial landscape" (paisaje industrial) or "post-industrial landscape" (paisaje postindustrial), and further screening was needed to identify the contents required for our analysis (Table 3). Sampling on the term "heritage" (patrimonio) provided more precise findings. The search was limited by a manual quantitative registration. Ranges, rather than exact figures, are provided in some cases: the lowest value being the one collected by manual registration and the highest value being the one indicated by the platform (usually 1000 in Facebook and 100 in Instagram).

Table 3. Results of sampling #1 with data relating to tags on the three social media platforms.

	Post-Industrial Landscape	Industrial Landscape	Industrial Heritage	Post-Industrial Heritage
Facebook	41–1000	103–1000	14,000	0
Twitter	34	109	191	9
Instagram	8	2327	26,995	0

More detailed data on the industrial landscape were recorded on the Instagram and Twitter profiles of the public administrations (Table 4). Dissemination was limited to specific sites with figures of cultural protection that are widely recognized; there is surprisingly little mention of hubs registered in the National Plan for Industrial Heritage or the 100 Spanish TICCIH Elements.

Apart from these profiles of cultural services within regional and local administrations, an important number of profiles dedicated to museums and other public entities should be considered. Some examples are *La fábrica de la Luz* in Ponferrada (Fundación Ciudad de la Energía, @museo_energia), *Fundació de la Comunitat Valenciana Patrimoni Industrial i Memòria Obrera de Port de Sagunt* (@fcvportdesagunt), and *Museu Nacional de la Ciència i la Tècnica de Catalunya* (MNACTEC, Generalitat de Catalunya, @MNACTEC).

Table 4. Results of sampling #1 for administrative profiles.

Instagram	Twitter
Ministry of Culture and Sport	
@culturagob (2248 publications)	@culturagob (23 million tweets)
Trade unionist Marcelino Camacho's archive Mining loading dock at Almeria—*Cable Inglés* (Enlisted) Ribeira Sacra reservoir (Enlisted) Image of planes from the National Film Archive San Isidro suspension bridge of Fraga (×2)	Médulas, Almadén, and UNESCO sites (Enlisted) Royal Glass Factory of Segovia (Enlisted) La Mancha windmill picture
Spanish Institute of Cultural Heritage (IPCE)	
No social media account	@ipcepatrimonio (9665 tweets)
	IPCE Industrial heritage conference Budapest Bridge San Sebastián-Donosti Channel of Castille (×2) (Enlisted) Contemporary industrial architecture of Asturias River Tinto Mines (Enlisted) Royal Glass Factory of the Segovia Royal Tapestry Factory (Enlisted) IPCE Dockyards and maritime heritage conference Vizcaya truss bridge (Enlisted) 20th century agricultural settlements—Docomomo Ibérico Cinema at Madrid—Palace of Music
Cultural Heritage Service of Castille & León Regional Government	
@patrimoniojcyl (202 publications)	@patrimoniojcyl (1886 tweets)
Chanel of Castille (×2) (Enlisted) Vallejo de Orbó company town (Photographs of Bustiello (Asturias) company town) Textile industry of Salamanca Las Médulas (×2) (enlisted)	1st meeting on mining landscapes of Castille and León Precautionary suspension of the demolition of Compostilla II power plant Sargentes de Lora Oil Museum (Enlisted) Barruelo de Santullán coal mine Mining heritage of León didactic units for scholars Lords of Eresma water mill (Enlisted) Aceñas (water mills) on Duero River in Zamora Channel of Castille (×2) (Enlisted) Navafria Ironworks (Enlisted) Rubagon collieries Castille & León mining monuments (Enlisted) Industrial Heritage Route in El Bierzo
Tourism Service of the Principality of Asturias Regional Government	
@ turismoasturias (3449 publications)	@ turismoasturias (29.1 million tweets)
Watermill (×6) Peñafura mines Pozu Espinu—coal mine shaft (Enlisted) Mining Museum of Asturias (×2) Pozu Samuño—coal mine shaft (×3) (Enlisted) Taramundi Ironworks (×5) (Enlisted) Pozo Sotón—coal mine shaft (Enlisted) Smith work Asturias Iron and Steel Museum Niemeyer Cultural Centre (×2)	Path route close to Pozo Espinu—coal mine shaft (×2) (Enlisted) Industrial warehouses of the Universidad Laboral de Gijón Rioseco company town and Texeo mines Mina Pozo Sotón—coal mine shaft (×3) (Enlisted) Industrial Heritage Route in Asturias Iron Heritage Route in Lluanco Museums and industrial tourism Mining Museum of Asturias (×2) Bustiello company town (Enlisted)

Table 4. *Cont.*

Instagram	Twitter
Department of Tourism, Culture, and Sport of the Andalucian Regional Government	
@turismoand (1170 publications)	@TurismoAND (40,900 tweets)
Industrial Tourism Innovation Forum Rodalquilar Mines (Enlisted) Riotinto Mines (Enlisted) Mining loading dock in Huelva	Riotinto minig park (Enlisted) Tuna fishing gear
Department of Culture, Tourism, and Sport of the Community of Madrid Regional Government	
@patrimoniocm (1112 publications)	@PatrimonioCM (17.4 million tweets)
Old taxi photography El Aguila brewery (×7) (Enlisted) Oil station Fulling mill of Colmenar Viejo El Gasco Dam Delicias Train Station (Enlisted) Undeground Services—Pacifico power station (Enlisted)	Atazar reservoir Old taxi photography El Aguila brewery (×7) Silo's photography in PhotoEspaña Contest Paper Factory—Photography of serie PELO Plaza Castilla water tank Petrol station on Aragon Road no. 388 (Enlisted) Delicias Station (Enlisted)—archived photography

Another important contribution comes from the following associations, with an unequal presence on the various social media platforms: *Hispania Nostra* (@HispaniaNostra); *Asociación Madrid, Ciudadanía y Patrimonio* (@madridcyp); *Asociación Vasca del Patrimonio Industrial y de la Obra Pública* (@AVPIOP); *Asociación INCUNA de Patrimonio Industrial* (@somos_incuna); etc. Standouts is the specific production dedicated to the enhancement of industrial heritage produced by the project *Patrimoniu Industrial de Asturias* (@Patrimoniu_Ind) alongside the dissemination work at the citizen level of the architect Diana Sánchez Mustieles PhD (@Patrindustrial). It is also worth highlighting the echo and dissemination of the elements inscribed on the Spanish Architectural Heritage Red List of endangered assets, carried out by Hispania Nostra (Figure 1).

Figure 1. Hispania Nostra's Instagram post about the demolition of the cooling tower of the Velilla del Río Carrión thermal power plant.

Actions against of the demolition of ENEL-ENDESA's thermal power plants in Andorra first (February 2021) and Compostilla later (December 2022) generated a significant volume of content related to the defense of distinctive elements of the landscape on individual accounts, NGOs/associations/clubs, or informal civic groups, with a notable echo on Facebook and Twitter. This brought together different profiles across media, public administrations, political parties and civic associations, public representatives, experts, and citizens. Additionally, the active role of Twitter is relevant in the dissemination of news and statements by the agents involved in the process of enlisting the cooling towers and chimneys of Compostilla II based on their landscape value, at the request of citizens' groups. There is also certain controversy on Twitter around the demolition of the Meirama thermal power plant (Cerceda, Galicia). The support shown by environmental groups such as Greenpeace Spain (@greenpeace_esp) in favor of the demolition had an important echo. Such a position is contrary to complaints from cultural heritage defense groups such as APATRIGAL (@apatrigal). Another example of citizen actions in favor of industrial heritage with an important echo on social media platforms is that of the *Plataforma en defensa de la Fábrica de Armas de la Vega* (@SalvemosVega). This is an initiative that calls for the modification of the urban integration project of this former weapons factory linked to INI, and the preservation of its values and heritage elements.

In addition to numerous discussion groups that can be found on Facebook, this social media platform has been identified as an important content niche for historical documentation and dissemination of regional community memory. There are numerous groups disseminating historical photographs, documentation, and information about industrial legacy or specific regional areas.

3.2. Sampling #2

Up to 50 tweets that incorporate photographs or videos were selected from conversations about industrial heritage. Table 5 shows some illustrative examples of various recognizable profiles. Each of the selected photographs (and/or comments) was labeled with a maximum of 4 variables among the 17 possible. In general, the esthetic–perceptive and social categories are widely represented; the attributes of collective memory, historicity, and heritage are the most frequent.

Table 5. Analysis of public publications (tweets) from Twitter.

Personal Profile	Academic Profile	Institutional Profile
@ALEJAND38485481, 17 May 2022 *If they put it in, why do they put it in? The megaphone is good for everything. Look, I see you chained yourself to a windmill when you were 90 years old.*	@McMulligan3, 2 June 2022 *This is yet another chapter in the destruction of Aragon's industrial heritage, in which neither Endesa nor the General Directorate of Heritage were able to keep up with their workers or the specialists who guaranteed their conservation and the assignment of new uses.*	@ENDESA. 13 May 2022 *This is the blasting of the 3 cooling towers of the Andorra thermal power plant, #Teruel. A historic step toward #FairEnergeticTransition in Spain.*
		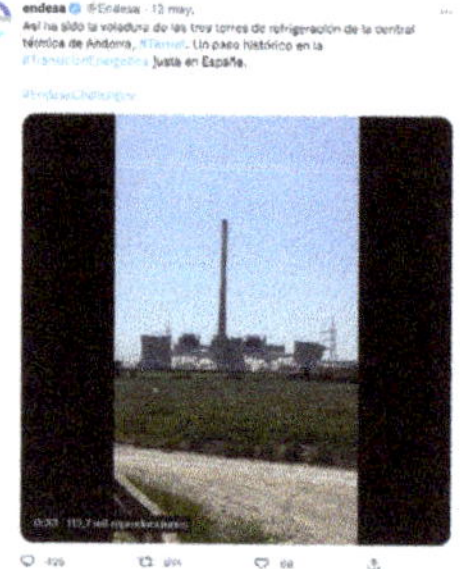

Table 5. *Cont.*

Personal Profile	Academic Profile	Institutional Profile
Assigned variables		
Critical reactions	Assessment of expert	Morphological analysis
Environmental impact	Perception of the local population	Infrastructures
Documentary contributions	Intangible heritage	Facilities
Collective memory	Collective memory	Environmental issues
Expert profile	**NGO/association/club profile**	**Institutional profile**

@javirevilla, 24 November 2022
This is excellent news and as such it is to be applauded. Heritage is defended and the institution responsible for it is the @jcyl. Congratulations on this decision that safeguards our #IndustrialHeritage.

@BierzoYa, 24 November 2022
Bierzo Ya requests to the Junta the declaration BIC of the towers and chimneys of Compostilla II [link] Are these elements difficult to forget or to protect? BIERZO YA

@patrimoniojcyl, 24 November 2022
The @jcyl agreed today the precautionary suspension of the demolition of the 4 towers of the Compostilla power plant, located in Cubillos del Sil #León, and initiated proceedings toward its declaration as an Asset of Cultural Interest with the category of industrial property.

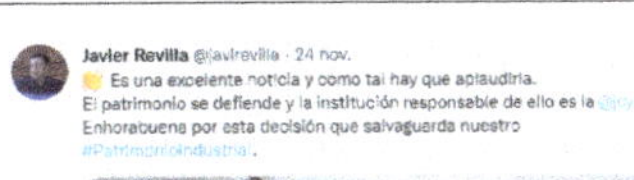

Personal Profile	Academic Profile	Institutional Profile
Assigned variables		
Assessment of expert	Perception of local populations	Assessment of expert
Heritage	Collective memory	Heritage
Historicity	Historicity	Esthetic interest
Scale	Functionality	Functionality
Personal profile	**NGO/association/club profile**	**Institutional profile**

@J_Merino_B, 6 February 2021
The channel and tower (former fire station) on the front page of ABC still exists. ENSIDESA veterans tell Hollywood movie stories about the explosion.

@Patrimoniu_Ind, 30 November 2022
Do you know who were the architects who designed the Llaranes settlement, a residential complex comprising more than 1000 dwellings inhabited by workers, foremen and other professionals from the ENSIDESA factory? Find out via this link

@UPCTnoticias, 11 November 2019
The Cloud Factory, an environmental regeneration project in a post-industrial landscape. The final degree project of the Asturian architect Daniel Suárez seeks to transform the urban façade of Avilés.

Table 5. *Cont.*

Personal Profile	Academic Profile	Institutional Profile
Assigned variables		
Collective memory	Assessment of expert	Urban design
Personal experiences	Architecture	Architecture
Historicity	Town planning	Functionality
Functionality	Esthetic interest	Esthetic interest
NGO/association/club profile	**NGO/association/club profile**	**Personal profile**
@greenpeace_esp, 21 December 2022 *This factory of climate change and health damage is already in the history. The demolition of the Meirama thermal power station will mark a before and after, as a symbol of coal burning disappears.*	@apatrigal, 21 December 2022 *Under the ideological flag of the false environmentalism of the @mitecogob (Ministry of Ecological Transition) our industrial heritage is being destroyed, causing an ecological footprint that we are not being told about.*	@devatrannquila, 24 December 2022 *One other example of an interesting intervention of historic industrial buildings. The intervention at La Vega factory could take reference from many of them.*

Assigned variables		
Morphological analysis	Heritage	Personal experiences
Environmental problems	Sense of identity	Architecture
Personal experiences	Collective memory	Town planning
Historicity	Critical reactions	Critical reactions

3.3. Sampling #3

The dissemination of content about hubs linked to INI, and the terms proposed at the beginning of this article as the basis of the case study, shows notable differences according to the type of social media platform (Table 6).

The National Institute of Industry does not offer too many results and barely has a presence on Instagram. It is only mentioned on Twitter and Facebook as a historical reference, or in conversations with a strong political meaning with hardly any allusions to landscape legacy or graphic references.

In the case of ENDESA (an acronym for the former National Electricity Company, active between 1944 and 1998), a detailed analysis was discarded given the high volume of content still associated with corporate advertising of the ENEL group in which it is integrated. Instead, we chose to explore two iconic production centers: the thermal power stations of Compostilla II (León) and Andorra (Teruel). Compostilla II offers different results depending on the social media platform; a notable echo about its landscape and heritage value is found especially on Twitter as a result of the controversy about its imminent demolition. Similar results were found in the case of Andorra (ENEL-ENDESA), with limited presence on Instagram compared with the broad debate on Facebook and Twitter (Table 7). Abundant content can also be found in inhabited urban hubs, such as the company towns of Compostilla, Llaranes, and Ciudad Pegaso.

Table 6. Results of sampling #2.

#	Facebook	Instagram	(# Similar)	Twitter
National Institute of Industry	12–1000	4	–	(>40) *
Parque Marqués de Suances	>100	23	8	90–100
ENSIDESA	68–1000	653	>23	162
Llaranes	79–1000	1000	>78	>500
ENASA	<1000	1429	–	60
Ciudad Pegaso	44–1000	594	9	80–100
Rodalquilar	>1000	21,000	2418	90–100
El Arteal	7–1000	37	–	5
ENCASUR	22–1000	11	–	77
Poblado Asdrúbal	(4) *	(14) *	–	(62) *
ENDESA	19,000	15,000	–	>100
Compostilla	200–1000	970	47	120

* No active hashtag, but term identified in conversations.

Table 7. Comparison of results between 2 ENEL-ENDESA thermal power plants subject to citizen-driven heritage processes.

	Twitter	Facebook	Instagram
CT. Compostilla II	>100	100–1000	7
CT. Andorra	>100	100–1000	14

Based on the specific study carried out on Instagram, a reduced presence of industrial production spaces located in rural areas outside tourist circuits is observed (Compostilla; Andorra; ENCASUR; Arteal; etc.). This limited presence contrasts with that of urban production facilities with recognizable urban profiles, such as ENSIDESA (Table 8). Spaces such as the Rodalquilar Mines, inscribed in tourist areas with a consolidated brand image and formal heritage recognition, enjoy notable diffusion. Much of the Instagram content related to the inhabited company towns of Compostilla, Ciudad Pegaso, and Llaranes could be aligned with landscape and heritage sensitivities (Figure 2).

Table 8. Results of sampling #3, with image numbers for each case study.

	#	(P)	(D)	(IP)	(RS)	(M)	(U)	(A)	(Di)	(RU)	(RN)	(H)	(O)	Visible
ENSIDESA	626	16	31	138	34	15	47	45		6		29	40	401
Llaranes	1012	21	10	8	5	1	164	42		1		12	315	579
ENASA	1429	6	123	2	1	650	21	18		2		2	181	1006
Ciudad Pegaso	586	13	47	1	1	6	72	22				35	142	339
Compostilla II	8			4		1		2						7
Rodalquilar	21,105	1	3	89		1	20	15		61		1	1642	1833
Compostilla	953	18	18	56	2	2	65	20	8	7	4	9	547	756
El Arteal	36			2			5			7			22	36
ENCASUR	12			4		1	1					4	2	12
P. Asdrúbal	14			2	1			1		3			7	14
Inst. Nac. Ind.	4		1			2		1						4
Parque Suances	23						5	1					17	23

(#) publications attached to each term; (Visible) public or consulted publications. (P) people, (D) document, (IP) industrial profile, (RS) railway stamp, (M) machinery and utensils, (U) urban stamp, (A) architecture, (Di) dismantling, (RU) ruin, (RN) renewable, (H) heritage event, or (O) others.

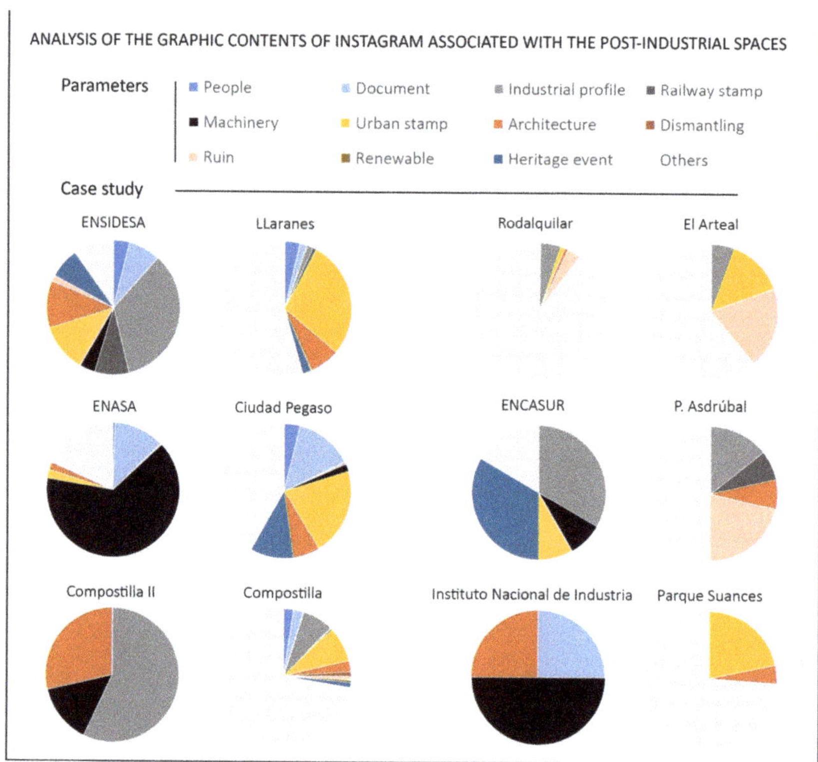

Figure 2. Pie charts relating to public content of the selected case studies found on Instagram.

3.4. Synthesis

The global results show that:

There is participation, via personal profiles, in debates about the loss of heritage and/or its degradation. The following factors tend to dominate: social aspects, dissemination of historical content, and oral memory.

NGOs/associations/clubs and expert profiles contribute to debates and complaints. They also disseminate information campaigns about industrial spaces or elements and examples of conservation and rehabilitation. Social and morphological aspects predominate, with an emphasis on community memory or complaints. Environmental and pollution issues are not usually featured in discussions about industrial heritage conservation; this is not the case for ecologist associations that support decommissioning programs.

Institutional profiles of cultural administration services are scarce and limited to discussing specific events. They are not particularly active in issues related to industrial heritage. Administrations responsible for heritage issues tend to focus on disseminating recurring content on landscapes or assets with a consolidated trajectory. Institutional attention to landscapes and industrial heritage on social media platforms is limited to the activity developed by entities specialized in the field, such as museums or interpretation centers.

4. Discussion

This quantitative study is limited by no-cost search tools: Facebook does not provide total data and Twitter is also limited. In addition, a manual count of the registers is not very operational; without a detailed register, partial data are not very representative. Therefore, manual counting is not a practical solution for obtaining data with these tools. Limitations due to the privacy constraints of numerous publications must be considered, yet the differential value can be discerned in cases such as Instagram. Apart from the difficulties in counting, the diversity of content in some cases does not correspond to the search field. For example, in "ENDESA", the sample relating to landscape that remains with respect to the rest of the message (corporate advertising, news, reviews, etc.) is negligible.

In sampling #1, the hashtags that received more interest are industrial heritage > industrial landscape > post-industrial heritage > post-industrial landscape. This could be interpreted as a greater acceptance of historical heritage values linked to industrial memory, over the recognition or understanding of the concept of post-industrial landscape.

In terms of content, several profiles could be distinguished whose content was aligned to sensitive approaches and distinctive elements of industrial landscapes. In personal profiles, especially on Twitter, complaints about the degradation and loss of heritage tend to predominate. Instagram mainly showcases esthetic images linked to travel circuits; however, the publications of heritage entities such as the NGO Hispania Nostra are likely to garner some debate. A particular role is played by personal profiles of experts in the given discipline who act on the different platforms as content disseminators and dynamizers of debates and complaints. This is the case for Diana Sánchez Mustieles (@Patrindustrial), J.J. Llera (@EspIndustrial), Diego Arribas (@McMulligan3), and Javier Revilla (@javirevilla), among others. Groups and associations are very active when it comes to disseminating news and informative campaigns about industrial spaces or elements, especially on Facebook and Twitter. Twitter and Facebook are identified as spaces for public debate and dissemination of citizens' actions that often do not have fixed digital platforms (websites or blogs). This is the case for the actions around the demolition of the cooling towers of the Compostilla and La Robla thermal power plants, when Twitter and Facebook served as a loudspeaker for civic heritage groups. Radio stations, local media, and political parties' profiles have also acted as disseminators of these types of content. In this sense, social media platforms serve as documentary sources for tracking dynamics, historical evolution, missing/obsolete functions, protection measures, and proposals for the recovery of industrial heritage.

The scarcity of content related to industrial memory in institutional profiles is striking. This phenomenon can be attributed to the lack of specific training of staff, the limited political weight attributed to these spaces, and the limitations of a reduced workforce. The dissemination of content by public administration is limited to very specific properties or landscapes and does not cover all areas subject to cultural protection or equipped with interpretation centers. There is a notable underrepresentation of industrial heritage and landscapes over other types of historical and artistic heritage. A specific study on the presence of content related to industrial heritage on social media profiles of competent administrations is a possible avenue for future analysis.

In sampling #2, the overall morphological, esthetic–perceptual, and social aspects are widely represented compared with the environmental aspect, which is directly mentioned in limited cases. Results show the state of abandonment, ruin, regional disarticulation of the industrial or post-industrial landscapes plus a growing demand for their patrimonialization. References related to environmental issues are limited. However, there are also socioeconomic issues in the background, and these should be analyzed separately. Problems related to the disappearance of heritage legacy of geographically disadvantaged regions could be related to findings presented by Barrio Rodríguez (2022) [43] regarding the digital campaigns of the NGO Hispania Nostra.

In sampling #3, the study cases show mixed results. There is a quite reduced echo of the industrial work attached to the National Institute of Industry as well as an uneven diffusion of the industrial landscape of the different companies. For example, in ENASA, the images of machinery stand out (64%), with the industrial and urban landscape of its factories being limited. ENSIDESA (Figure 3) shows the largest sample and less dispersion in content unrelated to the study; almost half of the analyzed content is related to the industrial profile (45.6%). Rodalquilar, although with a representative number of publications dedicated to the industrial profile or ruin (5%), returns content related to topics unrelated to the study. The situation with company towns is similar, with more than 25% of the analyzed content alluding to topics outside the scope of this study. This question allows us to raise the possibility that these enclaves have transcended their eminently industrial meaning, turning into enclaves or place names capable of alluding to other issues of life. In the case of ENDESA and Compostilla, it is worth mentioning that the "industrial profile" attributed to the industrial town of "Compostilla" and its former thermal power plant offers a greater number of publications than "Compostilla II", which has a limited echo on Instagram. However, it is possible that the volume of content associated with the landscape significance of Compostilla II increases while the controversy surrounding its demolition continues. Likewise, the inhabited towns of Llaranes and Ciudad Pegaso are the protagonists of an important volume of content attached to the analyzed topics. In the user-generated content studied in these towns are several references to civic actions of patrimonial recognition. On the contrary, the abandoned company towns of Asdrúbal or El Arteal have very limited diffusion in social media, and their presence would be explained by the esthetic component of their ruins.

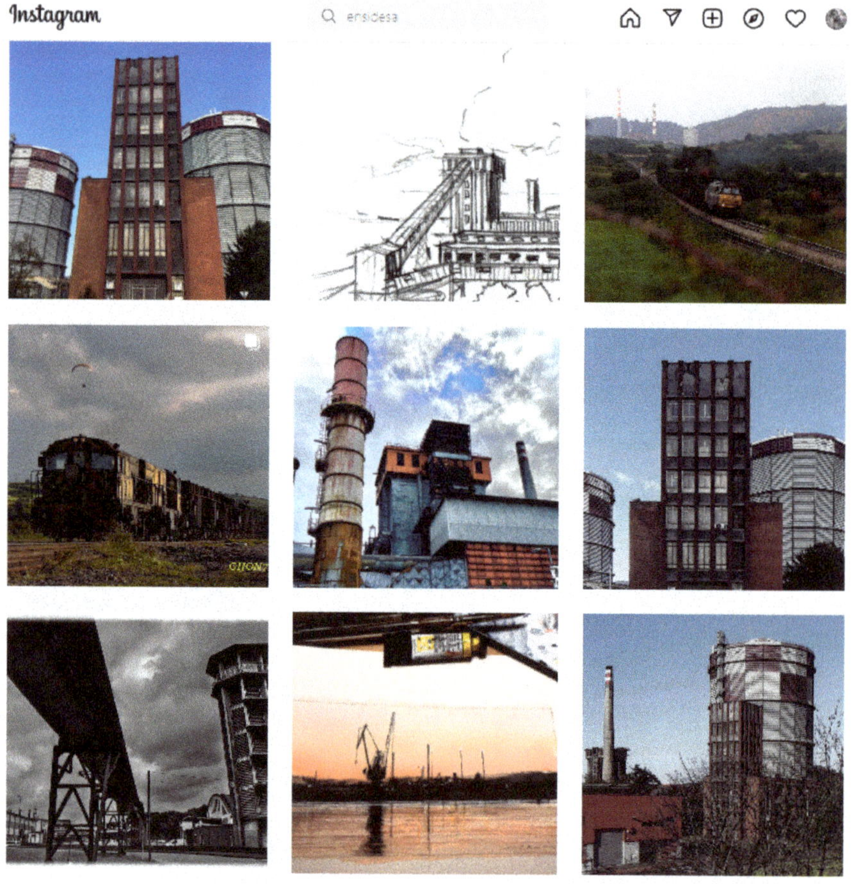

Figure 3. Some Instagram photos tagged with the hashtag ENSIDESA.

Social media platforms, in their most basic design, serve to relay the latest news. Undoubtedly, the great novelty of these digital media platforms is their interactivity and the possibility of producing a dialogue between users that can amplify and enrich the news. These platforms effectively constitute new forms, spaces, and times of social interaction as well as new dimensions of culture. Therefore, the information that flows through social media platforms somehow exceeds the mere means of communication, and is part of a new social configuration—a new way of understanding and interpreting culture [44].

Our results indicate that, at a general level, the information collected on social media platforms by individuals, groups, NGOs, and associations is likely to show the pulse of certain sites. The processes of valorization of obsolete industrial landscapes need a multidisciplinary study to which user-generated content contributes by highlighting only some of the characteristics of such landscapes. According to Akehurst (2009) [45], the credibility of user-generated digital content sources are trustworthy. In this sense, we can state that social media platforms respond to a social demand for the recognition of contemporary industry and the productive landscape as a community cultural heritage. However, it is difficult to quantify their scope and level of dissemination. Moreover, based on the social media platforms analyzed in detail (Twitter and Instagram), there is limited space for text and therefore not enough margin for a complete reflection.

As for civic debates on social media platforms, it is worth recognizing their current value in regard to the fact that they are practical cases of application of heritage considerations on elements that have not often been the subject of academic or press attention. In this sense, we could mention the novelty of the terms raised around the process of decarbonization that has opened up new avenues. The obtained results point to a remarkable echo of concrete heritage actions, such as those surrounding the demolition of Compostilla II or Andorra, with a significant influence. More accurate case studies would allow us to deepen this hypothesis. In addition, further studies could be oriented to the geolocation of specific sites or elements, and their echo on social media platforms. It would also be possible to identify recurrent elements over time.

In terms of its reach, Facebook stands out as the most used social media platform, with people interacting and keeping places of historical memory alive. This is in agreement with the general data, where Facebook is the second most visited website in Spain, only after YouTube [46], and the one with the highest number of users [35,47]. Findings of Liang et al. [26] also highlight Facebook being at the top of the rankings (30%), followed by Twitter (19%), and other websites (12%). This shows that platforms that are text-based are the most popular among the global audience. In our results, Twitter stands out as the most used platform to report cases of abandonment, building demolition, etc. Twitter is a platform that allows transmitting and maintaining the interest of an event, as pointed out by the Social Media and Events Report 2011 [48], since most tweets related to an event occur during its celebration (60%); a second peak arises thereafter due to the publication of material that users share, representing 35% [36]. In addition, the use of hashtags and their combination with others already consolidated on this platform allows fostering the dissemination of the event in social media platforms. Campos-Domínguez [49] highlights Twitter's social role in political communication.

Instagram, on the other hand, produces a call-to-action effect on photos uploaded from certain locations. Recently, researchers have noticed the so-called "Instagram effect", which implies that a point-of-interest becomes increasingly more popular when highlighted on social media [50], independent of its capacity to absorb visitors. Our results indicate that these are specialized content; industrial and post-industrial landscapes promote considerably less interest than other unrecognized places, in agreement with findings from Falk and Hagsten [32]. In this sense, Instagram can play a prominent role as one of the social media platforms that use visuals to raise awareness of cultural heritage [28].

Last but not least, when analyzing the role of institutions and other administrations, we found that the use of social media platforms to promote heritage is scarce and limited to the dissemination of information, and therefore not for user interaction. Hidalgo Giralt et al. [51] had already pointed out that the quality of information showcased on websites, and the dissemination of digital content in cultural spaces, only reach average values; the type of information they offer is basic, and user interaction is lacking.

5. Conclusions

The role that social media platforms play in Spain in relation to the conservation of industrial heritage and its revaluation has been studied for the first time. The sampling carried out here was small and very fragmented. Given the enormous volume of data from various platforms, we found that manual registration was difficult. Further screening is required due to the fact that the keywords used have meanings that go beyond the field of study. The use of hashtag tracking and metric analysis of social media platforms is required for a more precise quantitative study. The use of systems for automating tracking processes, such as Python scripts, could be considered for further analyses.

The findings of this study show that social media platforms can be used as tools to disseminate information about industrial landscapes. Although they are considered communication tools, their contribution is only partial in relation to the reality of post-industrial landscapes. However, it is worth recognizing the added value that their possibility of interaction brings as a reference for current civic debates. To this extent, concepts that can be incorporated into a theoretical discussion on the future of post-industrial landscapes can be identified in social media debates. In this way, social media could play a proactive role in the cultural reactivation of these landscapes. In addition, specific tracking analyses can be oriented toward ethnographic and documentary tasks that could complement inventory and cataloguing work.

Social media platforms have potential as essential tools for training, awareness of collective memory, and social cohesion around industrial and post-industrial landscapes. At the administrative and institutional level, launching social media platforms to encourage industrial heritage as a touristic resource is vital.

The contents of social media platforms could contribute toward mitigating the difficulties related to the recognition, comprehension, and protection of post-industrial landscapes.

Author Contributions: Á.L.R. elaborated the theoretical framework of the study and the methodological proposal. J.M.-M. carried out the data search, both authors developed the study, and J.M.-M. wrote the final manuscript. All authors have read and agreed to the published version of the manuscript.

Funding: This work is part of the research project "The image of the National Institute of Industry in the territory: Cartography and landscape of industry", funded by the Spanish Ministry of Science, Innovation and Universities. Ref. PGC2018-095261-B-C22. PI: Ángeles Layuno-Rosas.

Data Availability Statement: Not applicable.

Acknowledgments: This research has been supported by Ana Molina and Adrián Magaz. The guidance of Elena Marcos and Mario Peláez about social media tracking is much appreciated. The information about NGO Hispania Nostril's social media results provided by Beatriz Barrio is also gratefully acknowledged. The authors would also like to thank social media contributors involved in heritage and industrial landscape.

Conflicts of Interest: The authors declare no conflict of interest.

References

1. Council of Europe. European Landscape Convention; Council of Europe: Florence, Italy, 2000, October 20th. Signed by Spain in 2000, Ratified in BOE 31, 2008, February 5th. [BOE-A-2008-1899]. Council of Europe Treaty Series-No. 176. Available online: https://www.coe.int/en/web/conventions/full-list?module=treaty-detail&treatynum=176 (accessed on 5 December 2022).
2. UNESCO, Stockholm Conference. *Intergovernmental Conference on Cultural Policies for Development: Final Report*; CLT.98/CONF.210/5; UNESCO: Stockholm, Sweden, 1998.
3. Han, S.H.; Zhang, H. Progress and Prospects in Industrial Heritage Reconstruction and Reuse Research during the Past Five Years: Review and Outlook. *Land* **2022**, *11*, 2119. [CrossRef]
4. Roux, H.; Cestaro, G. Hostinng the Olympics through industrial regeneration and reuse: A comparative study of Turin 2006, London 2012 and Beijing 2022. In *Stati Generali del Patrimonio Industriale 2022*; Currà, E., Docci, M., Mennicchelli, C., Russo, M., Severi, L., Eds.; Marsilio Editori: Venezia, Italy, 2022.
5. Sun, M.; Chen, C. Renovation of industrial heritage sites and sustainable urban regeneration in post-industrial Shanghai. *J. Urban Aff.* **2021**. [CrossRef]

6.   Zhang, J.; Cenci, J.; Becue, V.; Koutra, S.; Liao, C. Stewardship of Industrial Heritage Protection in Typical Western European and Chinese Regions: Values and Dilemmas. *Land* **2022**, *11*, 772. [CrossRef]

7.   Biel Ibáñez, P. Unidad de producción térmica Teruel (Andorra). El final de un modelo energético y el inicio de un proceso de patrimonialización. *Rolde Rev. Cult. Aragon.* **2020**, *172–173*, 4–25.

8.   Clarke, J. 20th-Century Coal- and Oil-Fired Electric Power Generation. Introductions to Heritage Assets (Historic England Guidance). 2015. Available online: https://historicengland.org.uk/images-books/publications/iha-20thcentury-coal-oil-fired-electric-power-generation/ (accessed on 5 December 2022).

9.   Escudero, D.; de la O Cabrera, M.R. Energía y territorio: Inmersiones en un paisaje dinámico. *Ábaco Rev. Cult. Cienc. Soc.* **2015**, *86*, 23–32.

10. AAVV. *Industrial Heritage Re-Tooled*; Routledge: New York, NY, USA, 2012.

11. Orange, H. Industrial Archaeology: Its Place Within the Academic Discipline, the Public Realm and the Heritage Industry. *Ind. Archaeol. Rev.* **2013**, *30*, 83–95. [CrossRef]

12. Bangstad, T.R. *Defamiliarization, Conflict and Authenticity: Industrial Heritage and the Problem of Representation*; NTNU: Trondheim, Norway, 2014.

13. Strangleman, T. "Smokestack Nostalgia," "Ruin Porn" or Working-Class Obituary: The Role and Meaning of Deindustrial Representation1. *Int. Labor Work. Hist.* **2013**, *84*, 23–37. [CrossRef]

14. Zoido Naranjo, F. El paisaje, patrimonio público y recurso para la mejora de la democracia. *PH, Boletín Inst. Andaluz Patrim. Histórico* **2004**, *50*, 66–73. [CrossRef]

15. Jackson, J. *A Sense of Place, a Sense of Time*; Yale University Press: New Haven, CT, USA, 1994.

16. Lowentahl, D. Living with and looking at landscape. *Landsc. Res.* **2007**, *32*, 637–659. [CrossRef]

17. International Committee for the Conservation of the Industrial Heritage (TICCIH). Chart of Nizhny Tagil. In Proceedings of the TICCIH Congress 2003, Moscow, Russia, 17 July 2003.

18. ICOMOS—TICCIH. «The Dublin Principles». In Proceedings of the Principles for the Conservation of Industrial Heritage Sites, Structures, Areas and Landscapes, 17th ICOMOS General Assembly, Paris, France, 28 November 2011.

19. Arnaboldi, M.; Diaz Lema, M.L. Shaping cultural participation through social media. *Financ. Account. Manag.* **2022**, *38*, 299–321. [CrossRef]

20. Mihelj, S.; Leguina, A.; Downey, J. Culture is digital: Cultural participation, diversity and the digital divide. *New Media Soc.* **2019**, *21*, 1465–1485. [CrossRef]

21. King, L.; Stark, J.F.; Cooke, P. Experiencing the Digital World: The Cultural Value of Digital Engagement with Heritage. *Herit. Soc.* **2016**, *9*, 76–101. [CrossRef]

22. Secretaría del Patrimonio Cultura (Chile). 'Participación Digital en Patrimonio'; Santiago de Chile: Ministry of Cultures, Arts and Heritage of the Government of Chile, October 2022. Available online: https://www.cultura.gob.cl/publicaciones/participacion-digital-en-patrimonio/ (accessed on 7 December 2022).

23. Council of Europe. Convention on the Value of Cultural Heritage for Society; Council of Europe: Faro, Portugal, 2005, October 27th. Signed by Spain in 2018, Ratified in BOE 144, 2022, June 17th [BOE-A-2022-10041]; Council of Europe Treaty Series-No. 199. Available online: https://rm.coe.int/1680083746 (accessed on 7 December 2022).

24. González Sánchez, C. Comunicación y redes sociales en instituciones culturales. Introducción al debate. *Rev. PH* **2020**, *29*, 118–119.

25. Durán Salado, I.; Fernández Cacho, S. Una aproximación al paisaje cultural mediante etnografía virtual complementa el conocimiento experto de la percepción social. *Rev. PH* **2020**, *99*, 9. [CrossRef]

26. Liang, X.; Lu, Y.; Martin, J. A Review of the Role of Social Media for the Cultural Heritage Sustainability. *Sustainbility* **2021**, *13*, 1055. [CrossRef]

27. Barrientos, F.; Martin, J.; De Luca, C.; Tondelli, S.; Gómez-García-Bermejo, J.; Casanova, E.Z. Computational methods and rural cultural & natural heritage: A review. *J. Cult. Herit.* **2021**, *49*, 250–259.

28. Rosa, M.; Sunarya, Y. The Use of Social Media For Raising Awareness of Cultural Heritage And Promoting Industrial Heritage In Indonesia. In Proceedings of the IICACS: IV International and Interdisciplinary Conference on Arts Creation and Studies, Gdansk, Poland, 17 October 2020; Pascasarjana Institut Seni Indonesia: Surakarta, Indonesia, 2020; Volume 2, pp. 23–32.

29. Stuedahl, D.; Lowe, S. Experimentinng with culture, technology, communication: Scaffolding imagery and engagement with industrial heritage in the city. In Proceedings of the 9th International Conference in Culture, Technology, Communication: Celebration, Transformation, New Directions, Oslo, Norway, 18–20 June 2014; pp. 98–115.

30. Vukmirović, M.; Nikolić, M. Industrial heritage preservation and the urban revitalisation process in Belgrade. *J. Urban Aff.* **2021**. [CrossRef]

31. Naramski, M.; Herman, K.; Szromek, A.R. The Transformation Process of a Former Industrial Plant into an Industrial Heritage Tourist Site as Open Innovation. *J. Open Innov. Technol. Mark. Complex.* **2022**, *8*, 74. [CrossRef]

32. Falk, M.T.; Hagsten, E. Visitor flows to World Heritage Sites in the era of Instagram. *J. Sustain. Tour.* **2021**, *29*, 1547–1564. [CrossRef]

33. Sobrino Simal, J.; Sanz Carlos, M. Carta de Sevilla de Patrimonio Industrial 2018. Los Retos del Siglo XXI; Sevilla: Centro de Estudios Andaluces. 2019. Available online: https://www.centrodeestudiosandaluces.es/publicaciones/carta-de-sevilla-de-patrimonio-industrial-2018-los-retos-del-siglo-xxi (accessed on 5 August 2022).

34. Loures, L. Post-industrial Landscapes as Renaissance Locus:The Case Study Research Method. *WIT Trans. Ecol. Environ.* **2008**, *117*, 293–302.

35. We are Social & Hootsuite, 'Digital 2020, Report from Spain'. 2020. Available online: https://datareportal.com/reports/digital-2020-spain (accessed on 7 December 2022).
36. Campillo Alhama, C.; Ramos Soler, I.; Castelló Martínez, A. La gestión estratégica de la marca en los eventos empresariales 2.0. *Adres. ESIC Int. J. Commun. Res.* **2014**, *10*, 52–73.
37. Mariani, M.M.; Di Felice, M.; Mura, M. Facebook as a destination marketing tool: Evidence from Italian regional Destination Management Organizations. *Tour. Manag.* **2016**, *54*, 321–343. [CrossRef]
38. López-Sánchez, M.; Tejedor-Cabrera, A.; Linares-Gómez del Pulgar, M. Indicadores de paisaje: Evolución y pautas para su incorporación en la gestión del territorio. *Ciudad Territ. Estud. Territ.* **2020**, *52*, 719–738.
39. Magaz Molina, J.; Layuno Rosas, Á. Colonization and urbanization of the energy´s territory: National Institute of Industry company towns (1941–1975). In *Stati Generali del Patrimonio Industriale 2022*; Currà, E., Docci, M., Meninchelli, C., Russo, M., Severi, L., Eds.; Marsilio Editori: Venezia, Italy, 2022.
40. García García, R. Energías extinguidas, Centrales térmicas del periodo de la Autarquía. In Proceedings of the III Seminario del Aula de Formación, Gestión e Intervención sobre Patrimonio de la Arquitectura y la Industria, Madrid, School of Architecture of the Polytechnic University of Madrid, Madrid, Spain, 18–19 February 2016; pp. 163–185. Available online: http://gipai.aq.upm.es/index.php/publicaciones/actas-del-iii-seminario-gi_pai-energia/ (accessed on 5 December 2022).
41. García García, R.; Layuno Rosas, Á.; Llull Peñalba, J.; García Carbonero, M. El INI activador de la movilidad nacional: Redes de transporte, empresas e instalaciones de la primera épocca del Instituto (1941–1963). In *Hacia Una Nueva Época para el Patrimonio Industrial*; Álvarez Areces, M.Á., Ed.; INCUNA: Gijón, Spain, 2021; pp. 363–376.
42. Suárez Menéndez, G. Aportaciones a la Arquitectura del Movimiento Moderno desde el Patrimonio Industrial: La actividad de Cárdenas y Goicoechea en ENSIDESA. *Liño* **2011**, *17*. Available online: https://reunido.uniovi.es/index.php/RAHA/article/view/129 (accessed on 5 December 2022).
43. Barrio Rodríguez, B. El poder de las redes sociales para dar voz al patrimonio en peligro. In Proceedings of the II Simposio de Patrimonio Cultural ICOMOS España, Cartagena, Spain, 17–19 November 2022.
44. Vizer, E.; Carvalho, H. Las Tecnologías de la Información y la Comunicación como paradigmas de la Economía de la Información. *Rev. Soc.* **2007**, *39*, 5–17.
45. Akehurst, G. User generated content: The use of blogs for tourism organisations and tourism consumers. *Serv. Bus.* **2009**, *3*, 51–61. [CrossRef]
46. Asociación Para la Investigación de Medios de Comunicación—AIMC, 'Marco general de los Medios en España', Madrid. 2021. Available online: https://www.aimc.es/a1mc-c0nt3nt/uploads/2021/02/marco2021.pdf (accessed on 7 December 2022).
47. We are Social & Hootsuite, 'Digital 2022, Global Oveview Review. The Essential Guide to the World's Connected Behaviours.'. 2022. Available online: https://wearesocial.com/cn/wp-content/uploads/sites/8/2022/01/DataReportal-GDR002-20220126-Digital-2022-Global-Overview-Report-Essentials-v02.pdf (accessed on 7 December 2022).
48. Von Ferenczy, D.; Spiess, S.; Koch, L. *Social Media and Events Report*; AMIANDO: Munich, Germany, 2011.
49. Campos-Domínguez, E. Twitter y la comunicación política. *Prof. La Inf.* **2017**, *26*, 785–794. [CrossRef]
50. Miller, C. How Instagram Is Changing Travel. National Geographic, 26 Januray 2017. Available online: https://www.nationalgeographic.com/travel/article/how-instagram-is-changing-travel (accessed on 12 December 2022).
51. Hidalgo Giralt, C.; Palacios García, A.J.; Fernández Chamorro, V. La operatividad turística de los espacios culturales de origen industrial en Madrid. Un análisis de la oferta turística potencial mediante indicadores. *Cuad. Tur.* **2018**, *41*, 1989–4635. [CrossRef]

 *land*

Article

Postproduction in the Research on the Urban Cultural Landscape: From the Transfer of Results to the Exchange of Knowledge on Digital Platforms and Social Networks—The TRAHERE Project in Madrid

Graziella Trovato

Department of Architectural Composition, Escuela Técnica Superior de Arquitectura de Madrid, Universidad Politécnica de Madrid, 28040 Madrid, Spain; graziella.trovato@upm.es

Abstract: The urban scenarios outlined by the environmental and economic crisis have fostered, on one hand, the unstoppable gentrification of the most central neighborhoods of the cities; and on the other hand, a growing associationism committed to cultural and environmental values, which demands tools from academia to negotiate with the administration. An emblematic case is that of Arganzuela, one of the three districts of the Spanish capital affected by the rise and fall of industrialization (the freight railway in 1861 and the M-30 ring road in 1970). The burying of these infrastructures began in 1990 with the Green Railway Corridor (PVF) operation, a year after the inauguration in Paris of the Promenade Plantée on the disused railway lands, which allows us to foreshadow new scenarios. The TRAHERE project researches the state of abandonment and disaffection of the public spaces of the PVF using social networks as a connection platform with participatory channels to promote its regeneration. The challenge is to convert the concept of transfer of results into a more inclusive one of knowledge exchange, which implies a methodological change in research, with an integrating perspective that combines urban historical studies with artistic practices of production and postproduction for the dissemination of content on the networks.

Keywords: citizen participation; cultural landscape; industrial heritage; social media; postproduction

Citation: Trovato, G. Postproduction in the Research on the Urban Cultural Landscape: From the Transfer of Results to the Exchange of Knowledge on Digital Platforms and Social Networks—The TRAHERE Project in Madrid. *Land* **2023**, *12*, 31. https://doi.org/10.3390/land12010031

Academic Editors: Cecilia Arnaiz Schmitz, Nicolas Marine and María Fe Schmitz

Received: 30 November 2022
Revised: 16 December 2022
Accepted: 19 December 2022
Published: 22 December 2022

1. Introduction

The TRAHERE project is part of a little-explored line of landscape and urban studies that address documentary and historical research from a contemporary perspective with a clear vocation to become forums for citizen debate around the topics studied. As a national precedent, we point out "Aqueous Madrid", a documentary and artistic research that considers the emotional reconnection of the city of Madrid with its watery past [1]. A space for reflection and debate was curated by Malú Cayetano Molina in collaboration with the International Biological Station DueroDouro's "Water Project" [2]. Both projects share the aspiration to combine the exactitude and depth of scientific and urban historical studies with an artistic perspective, the consequence of a transversal approach, and a careful postproduction process of the results in order to provide emotional experiences that can be shared by citizens. In short, they seek to provide the didactic contents with an aesthetic, contemplative, and operational dimension at the same time.

As a starting point we ask ourselves:

Can the use of digital technology and social networks promote a process of reappropriation based on a shared knowledge of the urban landscape and its transformations? If that is the case, how can we carry it out from the universities?

If the concept of transfers of results entails "the dissemination of the results for their implementation in society through professional practice" [3], the exchange of knowledge through social networks implies an open attitude toward a shared knowledge within the reach of all citizens. In this sense, a change of attitude is proposed here that, in methodological terms, introduces the postproduction of results as a step after production and prior to dissemination through academic channels and social networks. Thus, a predominantly artistic practice is introduced into academic research. In fact, if the term "production" is synonymous with creation, elaboration, and original composition through the use of raw materials (from the Latin *producere*, in turn from *ducere*, meaning to drive toward something in particular), the concept of "postproduction" arose in 1989 in the digital culture as a set of operations that include dubbing, editing, mixing, and the eventual digital elaboration of sequences with special effects of videos, photos, or music tracks that precede the staging and/or marketing phase [4] (p. 2094).

The theoretical and conceptual framework is offered to us in this sense by the French historian and art critic Nicolás Bourriaud, who has theorized about this practice. According to Bourriaud, the contemporary creator acts as a disc jockey: rather than generating or composing new forms, he reprograms and combines those that already exist based on his information [5] (p. 25).

In this theoretical context, the objectives are, at a general level:

- Encourage a more democratic conservation of the landscape.
- Foster, at an academic level, a more democratic and open conception of research in the field of landscape.

At a specific level:

- To claim the care of the public space and its heritage, thereby promoting its knowledge in line with a new sensibility.
- To apply historical and heritage knowledge to more open and participatory social contexts through the use of storytelling techniques and the postproduction of results.
- Indicate strategies for the revitalization and enhancement of the public space by establishing physical, virtual, and cultural networks through social media.

Therefore, we are trying to give a contemporary response to the complexity of the contemporary landscape to an increasing presence of neighborhood associations in the political life of cities.

Our case study, the Pasillo Verde Ferroviario de Madrid (PVF), meets the requirements of a public space in a state of abandonment with a strong presence of neighborhood associations and that is very active on the social networks, which they use to share knowledge about the urban landscape and to call for actions for its improvement and conservation. Despite being the largest urban development operation carried out in the capital before Madrid Río, the residents of the area, and more generally of the city of Madrid, hardly know its history. Citizens have discovered their public spaces during the pandemic, when vegetation has taken over the city with an explosion of wild flora, especially around the tracks and monumental stations of Príncipe Pío and Delicias, which are located at the opposite ends of the area (Figure 1).

These starting conditions allow us to apply and validate our methodology through a study and rigorous analysis of the area and through a proposal for the future linked both to the river and to the postproduction of the results, which will in turn be disseminated to and exchanged with the public and the institutions through social networks. In this way, the aim is to turn the TRAHERE project into an operational tool for exchanging knowledge focused on the shared care of the urban landscape, which entails a change in the methodology of the research, as will be explained later (Figure 2).

We have chosen as an acronym of our project the word TRAHERE (*TRAin HERitage Experiences*) because it is a Latin verb from which derives, through the Norman and by successive adaptations, the word (French first and English later) *train*. From *trahere* (drag) derives the vulgar *traginare*, in Old French (*Middle* English) *trahiner* (push something), and from this the noun to which we refer, *train*, used in both French and English. TRAHERE could also be summarized as *TRAin HEritage REuse*, thereby emphasizing the concept of reuse as a postproduction process in which unused materials find, with few means and without the need for specialized intervention, a new life [6].

(a)

(b)

Figure 1. State of abandonment of the public space of the PVF: (**a**) surroundings of the Delicias station; (**b**) wild flora on the roads. Photos by Ramón Gómez of Herba Nova for the TRAHERE project.

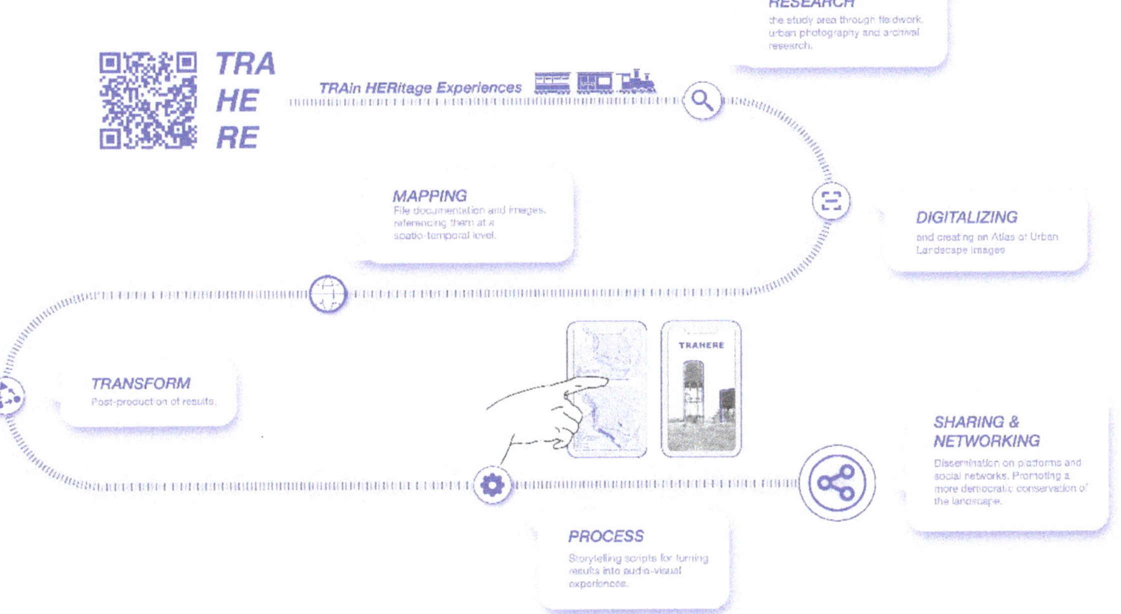

Figure 2. Initial objectives. Source: TRAHERE's own elaboration.

2. Case Study

2.1. Status of the Issue on the Area of Study

Studying the PVF means analyzing and measuring in qualitative and qualitative terms the impact of the irruption, rise, and decline of industrialization in the city of Madrid—an arc of two centuries whose analysis has allowed us to verify that in the case of the PVF, the relation between the generation of public spaces and the defense and care of the industrial landscape and its heritage is not entirely balanced.

The specific bibliography on the Green Railway Corridor is brief. The most relevant is formed by the publications of the Urban Consortium itself [7] and the subsequent ones of its technical director, Alfonso García Santos, who has detailed in several articles the experience of management and technical materialization of the operation [8]. There are brief useful reviews such as those from the guides of the Madrid City Council, in particular the one published in 2016 and coordinated by Ramón López de Lucio, Álvaro Ardura, José Javier Bataller, and Javier Tejera [9] (pp. 44–47); and the 2014 guide of the Historical Service of the Official College of Architects of Madrid (COAM) [10]. Both publications clarify, in a summarized way, the responsibilities, main legal procedures, phases, and results of the project. An important contribution is the study published in 1995 by geographers Dolores Brandis and María Isabel del Río, in which the repercussions of the project management and the successive modifications in buildability and uses in the affected area are explained in depth [11] (pp. 113–128).

Among the technical reports, we point out the evaluation of the Green Railway Corridor carried out by Testing Consulting in 2003 for the Ministry of Planning and Cooperation (MIDEPLAN) of Chile as an example of management along with seven other international case studies such as Puerto Madero (Buenos Aires, Argentina), San Diego Downtown (CA, USA), Paris Rive Gauche (Paris, France), and London Docklands (London, UK). The document is accessible through the digital library of the Inter-ministerial Office for Transport Planning (SECTRA). It thoroughly analyzes the management and financing model of the operation on the basis of data largely provided by the Urban Development Consortium [12].

The recently published book *The Urbanism of Transition: The Master Plan of Madrid 1985* by Carlos Sambricio and Paloma Ramos provides data and reflections of interest on aspects related to the PVF [13].

An important book to help understand the grounds of the operation is *Madrid Proyecto Madrid 1983–1987*, which collected the activity developed by the Municipal Management during the years indicated in the title and was based on a series of operations that laid the foundations of the Green Railway Corridor a few years before [14].

The original documentation of the Green Railway Corridor is distributed among multiple files. All the monthly reports of the Technical Assistance to the Works of the Green Railway Corridor of Madrid are kept in the deposit of the library of the Technical School of Civil Engineers of the Polytechnic University of Madrid donated by Alfonso García Santos, who informed us of its existence and location. It is a set of approximately 200 volumes (from the end of 1989, when the operation started, until the end of 1996) with detailed technical information that is especially related to the railway infrastructure. The bulk of the photographic, graphic, and documentary documentation is kept in the Regional Archive of the Community of Madrid (ARCM) and is organized in five volumes related to the five sections in which the operation was structured (A-B-C-D-E). The documentation relating to the new interchangers is kept in the General Archive of the Administration of Alcalá de Henares (AGA). The Museum of Science and Technology (MUNCYT) preserves documentation relating to the multiple files and projects for restructuring the Delicias station and its surroundings into the museum [15] (p. 153).

2.2. The Green Railway Corridor (PVF) Hypothesis and Objectives

The PVF was carried out between 1989 and 1996 through the 1st Modification of the General Plan with the creation of an Urban Consortium formed by representatives of the National Network of Spanish Railways (RENFE) and the City Council. The work involved the regeneration of an urban corridor of 8 km in length between Puente de los Franceses in the Moncloa district and the Delicias station in Cerro de la Plata. In other words, it was a radical transformation of the northwest–southeast arc of the capital parallel to the Manzanares River (Figure 3).

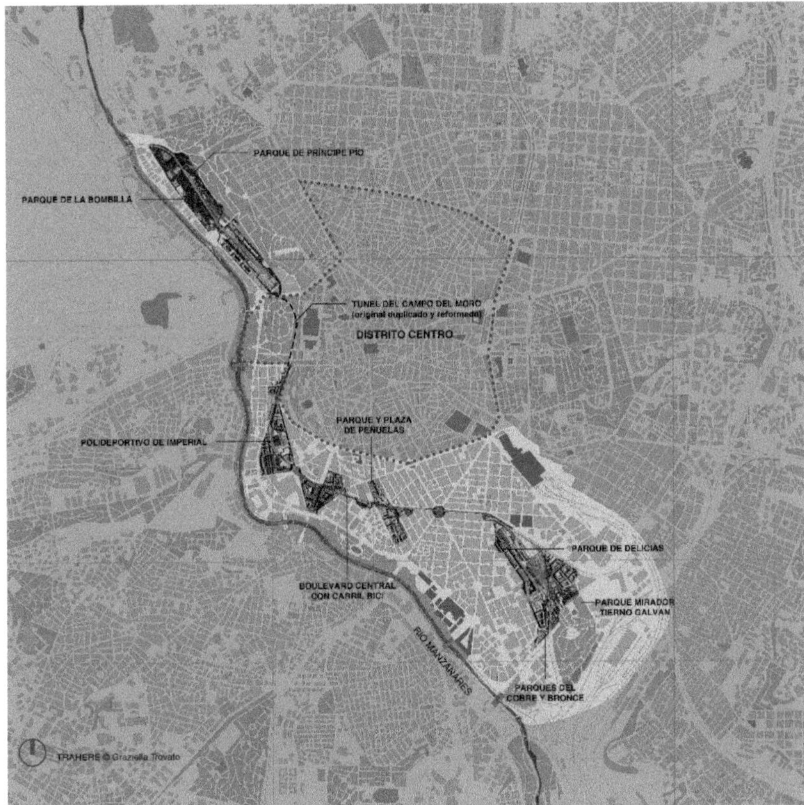

Figure 3. The PVF and the scope of action between the Central District of the capital and the Manzanares River. The parks of Príncipe Pío, Bombilla, Peñuelas, Delicias, Mirador de Tierno Galván, Cobre, and Bronce stand out, as well as the central boulevard with a bike lane. Source: TRAHERE's own elaboration.

The burial of the railway route was, in fact, the opportunity for the regeneration of a portion of the city that occupies 163 Ha in the districts of Moncloa-Aravaca, Centro (although tangentially), and Arganzuela. The area gained urban parks, intermediate spaces between landscaped blocks, bike lanes, a central boulevard, and non-residential buildings at no cost to the municipality. The cost in terms of industrial heritage, however, has been high: many factories disappeared, some of them of architectural, typological, and cultural interest; several buildings were reused without much sensitivity; and there is scarce tangible railway heritage left that is basically limited to the Delicias station, which was converted into the Railway Museum and the National Science and Technology Museum. In the case of the Príncipe Pío station, it is safe to say that it has been mistreated in many ways, both in the treatment of its surroundings and in a series of extensions unrelated to its initial use that have trivialized and altered the legibility of the original station.

It was the consequence of the decline of the industrial fabric and the obsolete character of the 19th-century single-track railway contour line that was intended for the transport of goods. The process began at the Planning Office of the Municipal Management in a climate of strongly politicized democratic construction. The burial of the railway route, in fact, was the opportunity for fairer growth that was in line with planning aimed at balancing the differences between a northern area of a privileged Madrid and a depressed and underdeveloped working-class south. In addition, the city discovered the difficulties of continuing to grow to the north according to the guidelines of the Zuazo-Jansen plan of 1929 due to, among other things, the existence of the neighborhoods of Fuencarral and Hortaleza. This highlighted the development opportunities of a wide swath of the city already immersed in a transformation phase and with abandoned plots and empty interstitial areas due to the decline of the industrial sector. The possibility of designing and shaping a new city with public parks, non-residential areas, and housing was foreshadowed in a new development in the southwest of the capital [16] (pp. 43–46).

It was a question of modernizing the transport network via the execution of a modern Network of Suburban and Metro Interchanges and at the same time regenerating the public space as derived from the restructuring of the disused railway fabric.

The residents of Arganzuela, one of the three districts affected by the operation and in turn composed of five neighborhoods of the capital, were responsible for the movement to demand the burial of the rails, which crossed our study area on the surface and frequently caused accidents [13] (pp. 250–255). This associative system was reactivated in 2002 to claim a public use of the old Municipal Slaughterhouse and achieved its conversion into a Cultural Center against the planned privatization, and finally in 2011, as a result of Operation Madrid Río, which is tangent to the one we are now dealing with, to defend the district from a growing gentrification [17] (Figure 4).

(a) (b)

Figure 4. Activism of neighborhood associations of Madrid after the Democratic Transition. (**a**) First protest of the neighborhood associations of Madrid in 1977 (Archivos de la Transición [18]); (**b**) an example of the presence of neighborhood associations on social networks: the current profile of the Pasillo Verde Imperial Association on Twitter.

However, since the PVF ended in 1996, it has remained in a state of total abandonment. Despite the beneficial effects of the operation, it was not publicized, probably as a result of a zero-cost financing model, which forced the disused land to be rezoned on several occasions to increase buildability and consequently its profitability. We found an urban landscape in a state of degradation due to lack of maintenance and unknown to a large part of the citizens. Publications, as we said, are scarce, and almost entirely driven by their own creators.

We ask ourselves: what are the causes of this ignorance and lack of care?

As a starting point we find:

- Loss of Industrial Memory and Disaffection.

The PVF today lacks a recognizable identity. In general, the set of railway facilities will be seriously compromised if the PVF chooses to replace the heritage landmarks with, according to the testimony of the then director of the Consortium, Manuel Ayllón, a symbolic framework of Masonic origin whose meaning is ignored by citizens and is not related to the city and its memory [7] (pp. 10–14). The structuring elements and landmarks of the PVF have a monumental character not related to the historical memory of the city, although fragments of its railway past appear occasionally that are disconnected from its original fabric and meaning. They are seen as solitary elements that are indifferent to the urban fabric and its inhabitants. All this led us to relate, in an initial hypothesis, the loss of neighborhood memory to disaffection.

- The Loss of Connection between the Green Railway Corridor and the River.

The PVF never sought a connection with the river despite being settled precisely on its old river terraces. In fact, "Terrazas del Manzanares" was the name given in the early 19th century to the archaeological area located on the old bank of the river at an approximate height of 600 m above sea level between El Pardo and Getafe. This area, which affects the entire Railway Green Corridor, was the first settlement in Madrid and has been logically close to the waters of the river since the Paleolithic, as evidenced by the discoveries preserved in the Madrid-based Museum of San Isidro and other samples in the Museum of History of the capital. The burial of the M-30 with Madrid Río in the first decade of the XXI century and the newly opened Plaza de España today allow us to imagine new scenarios for the future: an urban continuum that connects the Green Railway Corridor with the banks of the river through sustainable mobility and a greater biodiversity. Between the river and rails is, in short, a proposal for an integrated return to the river in a desirable circularity (Figure 5).

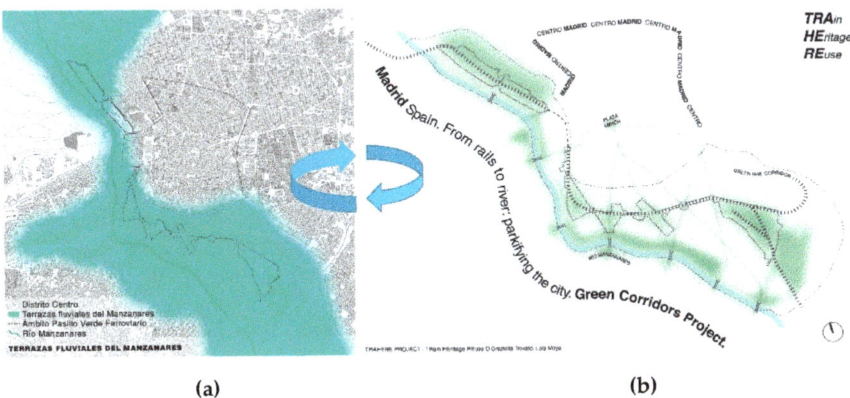

Figure 5. Two diagrams that synthesize the circularity proposal. (**a**) The scope of the PVF in the old fluvial terraces of the river. The following are distinguished: the downtown district, the area of the PVF, and that of the old terraces. (**b**) The proposal for the connection project between the PVF and Madrid Río "Towards a desirable circularity". Source: TRAHERE's own elaboration.

3. Materials and Methods, "Postproduction" vs. "Transfer", and Academia in a Process of Change

As has been explained, we distinguished in the research process between the "production" of results and the "postproduction" of them. The first focused on the transfer of results (book, book chapter, conference papers, and scientific articles) and the second on what we understand as dissemination and exchange of knowledge through digital platforms, physical and virtual networks and presence in public spaces. Both phases alternate and overlap so that production and postproduction of results are understood in a relationship of constant and mutual feeding.

In both phases the approach must be, in our opinion, holistic, transversal, and integrative. To achieve this, the first step is the formation of an interdisciplinary team with the ability to combine humanities with sciences and the digital world, as well as a quantitative analysis with a qualitative one, in functional, aesthetic, and patrimonial terms. A team composed of an architect (Graziella Trovato, main researcher), an urban planner (Luis Moya), a heritage manager (Melín Nava, granted for this project by the Carolina Foundation), urban photographers (Davide Curatola Soprana and Ramón Gómez; the latter specializes in wild flora and railway botany), and a digital design team (Diego Iglesias and Cristóbal Baños from Hyper Studio). The GIPC architects and researchers (Clara Álvarez García, Isabel Rodríguez de la Rosa, and Beatriz Salido) have also collaborated in graphic and audiovisual production as well as in data editing and documentary research. The Pasillo Verde Imperial Neighborhood Association, the largest of those present in the area, has also been involved, especially in fieldwork and the location of disused buildings for their possible transformation for public use.

3.1. Production of Results: Materials and Methods

As a basis for the landscape and urban analysis, the materials used in the first phase of production of results were:

- The General Plans: from the Castro Plan of 1860 until today, including the various modifications of the General Plan of 1985 with which the PVF was carried out. This documentation, which is kept in the Regional Archive, was unpublished because the modifications, with the exception of the first, responded to the need to rezone the soil in order to make bigger profits.
- Historical cartography: 15 plans in total that included land register and topographic plans and others related to plots of land from the years 1622 to 2020. In addition, it includes four orthophotos from 1975, 1991, 1999, and 2020. These documents were selected based on the degree of information provided to measure the impact of industrialization in the study area.
- A total of 850 documents that included various documentation from public and private archives; photo libraries; newspaper archives; and municipal, regional, and state libraries. They included unpublished and/or unreferenced documentation such as the original plans of the railway contour line project of 1861 as well as other materials that documented the creation and consolidation of the PVF in the 1990s.
- Material obtained from fieldwork carried out throughout the research process. As we stated previously, part of this process was carried out with the Pasillo Verde Imperial Association via actions organized through its social networks (mainly Twitter and Instagram) (Figure 6).

On this basis and with the use of graphic design programs (mostly InDesign, Auto-CAD, and Photoshop), the method consisted of:

- Developing different location diagrams of the study area accompanied by quantitative data on the of the operation in its entirety and the balances related to the area occupied by the railway and industrial heritage, housing, and public and non-residential spaces before and after the PVF.
- Tracing the evolution according to the stages of the area from its origins (river terraces) to the irruption of the railway (1861), the construction of the M-30 ring road (1970), the burial of the railway (PVF 1990), and the burial of the M-30 (Madrid Río 2011).
- Mapping the documentation of archives and locating and indexing it to historical plans and orthophotos. This allowed a spatio-temporal reading of the PVF understood as a cultural landscape and the measurement from the quantitative point of view of the heritage by identifying and differentiating between what has disappeared and what has been transformed through reuse operations.
- Photographing the public spaces and historical landmarks of the PVF and its area of influence. To the urban signature photography (Davide Curatola Soprana), we conferred an operative role in the critical sense that Zevi attributes to this word; that is

to say, propositional (for the quantitative and qualitative analysis of the operation) and at the same time contemplative. At a general level, we tried to capture and interpret the complex transformation phenomena that affect the Green Railway Corridor today, which we understand to be extensible to other areas of similar characteristics. These processes can be either natural or man-made in public policies and citizenship. In their interpretation, we understood that there was, in part, a solution: to indicate paths that point toward new and different singularities (Figure 7).

- Cataloguing the wild flora and railway botany in 142 files corresponding to the species registered during the pandemic in a time arc of 5 months. Spontaneous species have been one of the main protagonists of public spaces since the pandemic. The lack of maintenance due to COVID-19, in the case of the PVF, must be added to years of neglect by the administration. Far from constituting a threat to our environment, these species are a source of environmental wealth and biodiversity. Moreover, as the photos of the botanist Ramón Gómez show, they are beautiful and have a great richness of shapes and colors. The cataloguing includes data on the origin of the species, habitat, and ethnobotany, thus favoring the construction of multiple stories and tales. Gómez makes it clear: there are no bad species, and they all contribute to biodiversity. In our specific case, the railway corridors help channel biodiversity and, as demonstrated in this study, the largest number of species (with two of them on the path to extinction) have been registered around the stations of Delicias and Príncipe Pío and the remaining tracks on the surface. This allowed us to defend the protection of these areas as urban forests for public use at the height of a neighborhood protest calling for the temporary cession by ADIF on the land between the aforementioned Delicias station and the Regional Archive (the former El Águila Brewery) of the development of a privately managed leisure center.
- Interviewing the main actors of the operation, including urban technicians, engineers, and neighbors. The interviews were decisive for the research and highlighted the contributions of representatives of the Neighborhood Association with their testimony regarding the transformation of the neighborhood. Due to the pandemic, some of the interviews were conducted via Zoom, which allowed us to record without added costs and share previously selected graphic documentation on screen that enriched the discourse regarding heritage in the final video.
- Preparing, in a second phase, an Inventory and an Atlas of Heritage that included what still exists and what has disappeared as well as what is in use and/or reused. For its development, we considered the Green Corridor within a larger "area of influence" between the current boulevards (in the route of the old *Cerca de Felipe IV*) and the river. The methodology consisted of:
- Creating an inventory of 170 heritage elements in the corresponding files. Each of them included an image of the building that allowed it to be identified, the design authorship in the different phases and eventual modifications, and the list of uses and properties from a chronological perspective. There are 70 buildings that have maintained their original use while 44 have disappeared, of which 10% disappeared in the period in which the research was carried out (2019–2022). Additionally, three large groups were found in a state of abandonment, the most prominent being the former Museum of Military Pharmacy, which could be considered as the headquarters of the Arganzuela Neighborhood Space (EVA), which was recently evicted from the old Fruit and Vegetable Market near the Municipal Slaughterhouse by the current government.
- Sequencing heritage elements in an Atlas presented as a visual journey through the city "between river and rails". This created a spatio-temporal sequence through the area (both diachronic and synchronic) that was organized in the three sections in which the entire analysis of the operation was structured (Puente de los Franceses-Imperial, Imperial-Peñuelas, and Peñuelas-Delicias-Cerro de la Plata). The Atlas consists of 177 images from the archives that are alternated with contemporary photographs.

Figure 6. Fieldwork in the PVF. Action organized by the Pasillo Verde Imperial Association in the Parque del Gasómetro via Twitter. It reads: "We walked the #parquedelachimenea with the urban planning architects Luis Moya and Graziella Trovato @ETSAMadrid @La_UPM, with extensive experience designing urban parks, to assess recent actions #fencing and possible improvements #urbanregeneration #Arganzuela #Homeless".

Figure 7. Image by Davide Curatola Soprana from Urban Reports for TRAHERE. It intends to show the state of abandonment, squatting, and (at the same time) the potential of this urban space.

- Finally, we considered as part of the methodology the elaboration of a proposal for the future from the viewpoint of circularity and biodiversity to connect the PVF with Madrid Río: a proposal of return integrated to the Manzanares in a desirable circularity.

3.2. Postproduction of Results: Materials and Methods

The postproduction phase consisted of transforming the contents and documentation described above into an audiovisual format. It must be taken into account that if the world of theoretical production is linear, rigorous, organized, slow, distant, and timeless, digital postproduction is multiple, selective, fast, and constantly changing.

The phases of the process can be summarized as:

- Development of narrative scripts (storytelling) aimed at an audience not necessarily specialized but interested in understanding the city of Madrid, the transformations of its urban landscape, its potential, and its possibilities for future development.
- Selection of materials and techniques to transform the guides into different experiences of the study area (collages and 3D animations) by preparing virtual tours and augmented experiences that added temporalities to the physical space.
- Selection of physical and virtual platforms and appropriate social networks to share these experiences and improve results (Figure 8).

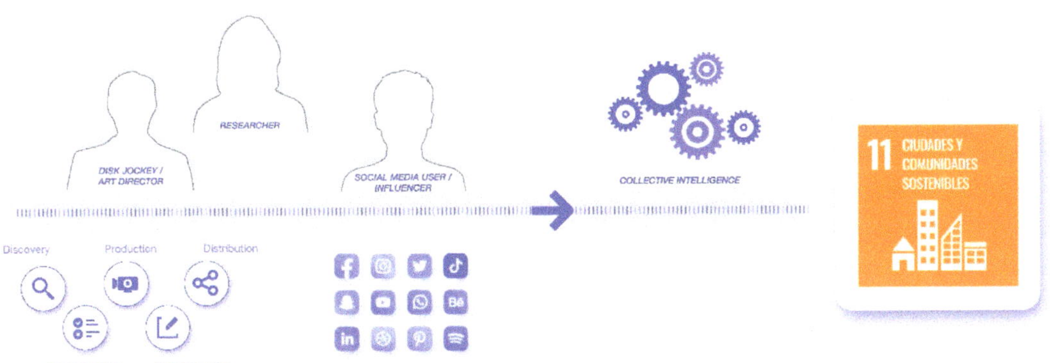

Figure 8. Conceptual diagram of the research method in which the postproduction work was combined with a presence on social networks. The purpose was collective intelligence and ultimately the Sustainable Development Goal 11, which refers to cities, within the 2030 Agenda. Source: TRAHERE's own elaboration.

4. Results

The storytelling scripts were outlined in the transformation of the material produced into audiovisual content in MP4 video format of variable durations and resolutions that ranged between 3 and 12 min and 720p and 1080p, respectively, depending on the destination network and whether it was recorded face-to-face or through a virtual platform. We highlight five derivative scripts in five results described below:

- The first virtual trip developed was a flight through the study area. It was a 3D animation made using Google Earth that covered an itinerary between Puente de los Franceses and Atocha. It was accompanied by indications related to the urban landmarks of the PVF and its urban spaces and by sounds of the urban environment (birds, tree leaves, and walkers) and the commuter train that currently runs through the area next to the metro that was previously registered.
- The second trip, which was of a documentary nature, was structured in three blocks:
- Background: rise and fall of the industrial fabric.
- The green railway corridor: the project, the works.
- Future prospects.

The tour immerses the viewer in the river terraces of the Manzanares: sequences of images and selected shots of the riverbed alternate with the previously elaborated graphic documentation to lay the foundation of the project proposal: if these were the terraces of the river, now they have to go back the way they were. In the background, the sound of the waters of the Manzanares and the birds that populate the area can be heard.

- The third trip, made by Curatola Soprana with sequences of images and shots taken in different areas of the PVF, portrays daily life through its inhabitants.
- The fourth journey, which is elaborated from the Atlas, is a space–time journey through the area. It shows the transformations of the urban landscape and its heritage landmarks. The sound was edited on the basis of the nostalgic theme "Just (After Song of Songs)" composed and performed by David Lang for the soundtrack of Paolo Sorrentino's *Youth* (2016). It is a song to lost loved ones that was used in reference to the demolished or abused heritage.
- The fifth trip, which is presented as a sensory and contemplative experience, was organized from the photographs of Ramón Gómez as a sequence of images of the plots occupied by the wild flora that alternates general shots with others of details of plants and flowers between rails, sidewalks, and pipelines.
- The sixth trip, "Between river and rails", consists of the design and curatorship of the exhibition currently installed in CentroCentro, Cultural Center and Exhibition Hall of the City of Madrid, housed at the Palacio de Cibeles [19] (Figure 9).

Figure 9. The exhibition "Between river and rails" announced and commented upon on LinkedIn by José María Ezquiaga, Head of Urban Planning of the Municipal Management in the 1980s and one of the first promoters of the PVF who today is an expert on urban planning.

For the dissemination of results, a TRAHERE YouTube channel was created and linked to the project website to seek the interconnection between the academic platform and the digital channels and networks using a QR code and a project profile on Instagram (username: entre_río_y_ra-iles.trahere).

The interviews conducted with the different actors involved in the PVF operation as well as those with international experts and representatives of neighborhood associations were edited, published on the TRAHERE YouTube channel, and referenced to the project website [20].

All of the documentation prepared was shared with the neighborhood associations that, in addition to collaborating in the previous phases as explained above, participated in various acts of the dissemination of the results. This included sharing the presentation of the book at the Ateneo de Madrid and the inauguration of the exhibition in CentroCentro mainly through their Twitter channel (PasilloVerdeImperial: @AVVerdeImperial), as well as materials and ideas: "if these used to be the terraces of the river, they must be so again" (Figure 10).

Figure 10. The presentation of the book *Madrid, Between River and Rails* [19] through a Twitter thread

of the Neighborhood Association. The images show a projection screen with the PVF study area, analysis, and proposals. The neighbors shared and made the information and the objectives of the research their own. The tweets read: " 'If these are the #ManzanaresTerraces, they must be so again' Graziella Trovato @ETSAMadrid #PasilloVerde" and " 'The #PasilloVerde, a great unknown, gained space for housing, but destroyed the #industrialheritage of which very little remains' Graziella Trovato @ETSAMadrid".

Through the exhibition, the TRAHERE project is present on the social networks of the Madrid City Council (CentroCentro), including Twitter (15,400 followers), Instagram (22,900 followers), and Flickr, where a complete photographic report of the room is available [21] (Figure 11).

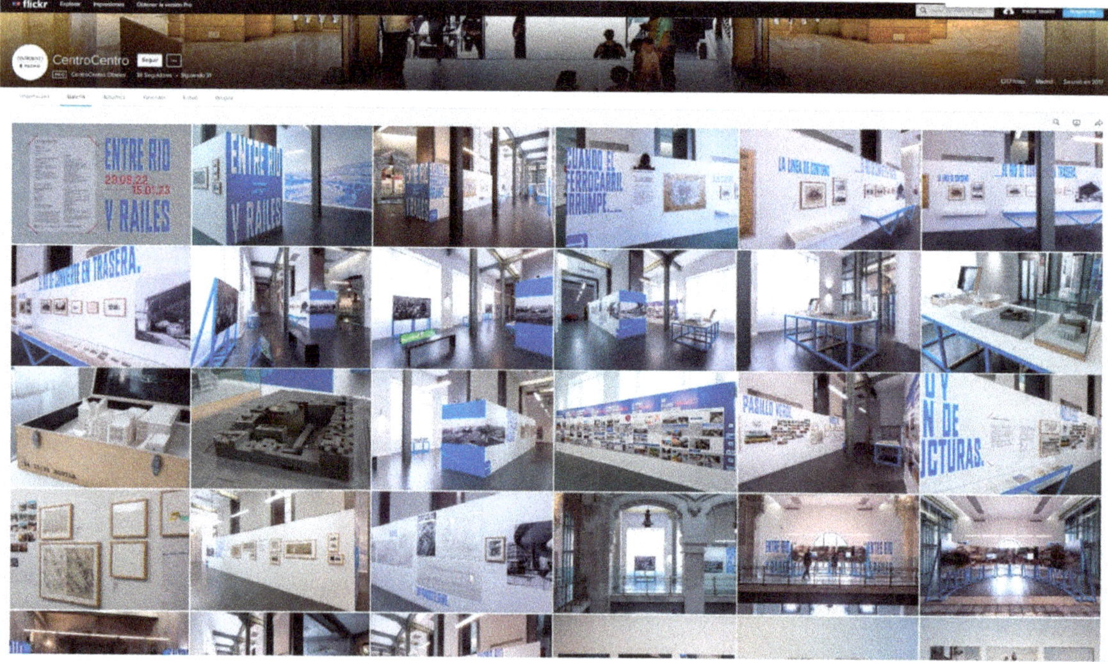

Figure 11. Image sequences of the exhibition "Between river and rails" on the Flickr profile of CentroCentro, an exhibition hall and cultural center of the Madrid City Council, at its headquarters in Plaza de Cibeles. Its curation was the result of a process of storytelling and postproduction of the results of the TRAHERE research.

All of the videos were uploaded to a YouTube channel specifically created by CentroCentro for the exhibition and shared through the aforementioned page, where it is also possible to download the brochure.

In addition to organizing face-to-face visits to the exhibition, there was a live one-hour presentation on Instagram within the Asymmetric Reading Club directed by María Fernández and Juan García Millán that included a tour and description of the different sections that composed it and all of its material (Figure 12).

The video, which can be downloaded through this social network on the Asymmetric Editions profile, has so far registered 428 visits [22].

The project has been shared on social networks by citizens. For some, it offers an opportunity to reflect on the city's past and its transformations with a certain nostalgia for its lost heritage. Sometimes it is carried out through play: this is the case of *Bico*, who poses riddles about singular patrimonial elements little known by the public. In his tweets, he provides data related to his recollections. In this instance, he shared one of the panels from the exhibition "Between the river and rails" with images of the old Colomina factory on Paseo de las Delicias that was transformed into a residential building in 1982 by Luis Moya and Guillermo Cabeza [23]. Bico specifies that Colomina manufactured liquid carbonic acid and was huge, facing two streets (Delicias and Embajadores) (Figure 13).

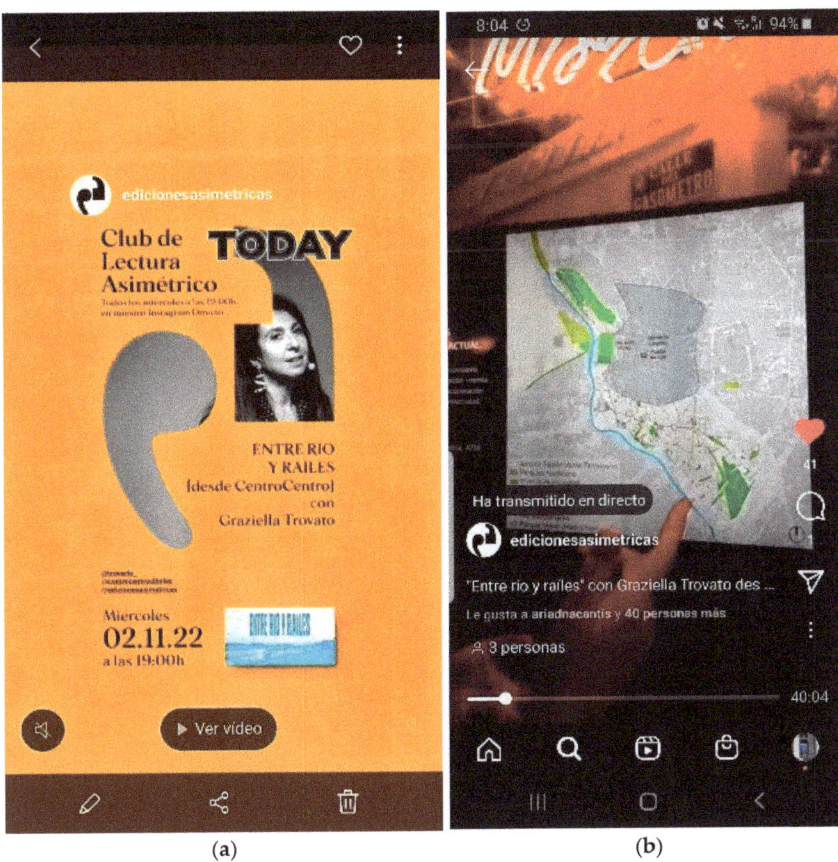

(a) (b)

Figure 12. One-hour Instagram Live guided tour showing all the results and including the videos. (**a**) The banner; (**b**) an image of a moment of the video documentary "Las terrazas del Manzanares".

For others, it is an occasion for critical debate as one of the topics on how the models for financing and management of public space can influence its conservation or abandonment (Figure 14).

(a)

(b)

Figure 13. Twitter thread with a photo of a panel from the "Between river and rails" exhibition with a riddle about the Colomina factory. Here, *Bico* shares his memories and presents new information about the building before the remodeling. (**a**) Twitter thread; (**b**) project of Moya and Cabeza from 1982.

Figure 14. Twitter thread by the Asociación Pasillo Verde Imperial that included photos of neighbors visiting the exhibition. A debate was generated around the "zero cost" management model and the state of abandonment. Two neighbors indicated the photo of the authorities in the presentation of the PVF model. Among the comments can be read the following: Naranch replying to @madridete: "The exhibition is very top. It captures very well the transformation of the downtown industrial belt +ffcc into urban fabric. And how in this case, brick prevailed over 'green'. Similar processes took place in Barcelona, Valencia or Bilbao."; Antonio Rguaz: "It is curious to see in the brochures of the time how they boasted that the operation did not cost the public treasury anything. Well, it did cost the transfer of a lot of land to build flats. An operation that Tierno (Enrique Tierno Galván) began to develop and that Manzano ended. No wonder it ended up degrading.".

5. Discussion

Fernando de Terán argued: "It is a very interesting and singularly satisfactory part of the not always pleasant history of modern urbanism which refers to the way in which the conceptual understanding of the action on the urban reality has been built (and the legal attention and social valuation derived from it) to protect, defend and care for what has been generically called cultural heritage. Until reaching the current situation, in which the idea itself has widened, evolving and transforming from a mainly objectual perspective to another that is fundamentally environmental, which implies and incorporates notions that come from ecology and an attention to the environment and leads to an intention of general protection of biodiversity. All of which configures a much broader and integrative understanding, which in reality supposes a certain redefinition of the conditions of human intervention on the existing, including of course and in the first place the inherited city" [15] (p. 8).

In our case, the study of the PVF gave us the opportunity to study and suggest a proposal for the future of an area equipped with a strong associative fabric that, approximately three decades ago, promoted a process of reuse of railway heritage that we continue to debate today.

In the final phase of the research, a discussion forum was established with international experts who endorsed our initial hypotheses and method of study and pointed toward the process of reuse as the future of urban and architectural heritage to respond to the crisis of resources in which we live. We cite, among the cases analyzed for their analogy with our case study, the Paris Rive Gauche (France) and The Seven Railway Scales of Milan 2030 (Italy), which is predicted to be one of the largest urban regeneration operations of the next decade [15] (pp. 230–270). Both *reuse* and *recyclage* are concepts that emerged, in fact, in the middle of the energy crisis in the 1970s as an expression of the need to curb the consumerist euphoria of the *good life* and promote the saving of natural resources. Reuse, in addition, is in itself participatory because, unlike restoration and other intervention techniques in heritage areas, it implies forms of appropriation that do not require technical specialization. Our era manufactures objects of all scales with a vocation of precariousness and temporality from a perspective of pure consumerism at the economic level and opportunism at the political level. This inevitably affects our immediate surroundings, our desks, our houses, our buildings, and our public spaces. Nowadays, thinking in terms of reuse, in our opinion, is better defined as an act that comes from a broader awareness of postproduction and inclusion rather than recycling. The prefix "post-", Bourriaud reminds us, does not indicate a conceptual overcoming but a zone of activity and an attitude [5].

The urgency of this change is evident, and the continuous aggressions toward the urban landscape confirm it. The TRAHERE project aspires, with its presence in public spaces such as CentroCentro and the social networks, to contribute to this change.

6. Conclusions

This research project aimed to discuss the postproduction of the urban environment through a process of the postproduction of academic results for their conversion into a forum for urban debate.

Converting the concept of the *transfer* of results into that of *knowledge exchange* implies, as we stated previously, a new approach to research. The digital age has imposed a paradigm shift in which any barrier between the theorist (the researcher/narrator) and their "spectator" has been erased. The critical distance imposed by the book as a physical object is blurred in the "copy-paste" world of networks where everything is shared and commented on and thus transformed. Hypertext, as McLuhan already warned us, resembles the pre-alphabetic oral world where stories were enriched by new stories and characters in their journeys from town to town [24]. Beyond the problems of regulation and control of copyrights that all this poses, as well as the future implementation of *blockchain* systems to defend institutional rights, the academic world faces new challenges that at the same time involve the possibility of turning scientific research into shared knowledge that is available to all citizens. After the publication of the book *Madrid, Between River and Rails. Past, Present and Future of the Railway Green Corridor* [15], the first result of the research, both the materials and the inspiring ideas of this project have been shared in physical (Madrid City Council) and virtual (Twitter, Instagram and other mentioned networks) forums of neighborhood associations and institutions (in addition to the City Council and the Spanish Association of Landscape Architects, we mention the Railway Museum, among others) apart from citizens who warn of modifications and changes and interact with the contents of the research (Figure 15).

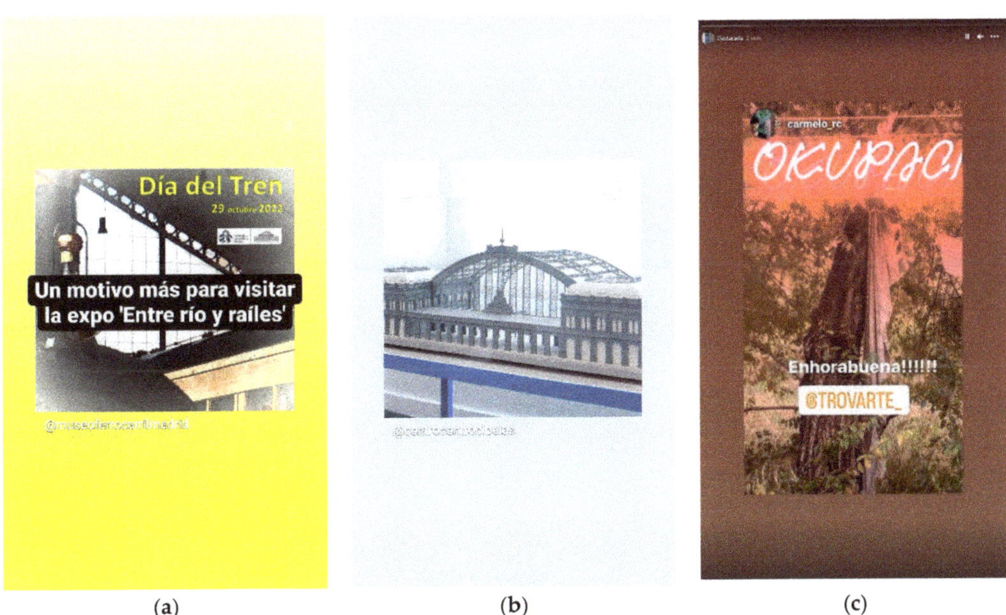

(a) (b) (c)

Figure 15. Presence of the research on Instagram through: (**a**) the profiles of the Railway Museum of the Spanish Railways Foundation (FFE); (**b**) the Madrid City Council; and (**c**) a citizen (the prominent architect and activist Carmelo Rodríguez, founder of *Enorme Estudio*).

All these stories in their various formats encourage new areas for reflection and debate on the conservation of the urban landscape. In addition, they open up new avenues for analyzing the current and future presence of academic research on social networks through postproduction processes and the dissemination of results. Their tweets and comments become part of the story and create new fields of reflection and avenues for research (Figure 16).

Figure 16. (**a**,**b**) Twitter thread from a citizen who made the story about the PVF his own and enriched it with his memories. (**a**) It reads: "The Exhibition Between River and Rails that can be seen in the Palacio de Cibeles is very interesting. It deals with the regeneration processes of the Arganzuela area between the Green corridor and Madrid Río. Surely some of you remember these panels" (. . .) "The whole area underwent a huge transformation. Here you can see, for example, the beginning of Pedro Bosch, Santa María de la Cabeza and Ferrocarril street, where the train used to go underneath a tunnel until it passed Paseo de las Delicias".

Funding: This work was supported by the Madrid Government (Comunidad de Madrid-Spain) under the Multiannual Agreement with Universidad Politécnica de Madrid in the line Excellence Programme for University Professors and in the context of the V PRICIT (Regional Programme of Research and Technological Innovation) and by the project LABPA-CM: CONTEMPORARY CRITERIA, METHODS AND TECHNIQUES FOR LANDSCAPE KNOWLEDGE AND CONSERVATION (H2019/HUM5692), funded by the European Social Fund and the Madrid regional government.

Institutional Review Board Statement: Not applicable.

Informed Consent Statement: Not applicable.

Data Availability Statement: Not applicable.

Acknowledgments: The project presented was based on the results of a line of research on which we have been working for years in collaboration with the Cultural Landscape Research Group (GIPC). We highlight in particular the participation in the recently completed research project, "The railway and the city at the crossroads: Urban landscape and industrial heritage in the surroundings of the stations of the Iberian Peninsula, 1850–2017. Digital Station" (main researcher Jordi Martí Henneberg) funded by the BBVA Foundation in the competitive Digital Humanities 2017 call. This project led to the creation of the interactive digital platform "The railway and the city in the Iberian Peninsula (1850–1950)". The work is a catalogue and a compilation of railway information of various types and formats for 91 Spanish and 25 Portuguese cities. In this project, we want to maintain the interdisciplinary approach to industrial heritage and the integration and convergence of humanities and digital technologies for knowledge aimed at a society without cognitive, physical and cultural barriers [25]. Thanks to Clara Álvarez, Davide Curatola Soprana, Rodrigo de la O, Ramón Gómez, Luis Moya, Melin Nava, Isabel Rodríguez de la Rosa, Beatriz Salidoand Diego Toribio for their collaboration in the research. We are also grateful to the "Archivos de la Transición" [Figure 4] and Milla Hernández Pezzi for the transfer of rights for the reproduction of images (Figure 14a,b).

Conflicts of Interest: The author declares no conflict of interest.

References

1. Madrid acuosa. LaPa Laboratorio de Paisaje. Available online: https://la-pa.es/madacuosa/ (accessed on 9 May 2022).
2. Proyecto Agua (Water Project). Available online: https://www.iagua.es/ (accessed on 11 December 2022).
3. RedOTRI Universidades. Las Oficinas de Transferencia de Resultados de Investigación (OTRI). Available online: http://www.redotriuniversidades.net/index.php/presentacion?id=272 (accessed on 11 December 2022).
4. Devoto, G.; Oli, G.C. *Dizionario della Lingua Italiana*, 1st ed.; Edumont Le Monnier: Florence, Italy, 2004; p. 2094.
5. Bourriaud, N. *Postproducción*, 2nd ed.; Adriana Hidalgo Editora: Buenos Aires, Argentina, 2004; p. 25.
6. Interferencias de Crítica: Arquitectura y Ciudad. Available online: https://interferenciasdecritica.com/trahere/ (accessed on 20 October 2022).
7. Consorcio Urbanístico Pasillo Verde Ferroviario de Madrid. *Folleto Informativo de la Operación*; Ayuntamiento de Madrid: Madrid, Spain, 1994.
8. García Santos, A. El Pasillo Verde Ferroviario de Madrid, una experiencia de gestión. *Planur-e Territ. Urban. Paisaje Sostenibilidad Diseño Urbano* **2015**, *25*, 1–32.
9. López de Lucio, R.; Ardura, A.; Bataller Erguix, J.J.; Tejera Parra, J. *Madrid 1979–1999: La Transformación de la Ciudad en Veinte Años de Ayuntamientos Democráticos*; Ayuntamiento de Madrid: Madrid, Spain, 2016; Volume 2, pp. 44–47.
10. Guía de Arquitectura de Madrid. Available online: https://www.coam.org/es/fundacion/servicio-historico/guia-arquitectura-madrid (accessed on 21 November 2020).
11. Brandis, D.; del Río, M.I. As grandes operaciones de transformación urbana. El Pasillo Verde Ferroviario de Madrid. *Ería* **1995**, *37*, 113–128.
12. SECTRA. Programa de Viabilidad y Transporte Urbano. Gobierno de Chile. Available online: http://www.sectra.gob.cl/biblioteca/detalle1.asp?mfn=999 (accessed on 12 July 2021).
13. Sambricio, C.; Ramos, P. *El Urbanismo de la Transición: El Plan General de Ordenación Urbana de 1985*; Ayuntamiento de Madrid: Madrid, Spain, 2019; pp. 378–381.
14. Ayuntamiento de Madrid. *Madrid Proyecto Madrid*; Ayuntamiento de Madrid: Madrid, Spain, 1997.
15. Trovato, G. *Madrid, Entre Río y Raíles. Pasado, Presente y Futuro del Pasillo Verde Ferroviario*; Lampreave: Madrid, Spain, 2022.
16. Sambricio, C. Madrid entre historia y actualidad. Estructura y forma de la ciudad. In *Madrid, Milano: Forma della Città e Progetto Urbano*; Caputo, P., Ed.; Electa: Milano, Italy, 1998; Volume 1, pp. 43–46.
17. Trovato, G. Entrevista a Maite Gómez, Presidenta de la Asociación Pasillo Verde Imperial. Available online: https://interferenciasdecritica.com/entrevista-a-maite-gomez/ (accessed on 21 November 2022).
18. Archivos de la Transición. Available online: https://archivodelatransicion.es/ (accessed on 21 November 2022).
19. Exhibition Between River and Rails—CentroCentro. Available online: https://www.centrocentro.org/exposicion/entre-rio-y-railes (accessed on 21 November 2022).
20. TRAHERE YouTube Channel. Available online: https://www.youtube.com/channel/UCFpiqlNQKbzqHBeOU5BugWA (accessed on 31 October 2022).
21. CentroCentro Cibeles. Available online: https://www.flickr.com/photos/centrocentrocibeles/ (accessed on 25 November 2022).
22. Instagram Live—Guided Tour @edicionesasimetricas "Entre Río y Raíles" con Graziella Trovato Desde @centrocentrocibeles. Available online: https://www.instagram.com/p/CkeCJ8XAGyo/?hl=es (accessed on 2 November 2022).
23. Cabeza, G.; Moya, L. Ampliación de la Casa de Delicias, 139. Madrid 1981–82; *Arquit. COAM* **1983**, *245*, 17–19.

24. Mantzou, P.; Trovato, G. Architecture in the Digital Machine Age. Critic All. In Proceedings of the I International Conference on Architectural Design & Criticism, Madrid, Spain, 12–14 June 2014. Available online: https://www.academia.edu/8975931/Architecture_in_the_digital_machine_age (accessed on 20 November 2022).
25. El Ferrocarril y la Ciudad en la Península Ibérica (1850–1950). Available online: http://ciudadyferrocarril.com/ (accessed on 31 October 2022).

 *land*

Article

Can We Foresee Landscape Interest? Maximum Entropy Applied to Social Media Photographs: A Case Study in Madrid

Nicolas Marine [1,*], Cecilia Arnaiz-Schmitz [2], Luis Santos-Cid [2] and María F. Schmitz [2]

[1] Department of Architectural Composition, Escuela Tecnica Superior de Arquitectura de Madrid, Universidad Politécnica de Madrid, 28040 Madrid, Spain
[2] Department of Biodiversity, Ecology and Evolution, Universidad Complutense de Madrid, 28040 Madrid, Spain; caschmitz@ucm.es (C.A.-S.); lusant02@ucm.es (L.S.-C.); ma296@ucm.es (M.F.S.)
* Correspondence: nicolas.marine@upm.es

Abstract: Cultural Ecosystem Services (CES) are undervalued and poorly understood compared to other types of ecosystem services. The sociocultural preferences of the different actors who enjoy a landscape are intangible aspects of a complex evaluation. Landscape photographs available on social media have opened up the possibility of quantifying landscape values and ecosystem services that were previously difficult to measure. Thus, a new research methodology has been developed based on the spatial distribution of geotagged photographs that, based on probabilistic models, allows us to estimate the potential of the landscape to provide CES. This study tests the effectiveness of predictive models from MaxEnt, a software based on a machine learning technique called the maximum entropy approach, as tools for land management and for detecting CES hot spots. From a sample of photographs obtained from the Panoramio network, taken between 2007 and 2008 in the Lozoya Valley in Madrid (Central Spain), we have developed a predictive model of the future and compared it with the photographs available on the social network between 2009 and 2015. The results highlight a low correspondence between the prediction of the supply of CES and its real demand, which indicates that MaxEnt is not a sufficiently useful predictive tool in complex and changing landscapes such as the one studied here.

Keywords: cultural ecosystem services; social media; geotagged photographs; maximum entropy models; MaxEnt

Citation: Marine, N.; Arnaiz-Schmitz, C.; Santos-Cid, L.; Schmitz, M.F. Can We Foresee Landscape Interest? Maximum Entropy Applied to Social Media Photographs: A Case Study in Madrid. *Land* **2023**, *11*, 715. https://doi.org/10.3390/land11050715

Academic Editor: Hannes Palang

Received: 20 April 2022
Accepted: 8 May 2022
Published: 10 May 2022

Publisher's Note: MDPI stays neutral with regard to jurisdictional claims in published maps and institutional affiliations.

1. Introduction

In the last ten years, there has been an increase in studies using social media geotagged photos to analyze both people's perception of their lived environment and their behavior in it [1–3]. These photographs, and their accompanying information, have opened up the possibility of quantifying landscape values that have been difficult to measure until now, especially those related to cultural ecosystem services (CES). CES can be defined as "non-material benefits people obtain from ecosystems through spiritual enrichment, cognitive development, reflection, recreation and aesthetic experiences" [4] (p. 58).

Despite their relevance, CES evaluation remains disregarded and poorly understood in comparison with the other material ecosystem services [5]. The socio-cultural preferences of the various stakeholders who enjoy a landscape are intangible and challenging to assess. Consequently, researchers usually resort to interviews, questionnaires, participatory mapping and focus groups [6]. Yet, the growth of social media's users and, particularly, platforms where people post geotagged photographs have provided us with a large amount of data on landscape perception. Numerous methods have unfolded to spatially define CES, as well as to characterize and to visualize them [7]. The more common tend to bring together quantitative and qualitative analysis, frequently combining land cover maps with image cluster analysis and automated image recognition [8–13].

Recently, a new line of research has been opened that consists of determining areas that can potentially provide CES from the current distribution of geotagged photos. For this purpose, distribution modeling software is used to identify degrees of significance of different environmental variables in relation to the photographs and to identify potential hotspot areas of CES. For example, well-known models, such as the Integrated Valuation of Ecosystem Services and Tradeoffs (InVEST), have developed specific applications to predict recreation and tourism hotspots and future patterns of use from social network photographs [14]. It is more common, however, to use the open-source software MaxEnt in this operation, a maximum entropy modeling software oriented, in principle, to biologists to predict the distribution of species from current presence data. The use of MaxEnt in relation to social appreciation of the environment is not new. The SolVES (Social Values for Ecosystem Services) program, developed by the U.S. Department of the Interior, already incorporates the use of maximal entropy modeling software to cross-reference social values (aesthetic, biodiversity or recreational) with explanatory environmental variables [15]. By crossing these data with environmental variables, it provides a map of suitable habitats where these species may be present [16]. In the present case, the species presence layer is replaced by that of georeferenced photographs to calculate the probability of a photograph being taken in a certain place. In short, the potential supply of certain recreational services.

In fact, so-called affective computing has been making use of maximum entropy models also in close relation to social networks and CES [17–19]. As a result, there has been a growing number of investigations using georeferenced photographs and MaxEnt to determine the potential supply of CES in a given geography. The advantage offered by this software is that the presence data are sufficient to model the potential distribution. Furthermore, by means of jackknife resampling, it us allows to establish the degrees of relationship between the presence of the photographs and the different environmental variables [20]. Specific applications of the study of geotagged social media photos using MaxEnt are varied, both in terms of task and procedure. For example, Richards and Friess [11] use it to study coastal mangrove forest habitats and potential visitor interest based on distances to access points, communication infrastructure and viewpoints. Yoshimura and Hiura [21] apply it to the island of Hokkaido by comparing an area of demand based on viewsheds using geotagged photos as a viewpoint and a supply provided by cross-referencing data with MaxEnt. Clemente et al. [22] apply it to a natural park in Portugal and study proximity indices at different variables, assuming that the greater the distance, the lower the attractiveness of a biophysical component or infrastructure. More recent studies have increased the scale of application to the whole European continent [23] or have added the location of historical and cultural sites to the environmental variables, which brings a heritage reading to the potential interest of the landscape [24].

These references coincide in that they obtain the georeferenced photographs from the Flickr social network, mainly because of the ease with which they can be downloaded from its API and because of the data it contains. The number of photos used in each study varies greatly, from 250 to almost 7 million. This depends on the size of the study area, which, as can be seen, is also very varied, ranging from small, protected areas to entire continents. The photos are usually classified according to different CES or according to the elements photographed, either manually or automatically (using, for example, Google Cloud Vision). The environmental variables with which they are crossed are generally the same: land use and land cover, geomorphology and co-communication infrastructures. Although other variables such as heritage assets are often added. The final result is highly dependent on these variables.

The references cited above use MaxEnt as a CES supply potentiality tool, but do not question the effectiveness of this software itself. In general, they all validate the quality of the models using the area under the receiver operating characteristic curve as a parameter [20]. If this area, or AUC, is close to 1, the prediction is considered perfect, and if it is below 0.5, the model prediction is considered to be as good as random. Therefore, this method is validated by a result provided by the software itself and not on the basis of external checks. A review of the literature shows that, although several of the references include in the discussion a critique of social media photographs as univocal representations of the population's interest [12,22,24], they do not criticize MaxEnt as a mechanism for predicting their potential distribution.

Based on these arguments, this paper takes as its starting point a criticism that The Natural Capital Project [14] had already made of predictive models: that they require assuming that people's responses to environmental variables will not change over time. That is, the use of MaxEnt presumes that, in the future, people will continue to be attracted or repelled by the same factors as today, and even that these factors will remain unchanged over time. A question arises here: how valid is MaxEnt really in predicting potential interest in specific CES? To answer it, this research consists of comparing the actual evolution of social media geotagged photos with the prediction that MaxEnt would have made based on passed information. To do this, we use the photographs uploaded to the Panoramio social media network between 2006 and 2015 in the Lozoya Valley, a complex landscape north of Madrid. The case study has been selected because it combines natural and cultural values, because it is of great tourist interest and because previous work has shown changes in the valuation of CES in recent decades [25].

Here, we question the validity of MaxEnt as a predictive tool for landscapes as complex as those of the Lozoya Valley. The objectives of the study are the following:

i.   To determine the validity of social media geotagged photos as a basis, and of MaxEnt as a tool, for predictive studies of future landscape users' behavior.
ii. To determine the differences between the quantification of the future spatial distribution of geotagged photographs and their real qualitative changes.
iii. To propose a comprehensive approach to the actual complexity of the photographs uploaded by users to social networks to assess future CES interest.

2. Materials and Methods

2.1. Research Area

The study area, selected within the Lozoya Valley, covers approximately 776 km², belonging to 25 of the 30 municipalities that constitute the valley. The Lozoya Valley (Figure 1), is located in the northern Sierra de Guadarrama in the Lozoya river basin (Madrid region, Spain). The area's main connection with the rest of the region is through a single highway (A-1 Route). The Lozoya Valley has its highest elevation at Peñalara peak (2428 m a.s.l.) and the lowest elevation in the adjoining area of the Lozoya rivers (100 m a.s.l.) and includes 30 municipalities. The Lozoya Valley is a heterogeneous landscape with forest, settlements, water bodies such as reservoirs and a mosaic of traditional land uses containing pastures, meadows, hedgerows, ash groves and riparian forests, all of which are well preserved in most cases [26]. Over the centuries, the valley has come under different land uses and rural activities which have shaped the landscape and their traditional way of living along the centuries that have resulted in a region of great socio-ecological value. Currently, this heritage landscape is under several categories of protection.

Thus, the Lozoya Valley is within the boundaries of the Sierra de Guadarrama National Park (established in 2013) and the Sierra del Rincón Biosphere Reserve (2005), and it also belongs to the European network of protected sites Natura 2000. Several areas inside the valley also fall under other types of recognitions: (i) The Montejo de la Sierra beech forest (Natural Site of National Interest, 1974, and subsequently UNESCO's World Heritage Site, 2017); (ii) The Monastery of El Paular (Historical-Artistic Monument of Spain, 1876), and (iii) the Neanderthals Valley (Cultural Interest Asset, 2004). Recently, the High Valley of the Lozoya River has been proposed as a model of Heritage Cultural Landscape to UNESCO by the Spanish National Plan of Cultural Landscape [27]. Because of this, the area is a touristic hotspot appreciated by visitors to the Madrid region [26]. However, recent studies have shown that increasing tourism and conflicting management legislation caused a rurality loss and an urban sprawl throughout the territory, transforming the ancient agropastoral landscape into a wilderness [28]. The comparison of surveys conducted with visitors in

2007 and 2017 shows that these changes have had an impact on their way of understanding the landscape of the Lozoya Valley: from valuing the cultural components more, to now valuing the "naturalistic" ones more [26].

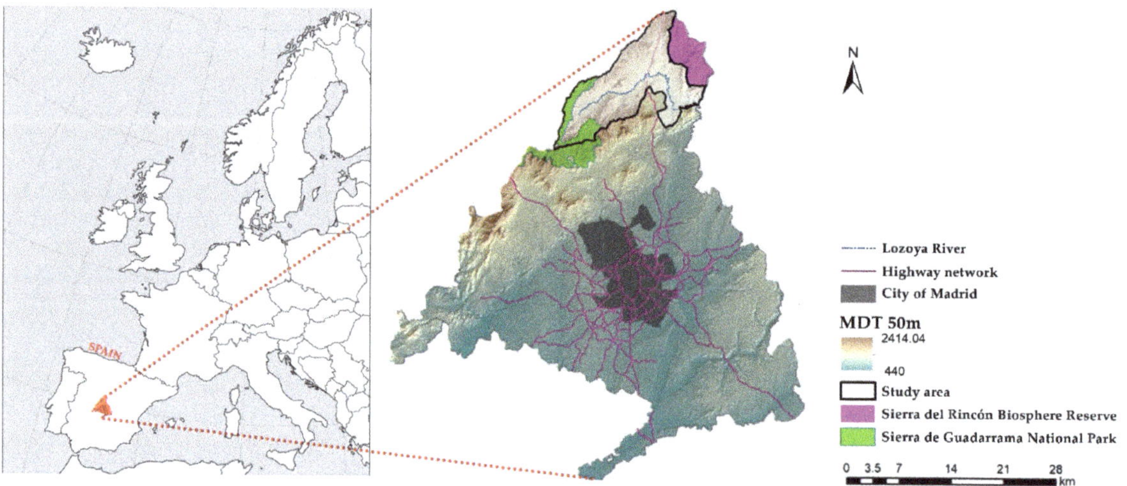

Figure 1. Map of the Madrid Region with the Lozoya Valley area of study in the north.

2.2. Materials

The basis of the study is the georeferenced photographs uploaded by users to the Panoramio website between 2007 and 2015. Panoramio was a website specialized in sharing georeferenced audiovisual material accessible as a layer in Google Earth and Google Maps. It was active between 2005 and 2016, when it closed, although the layer on Google Earth was available until January 2018. Panoramio is very similar to Flickr but, rather than a social network per se, it is considered to be a means of sharing photos and videos by users [24]. Sometimes both Panoramio and Flickr have been used to study the CES of a place [29] and have been found to provide similar patterns of landscape values, at least on the European continent [30]. Panoramio photos, as opposed to Flickr photos, are usually obtained manually [31], although they can also be obtained through APIs [32]. In the case of this article, the professional services of a company were hired to bulk download all the photographs uploaded to Panoramio in the Region of Madrid between 2007 and 2015.

As a result of this operation, a list of 54,956 photographs was obtained, half of which were located in the metropolitan area of Madrid. A selection of those located in the municipalities that make up the study area reduced this number to 3192 photographs located in the Lozoya Valley. Given that the references of studies of this type first make a classification of the photographs based on the elements that appear in them, only those photographs that could still be located on-line were taken for the sample. Before Panoramio disappeared completely, the Mapio website (https://mapio.net/, accessed on 7 May 2022) made an extensive transfer of its collection. Some of which can still be located today. An automated search made it possible to locate and download a total of 1728 photographs that serve as a starting point for the study. The attributes associated with the photos include the username, upload date, title and hashtags provided by the user, and the location of the photo. The biggest difference between the photos refers to the years in which they were uploaded (Figure 2). Subtracting 91 that do not contain a date, most of the photos in the sample were uploaded in 2007 (142), 2008 (752) and 2009 (442). The year with the least number of photos is 2014 (18).

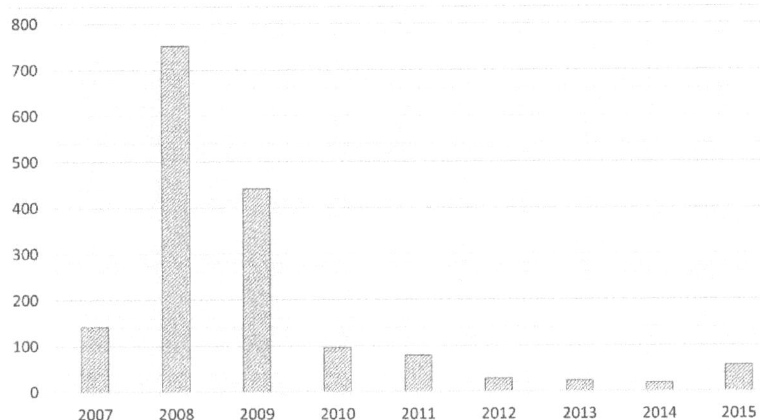

Figure 2. Number of photographs per year in the Panoramio sample.

2.3. Methodology

This research compares the prediction of potential demand of CES in an area with the actual demand. To do so, we use social media photos instead of species occurrences in a species distribution modeling software for understanding the spatial distribution of people preferences for CES. Actual demand is measured based on photo density. To correlate both datasets this methodology followed a four-step process: (1) database preparation and filtering, (2) variable selection and MaxEnt modeling, (3) modeling of the actual demand, and (4) elaboration of a correlation matrix.

2.3.1. Georeferenced Database of Social Media Photographs

First, the images are classified according to a series of categories linked to CES and the most represented elements (Figure 3). The classification is done manually and by two different researchers, and the differences are then compared by a third one. The method thus follows other manual classification methods used in similar studies [11,22]. In this way, each photograph is assigned one of the categories in Table 1.

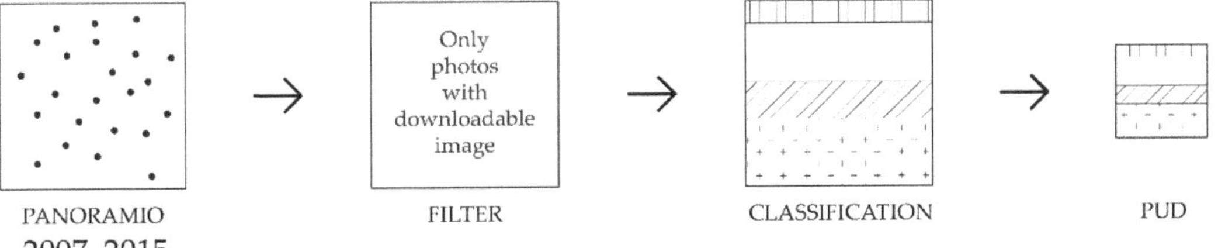

Figure 3. Diagram of photograph database processing and sampling.

Table 1. Photographs categories.

Category	Description
Natural system	Majority presence of flora and fauna in a wild state.
Urban system	Majority presence of architectural and urban elements
Rural system	Majority presence of agrosilvopastoral elements.
Water bodies	Majority presence of aquatic elements, very common in the zone.
Recreational activities	Presence of people engaged in sports or walking activities
Cultural activities	Presence of museums, monuments, food or typical products.

The categories are divided into systems (natural or naturalistic tendency, urban tendency and rural tendency) and activities linked to the CES (recreational and cultural). The presence of bodies of water was added as a category in its own right after checking the frequency with which they appeared in the photographs. The Lozoya Valley is characterized by a hydraulic system that combines rivers and artificial elements such as reservoirs. Since it is difficult to differentiate whether a user is photographing one of these bodies of water on the basis of whether it is considered natural or artificial, the presence of water is taken as a characteristic aesthetic value of the valley.

Once classified, a sieve is applied to the collection of images to remove unwanted tendencies. This is common in research related to social network photographs, since there are usually users who upload several photos on the same day, which can generate biased deductions. For this purpose, the PUD (Per User Day) method is used, a form of screening that avoids this problem based on randomly selecting, from the initial sample, one photo per user per day [7]. Given that there are also methodologies that consider this type of reduction in the original sample to be negative [33], and that this paper is oriented towards a critique of an established method, the analysis is carried out with two samples: the original, with 1728 photographs, and the PUD, with 709 photographs.

2.3.2. Data Processing and Confrontation

Once the photographs have been classified and screened, we proceed to their analysis using MaxEnt. To do this, we divided the sample of photographs into two periods: one used to develop the predictive model (Base Demand Sample) and the other with the actual distribution of the photographs (Actual Demand Sample). For the first period, photographs from 2007 and 2008 are selected, which account for approximately half of the photographs. For the second period, we take the photographs between 2009 and 2015. The purpose is to test how close the MaxEnt would have been to determine CES demand in the future. This division of the database makes it possible to develop a predictive model from the 2007–2008 data and check whether it matches the actual demand up to 2015 (Figure 4).

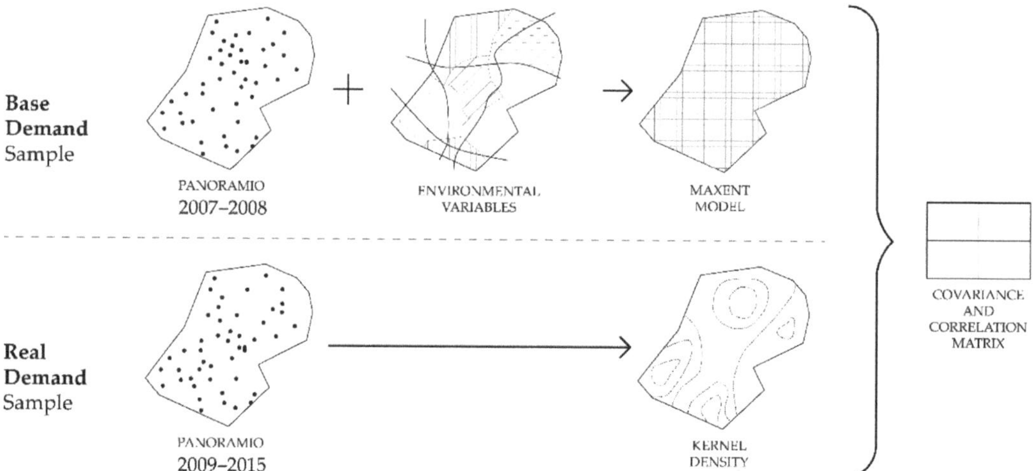

Figure 4. Modeling and correlation diagram.

Therefore, we use as species the point data from the Base Demand Sample and transfer it to MaxEnt 3.4.4 [20]. To this, we add a series of environmental variables: (1) 2006 land cover; (2) altitude; (3) average atmospheric temperature; (4) distance to roads; and (5) distance to cultural assets. Except for the last one, the data come from open sources (Table 2). All the variables are continuous, and the proximity indices have been calculated using the Kernel tool in ArcMap 10.5.1. The MaxEnt model provides an ACISS file that will be used for comparison. In addition, it provides a measure of goodness of fit that quantifies how closely the model is concentrated around occurrences.

Table 2. Environmental variables, source and processing.

Variable	Source	Processing
Land Cover	Corine Land Cover 2006 (https://land.copernicus.eu/pan-european/corine-land-cover, accessed on 7 May 2022)	Unificación de categorías
Altitude	SDI of Spain (https://www.idee.es/, accessed on 7 May 2022)	MDT as downloaded
Average atmospheric temperature	SDI of Spain	Kernel from medium temperature (station points)
Distance to roads	SDI of Spain	Kernel from road lines
Distance to cultural assets	Madrid Heritage Information System [34]	Kernel from cultural asset points

On the other hand, we study the real evolution of the photographs from the Actual Demand Sample (Panoramio photos uploaded between 2009–2015). Here, we conducted a photograph points density study using the Kernel tool in ArcMap 10.5.1. With these two layers (MaxEnt prediction from 2007–2008 data and density of photographs between 2009–2015) we performed a multivariate analysis using the Band Collection Statistics tool (ArcMap 10.5.1 Spatial Analyst). This tool allows us to confront the variation of two or more overlapping rasters. When requested, it computes covariance and correlation matrices. The result matrix presents the variances of all raster bands along the diagonal from the upper left to lower right and covariances between all raster bands in the remaining entries.

In our case, the final correlation matrix provides a correlation coefficient between the MaxEnt layer (prediction) and the density layer (actual demand). The proximity of this coefficient to 1 indicates that both layers vary similarly. That is, the potential and actual demand intentions not only coincide at certain points but are distributed equally. Therefore, we take the proximity of this coefficient to 1 as the measure of success of the MaxEnt predictive model. The process shown in Figure 4 is carried out with both the original sample and the PUD sample. From the one that is closest to 1 in the correlation matrix, we check the correlation of each of the categories.

3. Results

3.1. Photograph Samples

The classification of the photographs reveals large differences between categories in both the original and the unbiased or PUD sample (Table 3). In the former, photographs tend to be of elements related to natural systems (455) and bodies of water (426) followed by photographs related to urban environments (378). Photographs related to cultural activities are the least present (68). As for the PUD sample, water bodies are the most photographed (195), although closely matched by natural systems (190). After these categories, recreational activities (123) and urban systems (128) are closely matched. Cultural activities are again the least represented (29).

Table 3. Number of photographs in each sample and category.

Sample	Natural System	Urban System	Rural System	Water Bodies	Recreational Activities	Cultural Activities
Original sample	455	378	114	426	282	68
PUD sample	190	128	42	195	123	29

Percentage comparison of the two samples reveals very little difference between them (Figure 5). The representation of natural systems changes by only one percentage point when removing the bias of the original sample, and water bodies vary from 25% to

28%. Cultural activities, recreational activities and rural systems barely vary. The greatest variation from one sample to another occurs among the photos of urban systems (from 22% in the original sample to 18% in the PUD).

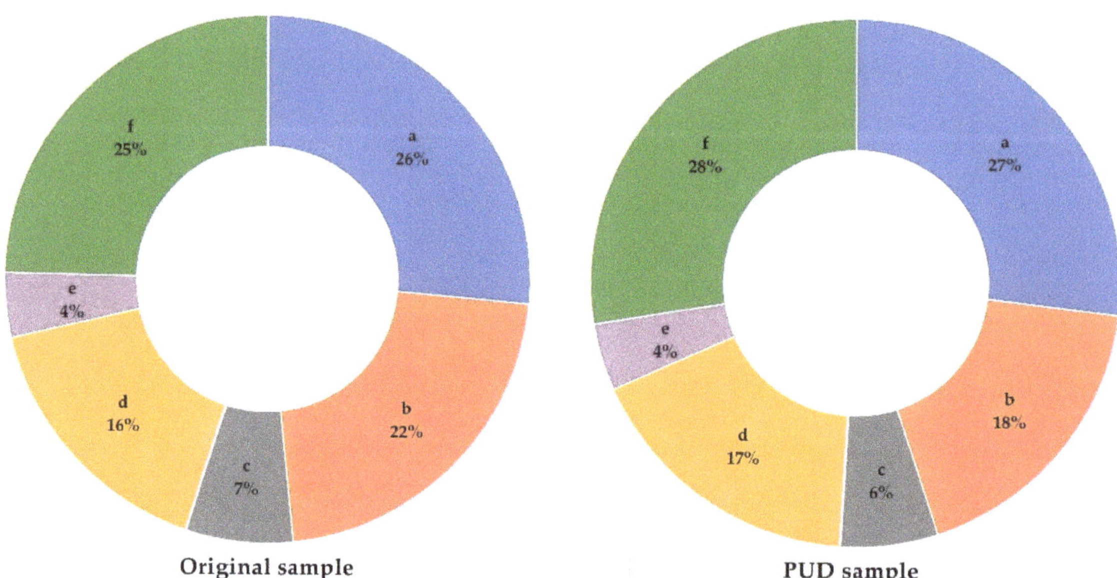

Figure 5. Percentage of each category in each sample: (**a**) natural systems, (**b**) urban systems, (**c**) rural systems, (**d**) recreational activities, (**e**) cultural activities, and (**f**) water bodies.

3.2. MaxEnt Modeling

Although there is little variation between the classes of photographs, the MaxEnt models performed with each sample do change more (Figure 6). The potential demand model from the original sample predicts a higher intensity in the southwest and northeast extremes. In between, the potential is also intensified by the different settlements located along the valley. The roads connecting them are also highlighted as points of potential interest. When the model is performed from the PUD sample, however, it places more emphasis on the southwestern part of the study area. The rest is shown as an area of low potentiality, except for the northeast zone, which is more intense than the center, but does not compensate for the more intense areas. The AUC, the standard measure of model reliability, is 0.755 in the case of the original sample and 0.908 in the case of the PUD sample. In both cases, it is much higher than 0.5, indicating reliability of the model. In the PUD sample, it is very close to 1.

The jackknife values also reveal several differences (Figure 7). In the MaxEnt model made from the original sample, the CES demand potential depends on the distance to roads and land uses. This corresponds to the map itself, where the surroundings of towns and infrastructure for road traffic are highlighted. The other variables (altitude, proximity to cultural assets and temperature) do not seem to have much influence on the prediction. In contrast, the model performed on the PUD sample shows a strong dependence on altitude and temperature. If we take into account that distance to roads and land use are the least influential variables, we observe that the change from the original sample to the PUD sample provides opposite models in MaxEnt.

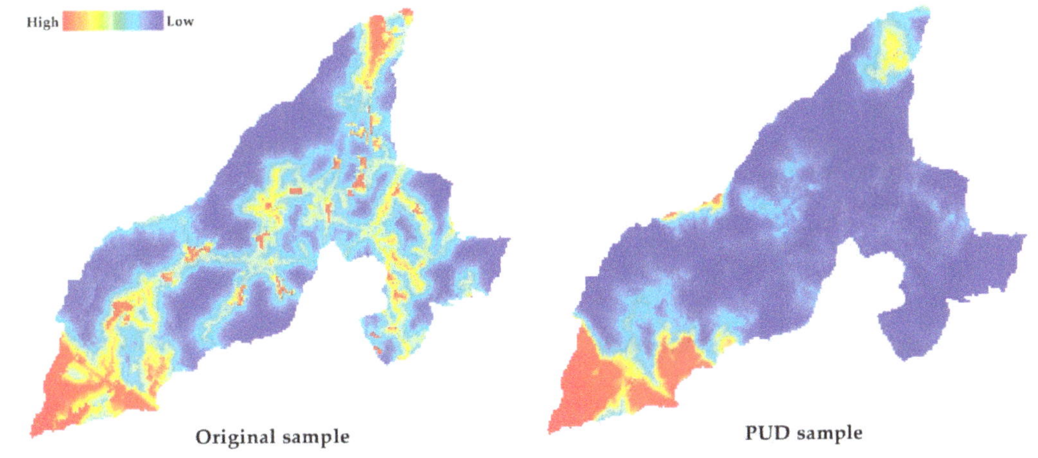

Figure 6. MaxEnt models for each sample.

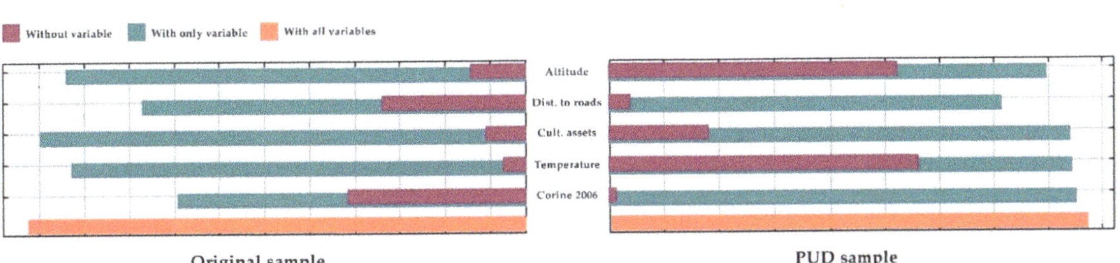

Figure 7. Jackknife test applied to environmental values for each of the samples.

3.3. Actual Demand and Correlation

The photo density from the actual demand provides similar results for both the original sample and the PUD (Figure 8). In both cases, there is a high concentration of photographs in the southwest area, although there is a proportionate distribution of lower concentrations throughout the rest of the Lozoya Valley. It is also noteworthy that in both samples, the concentration in the southeastern area is significant. The correlation matrix between these maps and their corresponding MaxEnt models gives a value of 0.30940 in the case of the original sample and 0.435 in the case of the PUD sample. That is, in both samples there is a low correlation (<0.5) between the actual demand and the potential demand determined by the MaxEnt. In other words, if MaxEnt had determined a potential evolution in 2008 from the information available in the Panoramio network, it would not have provided a model corresponding to the evolution of the photographs that actually took place.

Since the closest correlation occurs in the photographs of the PUD sample, we study the correlation of the photographs on the basis of the different categories (Table 4). It is observed here that the highest correspondence occurs in natural systems (65%) and outdoor activities (64%), being the only ones with a correspondence in the variation of more than 50%. Urban systems, however, maintain a negative correlation, meaning that the actual demand has more variation in intensity than the MaxEnt model.

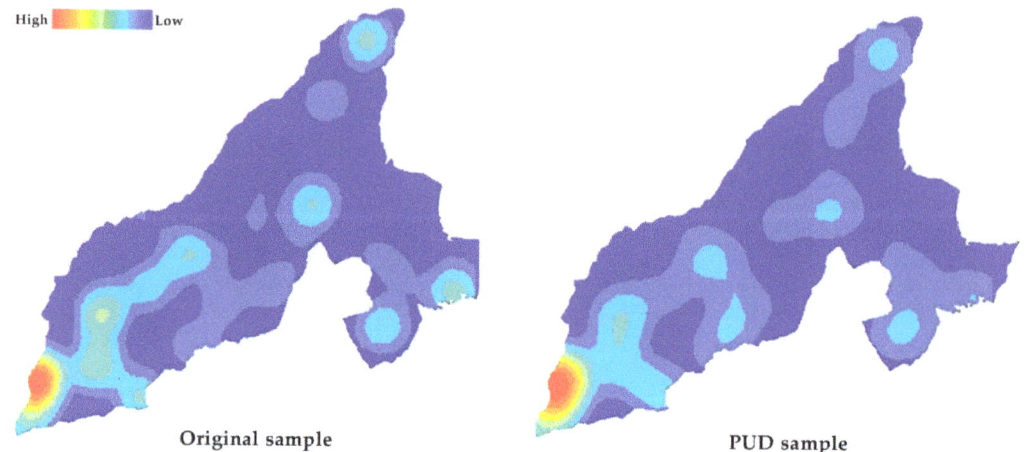

Figure 8. Density of photographs taken between 2009–2015 in both samples.

Table 4. Correlation between MaxEnt model and actual demand by category in PUD sample.

PUD Sample	Natural System	Urban System	Rural System	Water Bodies	Recreational Activities	Cultural Activities
Correlation Base Demand and Actual Demand	0.651	−0.07	0.204	0.246	0.642	0.244

4. Discussion

This study tests the effectiveness of MaxEnt predictive models, based on a machine-learning technique called maximum entropy approach, as a tool for land management to detect hot spots of CES. With this objective, we have evaluated this software through a set of photographs categorized according to their content in two different periods of time. From the photographs taken in the study area during the first period, the potential of the landscape for recreational use was estimated using MaxEnt. This estimate was correlated with the photographs of the second period, obtaining a series of results that are discussed below.

4.1. Photography Samples

The article presents the comparison of a predictive model of CES demand with the actual evolution of such demand. The demand is identified here with the georeferenced photographs uploaded to the Panoramio network by various users between 2007 and 2015. Several studies use MaxEnt to estimate the predicted future demand from social network photographs [1]. In contrast to the articles cited above, here we use photographs from the Panoramio network instead of Flickr, as both have been identified as similar [24]. In addition, some research defends the use of Panoramio over Flickr because it is a better measure of the aesthetic value of a place, since its contents were more focused on landscape and environment [35]. This is evident in the low number of photographs in our own sample devoted to cultural topics such as food, monuments or ethnographic elements. However, our sample of photographs reflects a high number of views of natural, urban and aquatic landscapes.

The fact that both the original sample of photographs and the PUD maintain a similar percentage classification by category indicates a certain consistency in the type of information uploaded by users to this type of network. This means that, at least thematically, the screening of photographs on the basis of user and day does not have an influence. However, the development of the MaxEnt model with one sample and with another provides opposite results. This means that spatially the PUD does have an impact. This allows us to refine certain explanations of this screening method [7], as it subtracts spatial bias, but not thematic bias. On the other hand, most of the photographs focus on natural or aquatic elements. This, in line with certain criticisms of samples made up of georeferenced photographs, could mean that, regardless of the type of sample, it will always be biased by an interest in photographing non-anthropized landscapes [12]. Specifically, in the study area, this is greatly influenced by the higher presence of photographs in the protected park areas (National Park in the southwest and Biosphere Re-reserve in the northeast).

4.2. MaxEnt Models

Neither of the two MaxEnt models developed bore any resemblance to the actual evolution of the distribution of the photographs except where a greater number of samples were concentrated. In spite of this, both models had AUC parameters above 0.5 and even close to 1. Interestingly, in most of the literature studied, the outcome of the models is closely linked to the environmental variables that are incorporated into the MaxEnt [11,21,24]. In our case, however, the dependence of the variables has changed greatly depending on the sample type. This contradicts the claim that MaxEnt results do not depend on the point sample size [24]. In reality, this is only true if the sample, when reduced, maintains a similar spatial distribution.

As can be seen in the two models developed (Figure 5), the PUD has reduced the number of photographs taken on the roads and settlements and, therefore, these factors are no longer important for the calculation of the prediction. What this shows is that MaxEnt is a program that is heavily influenced by high sample concentrations. A place, however small it may be in relation to the rest, will be decisive in the model if a very high number of points are located there. This is why some studies incorporate a percentage of points randomly distributed over the studied area [21]. The dependence of MaxEnt on the type of sample treatment is decisive. There are research methods that intentionally do not want to reduce the photographs to PUD, if, for example, researchers want to study the widest possible variety of images [33]. This means that different methods of spatial study from social media photographs would get different results in MaxEnt, since they would treat their samples differently.

Finally, several studies have shown certain changes in the public's interest in the values of the Lozoya Valley, with a tendency to value its wild aspects more highly [26,28]. Hence, the greatest overlap in correlation occurs in natural systems and recreational activities. However, the MaxEnt model predicts that there will be interest in infrastructure and settlements in the original sample. This model, therefore, errs in that it lacks sufficient complexity to adapt to changes in population interest. This is consistent with the criticism made by the authors of the InVEST model of predictive models in general [14]. In the case of the PUD sample, where MaxEnt predicts interest in locations at a certain altitude and temperature, the model falls short in its prediction, since it is based on a much smaller concentration of photographs.

5. Conclusions

In this article, we questioned the validity of MaxEnt as a predictive tool for landscapes as complex as those of the Lozoya Valley. The following conclusions can be drawn in line with our objectives:

i. Photographs from social networks are valid for predictive modeling as long as they are at sites that remain unchanged over time. If the configuration of the sites or the interest of the people changes, a present sample is invalid for determining future interest. On the other hand, MaxEnt is a program that allows us to determine with some accuracy to which spatial variables a certain sample of photographs is related, but it is very dependent on the concentration of these photographs. From the same sample, treated differently, it is possible to obtain models that are absolutely opposite. Comparison with the real evolution of the distribution of photographs shows that in complex and changing landscapes, MaxEnt is not useful as a predictive tool.

ii. There is a difference between the quantification of the future spatial distribution of geotagged photographs and their qualitative changes. MaxEnt establishes locations of potential interest of the photographs independently of the photographed element. The correspondence with the actual evolution of the photographs varies greatly depending on each category. For some categories, the model is closer in its prediction, but for others, the prediction is opposite to the actual evolution.

iii. This paper opens a comprehensive approach to the actual complexity of the photographs uploaded by users to social networks to assess future CES interest. Studies using MaxEnt to model potential demand can use other testers besides the AUC. For example, they can run the predictive model with a portion of the sample and use the most current photographs to check how accurate it is. They can also run the predictive models on a year-by-year basis, adjusting it according to the actual evolution of the photographs in the following year. Taking into account that most of the studies use time ranges of five or more years, this would allow us to establish rectification coefficients from one year to another to improve a global predictive model.

Author Contributions: Conceptualization, N.M. and C.A.-S.; methodology, N.M. and C.A.-S.; software, C.A.-S. and L.S.-C.; formal analysis, N.M., C.A.-S. and L.S.-C.; investigation, N.M., C.A.-S. and M.F.S.; writing—original draft preparation, N.M., C.A.-S., L.S.-C. and M.F.S.; writing—review and editing, N.M., C.A.-S. and M.F.S. All authors have read and agreed to the published version of the manuscript.

Funding: This research was funded by the project LABPA-CM: CONTEMPORARY CRITERIA, METHODS AND TECHNIQUES FOR LANDSCAPE KNOWLEDGE AND CONSERVATION (H2019/HUM-5692), funded by the European Social Fund and the Madrid regional government.

Institutional Review Board Statement: Not applicable.

Informed Consent Statement: Not applicable.

Data Availability Statement: Not applicable.

Acknowledgments: The authors would like to thank Ivan Nuñez Espinosa for his valuable help in image categorization.

Conflicts of Interest: The authors declare no conflict of interest.

References

1. Bubalo, M.; van Zanten, B.T.; Verburg, P.H. Crowdsourcing geo-information on landscape perceptions and preferences: A review. *Landsc. Urban Plan.* **2019**, *184*, 101–111. [CrossRef]
2. Toivonen, T.; Heikinheimo, V.; Fink, C.; Hausmann, A.; Hiippala, T.; Järv, O.; Tenkanen, H.; Di Minin, E. Social media data for conservation science: A methodological overview. *Biol. Conserv.* **2019**, *233*, 298–315. [CrossRef]
3. Karasov, O.; Heremans, S.; Külvik, M.; Domnich, A.; Chervanyov, I. On How Crowdsourced Data and Landscape Organisation Metrics Can Facilitate the Mapping of Cultural Ecosystem Services: An Estonian Case Study. *Land* **2020**, *9*, 158. [CrossRef]
4. Leemans, R.; De Groot, R.S. *Millennium Ecosystem Assessment: Ecosystems and Human Well-Being: A Framework for Assessment*; Island Press: Washington, DC, USA, 2003.
5. Cheng, X.; Van Damme, S.; Li, L.; Uyttenhove, P. Evaluation of cultural ecosystem services: A review of methods. *Ecosyst. Serv.* **2019**, *37*, 100925. [CrossRef]
6. Milcu, A.; Ioana, A.; Hanspach, J.; Abson, D.; Fischer, J. Cultural ecosystem services: A literature review and prospects for future research. *Ecol. Soc.* **2013**, *18*, 44. [CrossRef]
7. Zhang, H.; Huang, R.; Zhang, Y.; Buhalis, D. Cultural ecosystem services evaluation using geolocated social media data: A review. *Tour. Geogr.* **2020**, 1–23. [CrossRef]
8. Ghermandi, A.; Camacho-Valdez, V.; Trejo-Espinosa, H. Social media-based analysis of cultural ecosystem services and heritage tourism in a coastal region of Mexico. *Tour. Manag.* **2020**, *77*, 104002. [CrossRef]
9. Dunkel, A. Visualizing the perceived environment using crowdsourced photo geodata. *Landsc. Urban Plan.* **2015**, *142*, 173–186. [CrossRef]

10. Oteros-Rozas, E.; Martín-López, B.; Fagerholm, N.; Bieling, C.; Plieninger, T. Using social media photos to explore the relation between cultural ecosystem services and landscape features across five European sites. *Ecol. Ind.* **2018**, *94*, 74–86. [CrossRef]

11. Richards, D.R.; Friess, D.A. A rapid indicator of cultural ecosystem service usage at a fine spatial scale: Content analysis of social media photographs. *Ecol. Ind.* **2015**, *53*, 187–195. [CrossRef]

12. Richards, D.R.; Tunçer, B.; Tunçer, B. Using image recognition to automate assessment of cultural ecosystem services from social media photographs. *Ecos. Serv.* **2018**, *31*, 318–325. [CrossRef]

13. Abousaleh, F.S.; Cheng, W.H.; Yu, N.H.; Tsao, Y. Multimodal Deep Learning Framework for Image Popularity Prediction on Social Media. *IEEE Trans. Cogn. Dev. Syst.* **2020**, *13*, 679–692. [CrossRef]

14. Sharp, R.; Tallis, H.T.; Ricketts, T.; Guerry, A.D.; Wood, S.A.; Chaplin-Kramer, R.; Nelson, E.; Ennaanay, D.; Wolny, S.; Olwero, N.; et al. *InVEST User's Guide*; The Natural Capital Project: Stanford, CA, USA, 2014.

15. Sherrouse, B.C.; Semmens, D.J. *Social Values for Ecosystem Services, Version 4.0 (SolVES 4.0)—Documentation and User Manual*; US Department of the Interior, US Geological Survey: Washington, DC, USA, 2020.

16. Elith, J.; Phillips, S.J.; Hastie, T.; Dudík, M.; Chee, Y.E.; Yates, C.J. A statistical explanation of MaxEnt for ecologists. *Divers. Distrib.* **2011**, *17*, 43–57. [CrossRef]

17. Poria, S.; Cambria, E.; Bajpai, R.; Hussain, A. A review of affective computing: From unimodal analysis to multimodal fusion. *Inf. Fusion* **2017**, *37*, 98–125. [CrossRef]

18. Li, Y.; Fei, T.; Huang, Y.; Li, J.; Li, X.; Zhang, F.; Kang, Y.; Wu, G. Emotional habitat: Mapping the global geographic distribution of human emotion with physical environmental factors using a species distribution model. *Int. J. Geo. Inf. Sci.* **2021**, *35*, 227–249. [CrossRef]

19. He, S.; Su, Y.; Shahtahmassebi, A.R.; Huang, L.; Zhou, M.; Gan, M.; Deng, J.; Zhao, G.; Wang, K. Assessing and mapping cultural ecosystem services supply, demand and flow of farmlands in the Hangzhou metropolitan area, China. *Sci. Total Environ.* **2019**, *692*, 756–768. [CrossRef]

20. Phillips, S.J. *A Brief Tutorial on Maxent*; AT&T Research, American Museum of Natural History: New York, NY, USA, 2017; Available online: http://biodiversityinformatics.amnh.org/open_source/maxent/ (accessed on 9 May 2022).

21. Yoshimura, N.; Hiura, T. Demand and supply of cultural ecosystem services: Use of geotagged photos to map the aesthetic value of landscapes in Hokkaido. *Ecosyst. Serv.* **2017**, *24*, 68–78. [CrossRef]

22. Clemente, P.; Calvache, M.; Antunes, P.; Santos, R.; Cerdeira, J.O.; Martins, M.J. Combining social media photographs and species distribution models to map cultural ecosystem services: The case of a Natural Park in Portugal. *Ecol. Ind.* **2019**, *96*, 59–68. [CrossRef]

23. Long, P.R.; Nogué, S.; Benz, D.; Willis, K.J. Devising a method to remotely model and map the distribution of natural landscapes in Europe with the greatest recreational amenity value (cultural services). *Front. Biogeogr.* **2021**, *13*, 1. [CrossRef]

24. Arslan, E.S.; Örücü, Ö.K. MaxEnt modelling of the potential distribution areas of cultural ecosystem services using social media data and GIS. *Environ. Dev. Sustain.* **2021**, *23*, 2655–2667. [CrossRef]

25. Schmitz, M.F.; De Aranzabal, I.; Pineda, F.D. Spatial analysis of visitors preferences in the outdoor recreational niche of Mediterranean cultural landscapes. *Environ. Conserv.* **2007**, *34*, 300–312. [CrossRef]

26. Arnaiz-Schmitz, C.; Santos, L.; Herrero-Jáuregui, C.; Díaz, P.; Pineda, F.D.; Schmitz, M.F. Rural Tourism: Crossroads between nature, socio-ecological decoupling and urban sprawl. *WIT Trans. Ecol. Environ.* **2018**, *227*, 1–9.

27. Cruz, L.; Carrión, A. (Eds.) *Cien Paisajes Culturales en España*; Ministerio de Educación, Cultura y Deporte: Madrid, Spain, 2015.

28. Sarmiento-Mateos, P.; Arnaiz-Schmitz, C.; Herrero-Jáuregui, C.; Pineda, F.D.; Schmitz, M.F. Designing Protected Areas for Social–Ecological Sustainability: Effectiveness of Management Guidelines for Preserving Cultural Landscapes. *Sustainability* **2019**, *11*, 2871. [CrossRef]

29. Gülçin, D. Predicting visual aesthetic preferences of landscapes near historical sites by fluency theory using social media data and GIS. *Int. J. Geogr. Geogr. Educ. (IGGE)* **2021**, *43*, 265–277. [CrossRef]

30. Van Zanten, B.T.; Van Berkel, D.B.; Meentemeyer, R.K.; Smith, J.W.; Tieskens, K.F.; Verburg, P.H. Continental-scale quantification of landscape values using social media data. *Proc. Natl. Acad. Sci. USA* **2016**, *13*, 12974–12979. [CrossRef] [PubMed]

31. Orsi, F.; Geneletti, D. Using geotagged photographs and GIS analysis to estimate visitor flows in natural areas. *J. Nat. Conserv.* **2013**, *21*, 359–368. [CrossRef]

32. Lieskovský, J.; Rusňák, T.; Klimantová, A.; Izsóff, M.; Gašparovičová, P. Appreciation of landscape aesthetic values in Slovakia assessed by social media photographs. *Open Geosci.* **2017**, *9*, 593–599. [CrossRef]

33. Donaire, J.A.; Camprubí, R.; Galí, N. Tourist clusters from Flickr travel photography. *Tour. Manag. Perspect.* **2014**, *11*, 26–33. [CrossRef]

34. Bermudez, J. Sistemas de información geográfica y Administración pública: El sistema de información de Patrimonio Histórico inmueble de la Comunidad de Madrid. In *Manual de Tecnologías de la Información Geográfica Aplicadas a la Arqueología*; en Minguez, M.C., Capdevila, E., Eds.; Comunidad de Madrid, Museo Arqueológico Regional: Madrid, Spain, 2016; pp. 399–424.

35. Casalegno, S.; Inger, R.; DeSilvey, C.; Gaston, K.J. Spatial Covariance between Aesthetic Value & Other Ecosystem Services. *PLoS ONE* **2013**, *8*, e68437.

MDPI

St. Alban-Anlage 66

4052 Basel

Switzerland

www.mdpi.com

Land Editorial Office

E-mail: land@mdpi.com

www.mdpi.com/journal/land

www.ingramcontent.com/pod-product-compliance
Lightning Source LLC
LaVergne TN
LVHW071301170726
843515LV00006BA/1461